Sol-Gel Chemistry Applied to Materials Science

Sol-Gel Chemistry Applied to Materials Science

Special Issue Editor

Michelina Catauro

MDPI • Basel • Beijing • Wuhan • Barcelona • Belgrade

Special Issue Editor
Michelina Catauro
Second University of Naples
Italy

Editorial Office
MDPI
St. Alban-Anlage 66
4052 Basel, Switzerland

This is a reprint of articles from the Special Issue published online in the open access journal *Materials* (ISSN 1996-1944) from 2017 to 2018 (available at: https://www.mdpi.com/journal/materials/special_issues/sol_gel_chemistry_materials)

For citation purposes, cite each article independently as indicated on the article page online and as indicated below:

LastName, A.A.; LastName, B.B.; LastName, C.C. Article Title. *Journal Name* **Year**, *Article Number*, Page Range.

ISBN 978-3-03921-353-5 (Pbk)
ISBN 978-3-03921-354-2 (PDF)

Contents

About the Special Issue Editor

Michelina Catauro Michelina Catauro was enrolled at the University of Naples "Federico II" at the Faculty of Mathematics, Physics and Natural Sciences to study Chemistry in 1984. She was awarded an Erasmus Program Scholarship for the academic year 1989/90 sponsored by the University of East Anglia's School of Chemical Sciences in Norwich (Great Britain) under the joint supervision of Professor Vincenzo Vitaliano of the Chemistry Department of the University of Naples "Federico II" and Professor Brian H. Robinson of the UEA. During her stay abroad, she began the experimental work towards her thesis, entitled "Lipase in AOT Microemulsion-Based Gels". In 1991, she graduated with honors from the University of Naples "Federico II". She was awarded another scholarship sponsored by the University of East Anglia's School of Chemical Sciences in Norwich, Great Britain, under the supervision of Professor L. Salerno. She was enrolled in a PhD course on Materials Technology and Industrial Plants (VII) at the Department of Materials Engineering and Production of the University of Naples "Federico II" under the supervision of Professor Alberto Marotta and Professor Alberto Buri. She successfully completed her PhD course and presented a dissertation on glasses materials and ceramic glass materials prepared from sol–gel processing.

Having passed the examination in November 1995, she started to work as a University Researcher in Chemistry at the Faculty of Engineering of the University of Naples "Federico II". In 1996, she became a member of the Department of Materials Engineering and Production of the Faculty of Engineering of the University of Naples "Federico II". In 1998, she was confirmed in her role as Researcher with the University of Naples at the Faculty of Engineering of the University of Naples "Federico II". She passed the selection for Associate Professor, subject group CHIM/07 (Chemistry).

She has been Associate Professor at the Department of Engineering, University of Campania "Luigi Vanvitelli" since her appointment in 2005.

Professor Catauro is the author of 142 publications currently indexed by Scopus (h-index = 30).

Preface to "Sol-Gel Chemistry Applied to Materials Science"

Though the human body can be considered a beautifully designed machine, it has limited capacity to repair its main tissues and organs when serious damage, such as trauma or various diseases, occurs. In recent decades, it has been possible to overcome these problems by developing new materials that can replace tissues or parts of the body. Bioactive glasses constitute a promising class of bioactive materials for bone repair and substitution. They have the capability of bonding with living bone by forming a hydroxyapatite layer on their surface that has a composition equivalent to that of the mineral phase of bones. The biological properties of bioglasses are influenced by their composition and, also, by the synthesis method used. An ideal method to prepare a bioglass is the sol–gel technique, a versatile synthesis process. The process involves the transition of a system from mostly colloidal liquid ('sol') into a solid 'gel'. The sol–gel method has many advantages, such as product purity and, among the most important ones, the possibility to incorporate thermolabile molecules. In fact, it is possible to obtain organic–inorganic hybrid materials, in which the organic and inorganic phase are bonded together at the nanometer to submicrometer scales. The starting point for synthesizing sol–gel materials is to understand the chemistry behind this method. In the synthesis process, molecular precursors, such as metal alkoxides, are used, which are involved in two important reactions: hydrolysis and polycondensation. The drying stage is a critical part of the whole sol–gel process, which involves removal of the liquid phase from the wet gel. As evaporation occurs, drying stress can cause the cracking of bulk materials. During the drying process, the gel shrinks by the volume that was previously occupied by the liquid, which flows from the internal of the gel body to its surface. Upon shrinkage, OH groups at the internal surface approach each other and can react with each other. As drying proceeds, the network becomes increasingly stiffer and the surface tension in the liquid correspondingly increases due to pore radii becoming smaller. Furthermore, it is possible to obtain a xerogel or an aerogel, respectively, using different drying conditions at ambient pressure or in supercritical conditions. The sol–gel process is preferred due to its economic feasibility and that the low-temperature process allows control over the composition of the product that is achieved. The sol–gel technique intends to desirably control the dimensions of a material on a nanometric scale from the initial stages of processing. Chemical processing, highly controlled purity, and improved homogeneity can be used to improve material properties. This low-temperature processing technique is a great advantage over conventional synthesis techniques, and a very wide range of materials are actually fabricated using this method. This includes materials with optical and photonic functions, electronic functions, thermal functions, mechanical functions, chemical functions, and biochemical and biomedical functions. Most of the materials prepared by the sol–gel method are advanced materials needed for the development of advanced technologies.

Michelina Catauro
Special Issue Editor

Article

Synthesis of Bioactive Chlorogenic Acid-Silica Hybrid Materials via the Sol–Gel Route and Evaluation of Their Biocompatibility

Michelina Catauro [1,*] and Severina Pacifico [2]

[1] Department of Industrial and Information Engineering, University of Campania "Luigi Vanvitelli", Via Roma 29, 81031 Aversa, Italy

[2] Department Environmental, Biological and Pharmaceutical Sciences and Technologies, University of Campania "Luigi Vanvitelli", Via Vivaldi 43, 81100 Caserta, Italy; severina.pacifico@unicampania.it

*   Correspondence: michelina.catauro@unicampania.it; Tel.: +39-081-501-0360

Received: 14 June 2017; Accepted: 17 July 2017; Published: 21 July 2017

Abstract: Natural phenol compounds are gaining a great deal of attention because of their potential use as prophylactic and therapeutic agents in many diseases, as well as in applied science for their preventing role in oxidation deterioration. With the aim to synthetize new phenol-based materials, the sol–gel method was used to embed different content of the phenolic antioxidant chlorogenic acid (CGA) within silica matrices to obtain organic-inorganic hybrid materials. Fourier transform infrared (FTIR) measurements were used to characterize the prepared materials. The new materials were screened for their bioactivity and antioxidant potential. To this latter purpose, direct DPPH (2,2-diphenyl-1-picrylhydrazyl) and ABTS (2,2′-azinobis-(3-ethylbenzothiazolin-6-sulfonic acid) methods were applied: radical scavenging capability appeared strongly dependent on the phenol amount in investigated hybrids, and became pronounced, mainly toward the ABTS radical cation, when materials with CGA content equal to 15 wt% and 20 wt% were analyzed. The in vitro biocompatibility of the synthetized materials was estimated by using the MTT assay towards fibroblast NIH 3T3 cells, human keratinocyte HaCaT cells, and the neuroblastoma SH-SY5Y cell line. As cell viability and morphology of tested cell lines seemed to be unaffected by new materials, the attenuated total reflectance (ATR)-FTIR method was applied to deeply measure the effects of the hybrids in the three different cell lines.

Keywords: sol–gel method; organic-inorganic hybrids; chlorogenic acid; cytotoxicity; biocompatibility

---

1. Introduction

The growing interest in plants' secondary metabolites is due to their ability to be bioactive compounds with pharmacological or toxicological effects in humans and animals. Indeed, these naturally-occurring substances, also known as phytochemicals, are still the main source of lead molecules in modern drug discovery and development, and their evidenced health promoting benefits in counteracting chronic and degenerative pathologies (e.g., cancer, cardiovascular, and neurodegenerative diseases), make the natural products research an endless and intriguing research field with multidisciplinary approach [1]. Among phytochemicals, phenol, and polyphenols have gained a great deal of importance due to their antioxidant capability, which allows them to exert preventive and protectant effects in human cells [2,3]. Indeed, the highly-acclaimed chemopreventive role of these substances appear to be related to their ability to induce dose-dependent oxidative stress, DNA damage, and apoptosis in tumor, but not normal tissue [4,5].

5-O-caffeoylquinic acid, better known as chlorogenic acid (CGA), is a hydroxycinnamoyl derivative, whose structure consists of a caffeic acid moiety esterified with (−)-quinic acid (Figure 1).

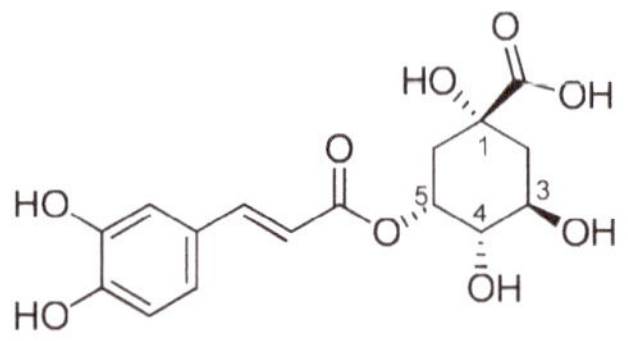

Figure 1. Chemical structure of chlorogenic acid (CGA).

This dietary metabolite, broadly distributed in edible plants, possesses many health-promoting properties [6]. Accumulating evidence demonstrated that CGA possesses antibacterial, anti-inflammatory, and anti-oxidant activities [7], as well as appearing to be an effective chemopreventive agent. The growing interest in the dietary supplementation of CGA as a nutraceutical agent in food formulations, due to its various medicinal properties, its low bioavailability and stability, addressed the encapsulation of CGA in a variety of polymers and the synthesis of CGA loaded chitosan nanoparticles with preserved antioxidant activity [8]. Biomaterial science was further fascinated by this molecule and recently CGA-gelatin was prepared as a coating for the preservation of seafood [9]. Moreover, highly-adhesive bioinspired polyurethanes based on CGA were prepared from 4,4′-methylenebis (cyclohexyl isocyanate) and polyethylene glycol 200 providing biocompatible-adhesive bioinspired polyurethanes, which appeared to be good candidates for medical applications as a tissue adhesive material. The sol–gel technique was employed for the preparation of carbon composite electrode modified with electroless deposition of chlorogenic acid, which were evaluated for their stability and electrochemical properties was [10]. The high versatility of sol–gel routes for the formulation of organic-inorganic hybrid materials [11–13], together with data from our recent researches aimed at entrapping another natural antioxidant compound, such as quercetin, in a silica matrix [14–16], intrigued us to investigate the possibility of synthesizing new materials having chlorogenic acid as the organic component. Thus, silica-based materials, differing in their CGA amount (5 wt%, 10 wt%, 15 wt%, and 20 wt%), were synthetized and characterized by Fourier transform infrared spectroscopy (FTIR) and UV-VIS spectroscopy. The preservation of CGA antioxidant chemical features was demonstrated by applying DPPH and ABTS tests. Bioactivity was studied by soaking the samples into a simulated body fluid (SBF) and evaluating the formation of a hydroxyapatite layer on their surface by FTIR spectroscopy and scanning electron microscopy (SEM) after 21 days of exposure. Biocompatibility was assessed by the MTT direct contact test using murine fibroblast NIH 3T3 cell line, human keratinocyte HaCaT cells and neuroblastoma SH-SY5Y cell line. The choice of cell lines was deliberate. Fibroblasts are cell types that interact with proteins on biomaterials surfaces, playing important roles in biomaterials rejection and implant failure. The HaCaT cell line, although obviously immortal, is a non-tumorigenic cell line [17]. On the other hand, considering the ability of CGA to act as an antioxidant at low doses and pro-oxidant at high doses, SH-SY5Y cells, particularly sensitive to oxidative stress onset, were also used. In fact, the brain is highly vulnerable to oxidative stress due to its high O_2 consumption, its modest antioxidant defenses and its lipid-rich constitution. Attenuated total reflectance (ATR)-FTIR analyses were also applied to deeply measure the effects of the hybrids in the treated cell lines.

2. Results and Discussion

Recently, in the search for new biocompatible biomaterials able to provide antioxidant functionality and to not exacerbate the body's normal oxidant and inflammatory response, our research group has optimized the synthesis of novel intrinsically-antioxidant quercetin-based biomaterials, which could be employed in dentistry, as components of glass ionomer cement, and in medicine, as replacements for bone implants. With the aim at preparing new bioactive and biocompatible organic-inorganic hybrids, our attention is turned to chlorogenic acid, a small phenol compound, broadly investigated for its several bioactive and health-promoting properties, which appears as an interesting compound to be incorporated into pharmaceutical, cosmetic, or food products.

2.1. Characterization of Synthetized Organic-Inorganic Hybrid Materials

Figure 2 shows the spectra of the SiO_2/CGA hybrids (curve from b to e) compared to the spectra of the pure SiO_2 (curve a) and CGA (curve f). The spectrum of the pure SiO_2 (curve a) shows all the typical peaks of the silica sol–gel materials [18–20].

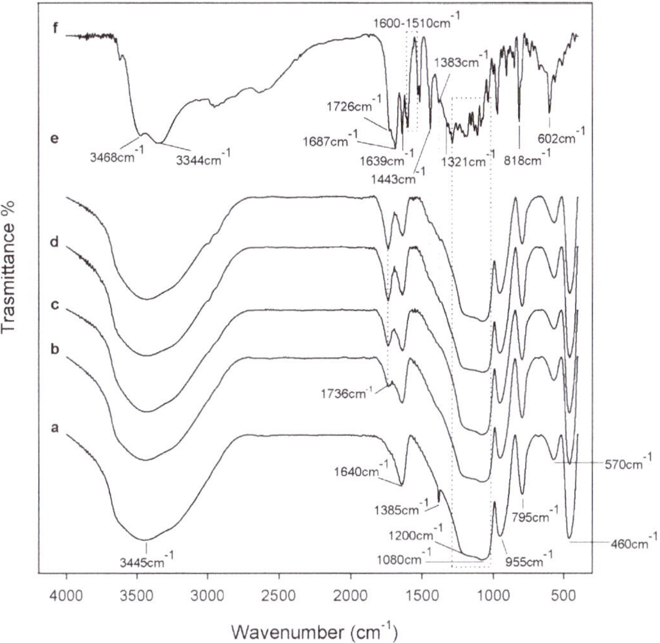

Figure 2. FT-IR spectra of (**a**) pure SiO_2; (**b**) SiO_2/CGA, 5 wt%; (**c**) SiO_2/CGA, 10 wt%; (**d**) SiO_2/CGA, 15 wt%; (**e**) SiO_2/CGA, 20 wt%; and (**f**) pure CGA.

The broad intense band at 3445 cm^{-1} and the peak at 1640 cm^{-1} are due to –OH stretching and bending vibrations in the hydration water. The bands at 1080 cm^{-1} and 795 cm^{-1} and the shoulder at 1200 cm^{-1} are associated to the asymmetric and symmetric Si–O stretching vibrations. The signal at 460 cm^{-1} is due to the bending of the Si–O–Si bonds. Moreover, three peaks generally observed in alkoxy-derived silica gels are visible at 1385 cm^{-1}, 955 cm^{-1} and at 570 cm^{-1}, which are due to residual nitrate anions [21], Si–OH bonds, and four-fold siloxane residual cyclic structures in the silica network, respectively [18,19,22]. The chlorogenic acid IR spectrum shows OH groups stretching at 3468 and 3344 cm^{-1}, whereas OH bending of the phenol function was at 1383 cm^{-1}. The band assigned to the stretching C=O vibration of the carboxylic group is located at 1726 cm^{-1}, whereas the band at 1687 cm^{-1} is due to the stretching C=O vibrations of the ester group [23]. The stretching vibration of the C=C fragment is at 1639 cm^{-1}, whereas the bands derived mainly from stretching vibrations of the aromatic ring are in the range of 1600–1510. The band at 1443 cm^{-1}, which can be assigned to a phenyl ring stretch, was previously attributed to the C_3–O–H group, whose contribution was found equal to 64% [24]. The in-plane bending band of C–H in the aromatic hydrocarbon is detectable at 1321 cm^{-1}, and out-of-plane bending bands are at 818 and 602 cm^{-1}. The spectra of the SiO_2/CGA hybrids show all the described SiO_2 peaks, whereas it lacks nitrate signals. Indeed, a broadening of the SiO_2 strong band at 1080 cm^{-1} occurs, together with a marked increase in the intensity of the shoulder at 1200 cm^{-1}, as a result of the contribution of the several, intense signals of the phenyl ring and C–O–C bonds in CGA (see curve f), which are present in this spectral region. The observation of two weak bands at 1447 and 1373 cm^{-1} in the spectrum of all the hybrid samples, and are clearer in those containing 15 wt% and 20 wt% of CGA (curves d, e), corresponding to 1443 cm^{-1} and 1383 cm^{-1} in the CGA spectrum, which seemed to confirm this hypothesis. In fact, these signals, whose wavenumber are displaced by 4 cm^{-1} and 10 cm^{-1} with respect to those detected in CGA, are ascribable to phenyl ring stretching and phenol bending vibrations [9]. The displacement of the CGA carboxylic group C=O stretch by about 10 cm^{-1} (1736 cm^{-1}) and the observation of a weak shoulder at a lower wavenumber, attributable to the ester C=O stretch vibration, allowed us to hypothesize the establishment of H-bonds with the SiO$_2$ inorganic matrix. FTIR data suggested the synthesis of materials in which chlorogenic acid was embedded in the silica network. UV-VIS spectra, recorded on extracts obtained by swelling powders of investigated materials in water, strengthen this hypothesis (Figure 3). In SiO$_2$ spectrum, two absorption bands were also observed, the first one located at 204 nm corresponds to electronic transitions exhibited by sol–gel SiO$_2$ materials [25], while the weak other one (260 nm) could be the result of the applied acid-catalyzed sol–gel route (Figure 3a). For CGA, maximum absorbances occurred at 217 nm (with shoulder at 240 nm) and at 324 nm (with shoulder at 296 nm), whereas the minimum point was at 262 nm (Figure 3b). The UV spectra of hybrids were in accordance with silica-induced changes of the chlorogenic acid skeleton, which modified the characteristic different electronic transitions of the caffeoyl quinic acid (Figure 3c,d).

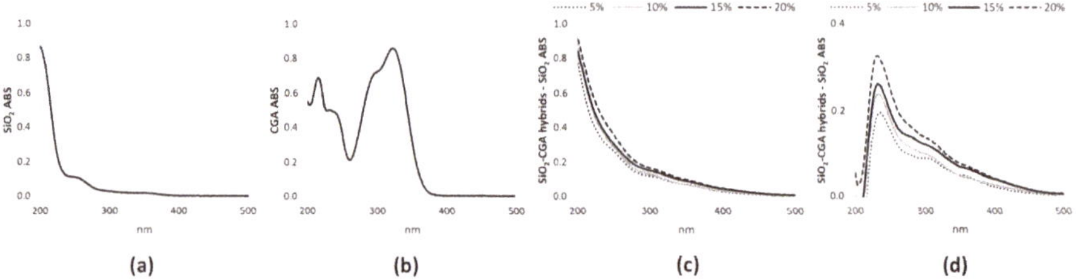

Figure 3. UV-VIS spectra of (**a**) pure SiO$_2$; (**b**) pure CGA; (**c**) SiO$_2$/CGA hybrids. Panel (**d**) shows different UV-VIS spectra obtained by importing the acquired data into Excel and subtracting SiO$_2$ signals from the SiO$_2$ scan to those of each hybrid.

2.2. Bioactivity Test

After 21 days of soaking into the SBF solution, both the sample powders and sample disks were air dried and their ability to induce the nucleation of a hydroxyapatite layer on their surface was evaluated by FTIR (Shimadzu, Tokyo, Japan) and SEM (Quanta 200, FEI, Eindhoven, The Netherlands) analyses, respectively. Comparing FTIR spectra of the sample powders before (Figure 2) and after the exposure to SBF (Figure 4), a new peak at 630 cm^{-1} and the split of the band at 570 cm^{-1} in two new peaks at 575 cm^{-1} and 560 cm^{-1} were observed. Those spectra modifications could be ascribable to the formation of the hydroxyapatite precipitate and, in particular, to the stretching of the hydroxyapatyte –OH groups and the vibrations of PO$_4$$^{3-}$ groups, respectively [26,27]. Moreover, a slight up-shift of the Si–OH band (from 955 cm^{-1} to 960 cm^{-1}) suggest the interaction of the hydroxyapatite layer with the –OH groups of the silica matrix.

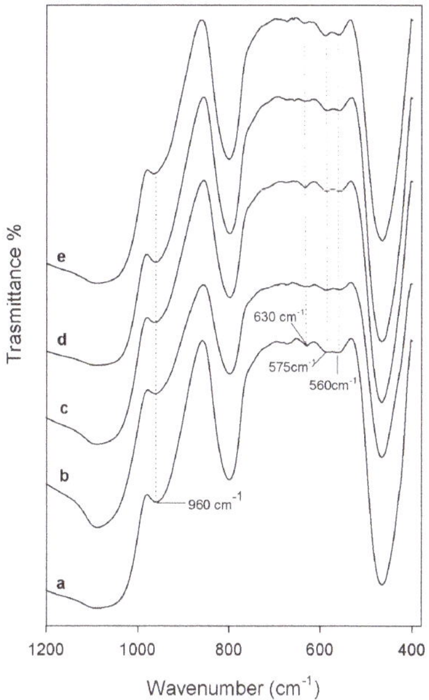

Figure 4. FT-IR spectra of (**a**) pure SiO_2; (**b**) SiO_2/CGA, 5 wt%; (**c**) SiO_2/CGA, 10 wt%; (**d**) SiO_2/CGA, 15 wt%; and (**e**) SiO_2/CGA, 20 wt% after 21 days of exposure to SBF.

The formation of the hydroxyapatite precipitate on the sample surfaces was confirmed by SEM imagery (Figure 5). After 21 days of exposure to SBF, indeed, the surfaces of all samples appear covered by a precipitate with the globular shape typical of hydroxyapatite [28]. No difference was detected in the distribution and amount of precipitate, as the whole surface of the samples is covered by the globules. Therefore, only representative SEM micrographs of the SiO_2 and SiO_2/CGA systems are reported (Figure 5a,b). The energy-dispersive X-ray (EDX) (Quanta 200, FEI, Eindhoven, The Netherlands) microanalysis (Figure 5c) confirmed that the globules consist of Ca and P in an atomic ratio equal to 1.67.

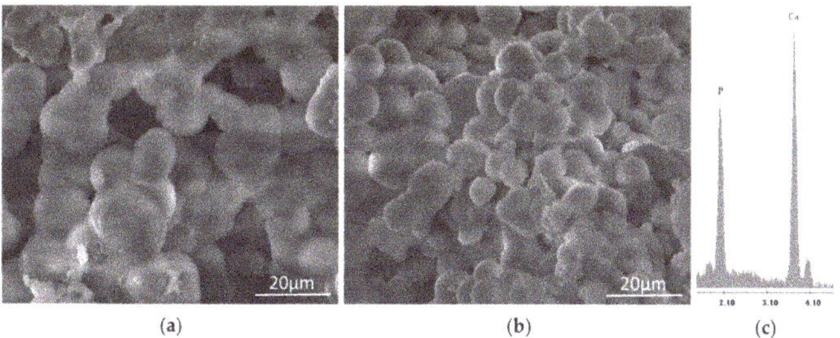

Figure 5. SEM micrographs of (**a**) pure SiO_2 and (**b**) a representative SiO_2/CGA hybrid; and (**c**) EDX analysis.

2.3. Antiradical Capability of SiO_2-CGA Hybrids

DPPH$^\bullet$ and ABTS$^{\bullet+}$ methods, which use two radical probes which may be neutralized by the transfer of an electron and/or a hydrogen atom, allowed us to evaluate the radical scavenging capacity of the synthetized hybrids and to compare it to that exercised by pure chlorogenic acid. Caffeoyl quinic acid silica-based materials were able to exert an anti-radical power strongly dependent on the phenol concentration therein, reaching its maximum effect when the highest chlorogenic acid dose level (20 wt%) was embedded (Figure 6). Analogously, the amount of hybrids placed in contact with the probe solutions seemed to affect the antiradical response, which appeared far below that elicited by pure chlorogenic acid. This latter, which is known to display notable free radical scavenging effects, showed ID_{50} values of 0.53 and 6.06 µg/mL vs. DPPH$^\bullet$ and ABTS$^{\bullet+}$, respectively [29]. The scavenging efficiency of pure CGA, as well as of other phenol compounds exhibiting similar structural features, is commonly ascribed to the two exchangeable hydrogen atoms (those of catechol moiety), whose presence makes phenol compounds biologically-reactive molecules capable of exhibiting both anti- and pro-oxidant behavior.

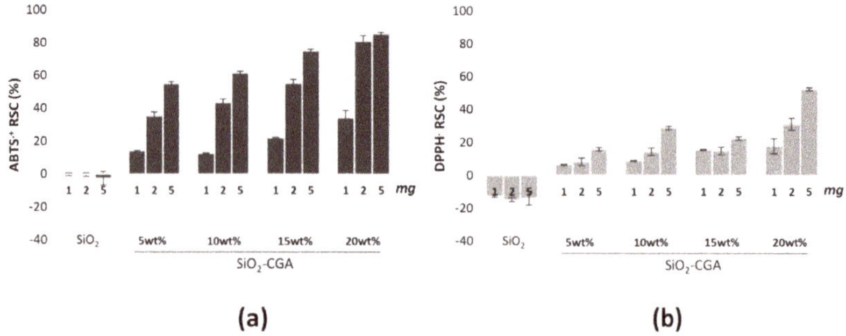

Figure 6. Radical Scavenging Capacity (RSC, %) of different amounts of SiO$_2$-CGA hybrids, and SiO$_2$ samples towards (**a**) ABTS$^{\bullet+}$ and (**b**) DPPH$^\bullet$. Values, reported as percentage vs. a blank, are the mean ± SD of measurements carried out on three samples (n = 3) analyzed three times.

2.4. Cytotoxicity of SiO$_2$-CGA Hybrids

In order to assess the influence of the synthesized hybrid materials on morphology and cell proliferation, NIH-3T3 murine fibroblast, HaCaT human keratinocyte, and SH-SY5Y human neuroblastoma cell lines were grown in the presence of powders of the investigated materials. After 48 h exposure, the MTT cytotoxicity assay was performed. In Figure 7 morphological changes detected directly from NIH-3T3 culture plates with a phase-contrast microscope are reported. The synthetized materials did not seem to affect NIH-3T3 cell morphology.

The proliferation of the embryonic fibroblast cells was observed to increase depending on the content of the embedded phenol (Figure 8a). In particular, it reached its maximum percentage value when SiO$_2$-CGA, 10 wt% was tested, whereas a weak decrease in cell viability was observed in cells treated with SiO$_2$-CGA, 15 wt%, and SiO$_2$-CGA, 20 wt% samples.

Similar behavior was observed for HaCaT cells, which strongly preserved the morphology after the treatment with the investigated hybrids. As for NIH-3T3 cells, MTT data were in accordance with a mild in vitro suppression of HaCaT cells' mitochondrial redox activity to levels that would be acceptable based on standards used to evaluate alloys and composites (<25% suppression of dehydrogenases activity; Figure 8b) [30]. Thus, SiO$_2$/CGA hybrids were found biocompatible towards non-tumorigenic NIH-3T3 and HaCaT cell lines, highlighting that the adopted synthesis strategy provided materials in which the establishment of a network between the phenol compound and the silica matrix was conducive to maintaining antioxidant functionality of the organic component, inhibiting the dose-dependent anti-proliferative efficacy, commonly observed when high doses of chlorogenic acid were tested.

Recently-published findings from pre-clinical experimental and phase I clinical studies have shown that treatment with CGA has shown therapeutic effects in breast cancer, brain tumors, lung cancer, colon cancer, and chronic myelogenous leukemia [31]. The ability of chlorogenic acid to exert an anti-tumor effect in multiple malignant tumors appeared to be shared by synthetized hybrids towards neuroblastoma SH-SY5Y cells, which seemed to change their phenotype, exhibiting a decrease in proliferation dependent on both the phenol amount embedded and the dose of hybrid directly placed in contact with them (Figure 8c).

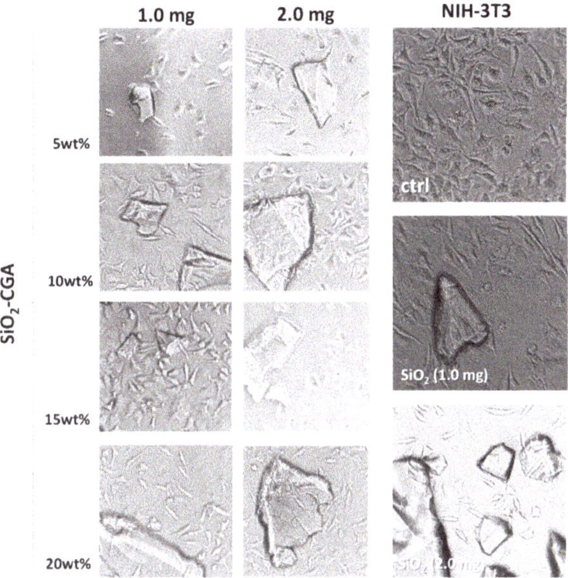

Figure 7. Morphological changes in hybrids- and SiO_2-treated NIH-3T3 cells. Representative images were acquired by an inverted phase contrast brightfield Zeiss Primo Vert Microscope. Ctrl = untreated cells.

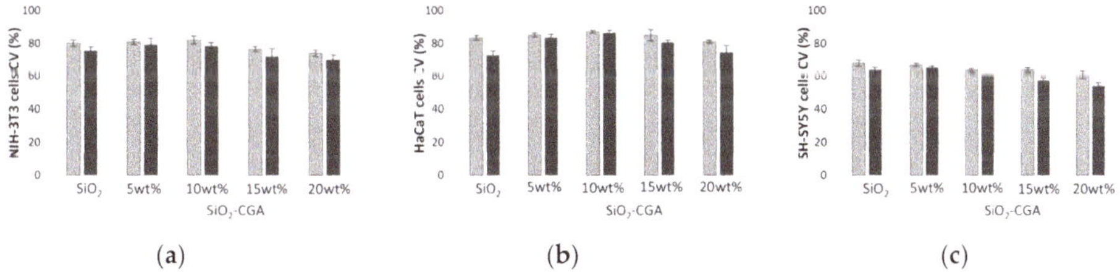

Figure 8. Cell Viability (CV, %) of (**a**) NIH-3T3, (**b**) HaCaT, and (**c**) SH-SY5Y cells treated with ■ 1.0 mg, and ▨ 2.0 mg of SiO_2-CGA hybrids, after 48 h exposure time by means of MTT test results. Values, reported as percentage vs. an untreated control, are the mean ± SD of measurements carried out on three samples (*n* = 3) analyzed six times.

In fact, when a dose equal to 2.0 mg of SiO_2-CGA, 20 wt% was tested, mitochondrial redox activity was inhibited by 49.9%. A marked dose-dependent anti-proliferative activity was found for pure chlorogenic acid, which was able to inhibit SH-SY5Y cell viability by 50% at a concentration level equal to 31.1 μg/mL. Thus, the embedment of high doses of chlorogenic acid in silica matrix, while massively preserving the cell growth of treated cells with respect to the pure compound, seemed to provide a material able to exert pro-oxidant activity. This hypothesis was supported by the experimental data of several studies, which highlights that dietary phenols and polyphenols can potentially confer additional benefits, but high-doses may elicit toxicity, thereby establishing a double-edged sword in their use as supplements [32].

In order to unravel the mechanism underlying the observed cytotoxicity, ATR-FTIR analyses were carried out [33]. The spectra, acquired in the 650–4000 cm^{-1} region of cell suspensions untreated or previously treated with synthetized hybrids dose are depicted in Figure 9. ATR-FTIR spectra of viable, apoptotic, and necrotic cells are dominated by bands assigned to protein absorption modes: the amide I band is the most intense, centered near 1640 cm^{-1}, which corresponds to the C=O stretching vibration coupled to the N–H bending and to C–N stretching modes of peptide bonds [33]. The amide II band at 1539 cm^{-1} is due to vibrational modes involving the C–N–H bending and C–N stretching of the peptide bonds [34]. The spectral analysis showed some significant differences between viable and apoptotic cells. The spectra of apoptotic cells, compared to vital ones, showed a significant decrease in the region between 900 and 1300 cm^{-1}. The complexity in this spectral region was due to the contribution of nucleic acids (DNA, RNA), carbohydrates, and phosphates. The band centered at 1080 cm^{-1} was assigned to the symmetric stretching mode of phosphodiesteric bonds in nucleic acids, whereas the band at 1230 cm^{-1} originated from the asymmetric stretching of the same bonds. The occurrence of apoptosis defined an important decrease of the intensity of the region assigned to nucleic acids, compared to that of the amide bands. In particular, the calculated nucleic acids/amide II area ratio, which was equal to 1.07 for untreated cells, was massively reduced to 0.77, 0.65, and 0.45 in cells directly placed in contact with SiO_2 powder, SiO_2-CGA, 5 wt%, and SiO_2-CGA, 15 wt%, respectively. Obtained data were consistent with those reported by Gasparri and Muzio [33] who assessed that the area between 1000 and 1140 cm^{-1}, relative to nucleic acids, was the one with the most prominent differences and was, therefore, indicated as a marker of apoptosis. The second detected difference was the increase of the ratio between the area of amide I and amide II peaks. The increase could be due to a cleavage of cellular proteins by different caspases, to the modulation of chaperone activity and of proteasome function or to cytoplasmic acidification, processes that occur during the whole process of apoptosis.

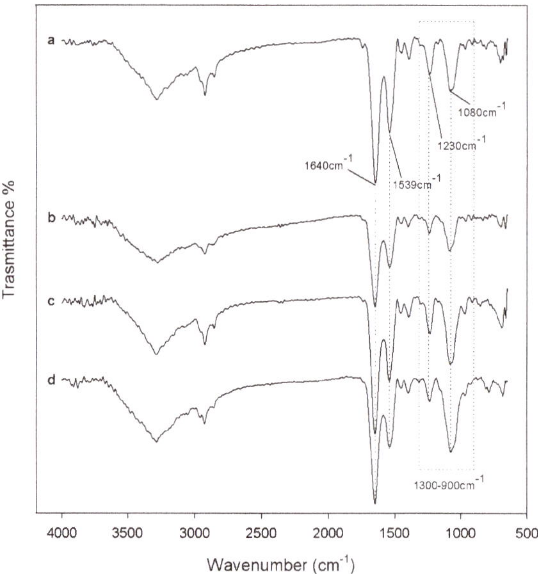

Figure 9. Representative ATR-FTIR spectra of cell suspensions (**a**) untreated or previously treated with (**b**) pure SiO_2; (**c**) SiO_2/CGA, 5 wt%; and (**d**) SiO_2/CGA, 15 wt%.

3. Materials and Methods

3.1. Sol–gel Synthesis

The organic-inorganic hybrids materials, consisting of a SiO_2 inorganic matrix and different content (5 wt%, 10 wt%, 15 wt%, and 20 wt%) of organic chlorogenic acid (CGA), were synthesized by means of a sol–gel route. A solution of tetraethyl orthosilicate (TEOS, reagent grade, 98%, Sigma Aldrich, Milan, Italy) was used as precursor of the SiO_2 inorganic matrix. Water was added drop by drop to a solution of TEOS and nitric acid (solution 65%, Sigma Aldrich, Milan, Italy) in ethanol 99% (Sigma Aldrich, Milan, Italy) under stirring. The molar ratio among the reagents in the obtained solution are: EtOH/TEOS = 6, TEOS/HNO_3 = 1.7, H_2O/TEOS = 2. After 20 min under stirring, a solution of CGA in pure ethanol was added drop by drop to the prepared TEOS solution under stirring. After 20 min the stirrer was stopped and the prepared solutions were left to gel at room temperature. The gels, then, were put into an oven at 40 °C to allow the removal of the solvent residue avoiding the thermal degradation of the drug. A flowchart of the sol–gel process is reported in Figure 10.

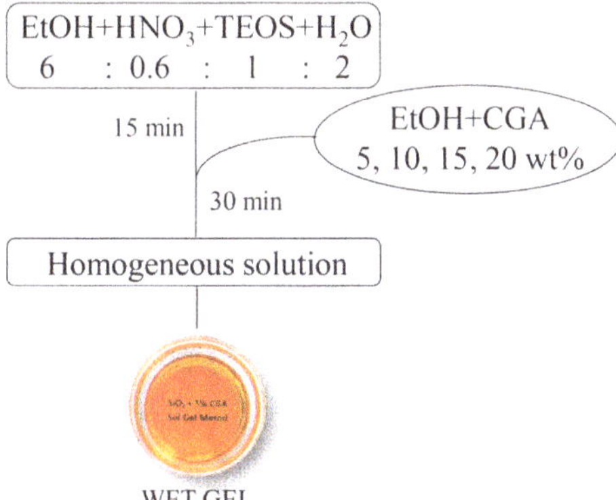

Figure 10. Flowchart of the sol–gel process used to synthesize the SiO_2-CGA hybrids.

3.2. Materials Characterization

The chemical structure of the synthetized materials was investigated by FTIR. A Prestige 21 (Shimadzu, Tokyo, Japan) system, equipped with a DTGS KBr (deuterated tryglycine sulphate with potassium bromide windows) detector allowed us to record the transmittance spectra in the 400–4000 cm^{-1} region, with resolution of 2 cm^{-1} (45 scans). 2 mg of sample powder, mixed with 198 mg of KBr, was then compacted into discs under a pressure of 7 t by using a hydraulic press (Specac, Ltd., Orpington, UK). The FTIR spectra were elaborated by Prestige software (1.30 IRSolution, Shimadzu, Tokyo, Japan).

The UV–VIS spectra of extracts form hybrid materials were also recorded. To this purpose, 1.0 mg of powder of each investigated material underwent ultrasound assisted maceration (Advantage Plus model ES, Darmstadt, Germany) for 1 h using distilled water (2.0 mL) as extracting solvent. Aqueous extracts were then centrifuged at 4500 rpm for 5 min. Supernatants were collected and their spectra acquired in the range 200–500 nm using a UV-1700 spectrophotometer from Shimadzu (Kyoto, Japan).

3.3. Bioactivity Test

The bioactivity of the synthesized materials was investigated by evaluating their ability of inducing the hydroxyapatite nucleation when soaked in a simulated body fluid (SBF) solution. SBF is a solution with ion concentrations nearly equal to those in human blood plasma [28]. This was prepared by dissolving NaCl, $NaHCO_3$, KCl, $MgCl_2·6H_2O$, $CaCl_2$, Na_2HPO_4, and Na_2SO_4

(Sigma-Aldrich, Milan, Italy) in ultra-pure water buffered at pH 7.4 using 4-(2-hydroxyethyl) piperazine-1-ethanesulfonic acid hemi-sodium salt (HEPES, Sigma-Aldrich).

The dried gels were grinded in a mortar to obtain powders. A part of those powders were pressed by a hydraulic press (Specac, Orpington, UK) to obtain disks with a diameter of 13 mm and a thickness of 2 mm.

The disks and the powders were soaked in SBF within polystyrene bottles, which were placed in a water bath at 37.0 ± 0.5 °C. Taking into account that the ratio of the exposed surface to the volume solution influences the reaction, a constant ratio was maintained as reported in literature [26,28]. The solution was replaced every two days to avoid depletion of the ionic species in the SBF due to the nucleation of biominerals on the samples. After 7, 14, and 21 days of exposure to the SBF solution, the samples were gently rinsed and dried in a glass desiccator. The ability of forming an apatite layer on the surface of both powders and disks after each time of exposure to SBF was investigated by FTIR analysis and using a Quanta 200 SEM (FEI, Eindhoven, The Netherlands), equipped with energy-dispersive X-ray (EDX), respectively.

3.4. Determination of DPPH Scavenging Capacity

In order to estimate the DPPH$^{\bullet}$ (2,2-diphenyl-1-picrylhydrazyl) scavenging capability, investigated matrices (1.0, 2.0, and 5.0 mg) were directly added to a DPPH$^{\bullet}$ methanol solution (9.4×10^{-5} M; 1.0 mL final volume) at room temperature. After 30 min, the absorption at 515 nm was measured in reference to a blank by a Shimadzu UV-1700 spectrophotometer. The results were expressed in terms of the percentage decrease of the initial DPPH$^{\bullet}$ radical absorption by the test samples [14–16].

3.5. Determination of ABTS$^{\bullet+}$ Scavenging Capacity

The determination of ABTS$^{\bullet+}$ solution scavenging capacity was estimated as previously reported [14–16], with slight modifications. The ABTS radical cation was generated by reacting ABTS (2,2′-azinobis-(3-ethylbenzothiazolin-6-sulfonic acid); 7.0 mM) and potassium persulfate (2.45 mM). The mixture was allowed to stand in the dark at room temperature for 12–16 h. Thus, the ABTS$^{\bullet+}$ solution was diluted with PBS (pH 7.4) in order to reach an absorbance of 0.70 at 734 nm. Powders of investigated materials (1.0, 2.0, and 5.0 mg) were directly added to the diluted ABTS$^{\bullet+}$ solution (1.0 mL final volume). After 6 min of incubation, the absorption at 734 nm was measured by a Shimadzu UV-1700 spectrophotometer in reference to a blank. The results were expressed in terms of the percentage decrease of the initial ABTS$^{\bullet+}$ absorption by the test samples.

3.6. Cell Culture and Cytotoxicity Assessment

Cytotoxicity was measured via the MTT (3-(4,5-dimethyl-2-thiazolyl)-2,5-diphenyl-2H-tetrazolium bromide) cell viability assay using the murine fibroblast NIH-3T3 cell line, human keratinocyte HaCaT, and neuroblastoma SH-SY5Y cell lines. All the cell lines were grown in Dulbecco's Modified Eagle Medium supplemented with 10% fetal bovine serum, 50.0 U/mL penicillin, and 100.0 µg/mL streptomycin, at 37 °C in a humidified atmosphere containing 5% CO_2. Powders of each synthesized material (1.0 and 2.0 mg) were placed in 24-well plates, and the cells were seeded (1.0×10^5 cells/well). After 48 h of incubation, cells were treated with MTT (500 µL; 0.50 mg/mL), previously dissolved in culture media, for 2 h at 37 °C in a 5% CO_2 humidified atmosphere. The MTT solution was then removed and DMSO was added to dissolve the originated formazan. Finally, the absorbance at 570 nm of each well was determined using a VictorIII Perkin Elmer fluorescence and absorbance reader.

Cell viability was expressed as a percentage of mitochondrial redox activity of the cells directly exposed to powders compared to an unexposed control [15]. Tests were carried out by performing six replicate ($n = 6$) measurements for three samples of each extract (in total: 6×3 measurements).

3.7. ATR-FTIR Analysis

Treated and untreated cells underwent spectroscopic analysis by ATR-FTIR [33]. Since the degree of hydration influences the spectral characteristics of the main cellular components, cells were in the form of an anhydrous bio-film favoring the recording of the absorption spectra in the 700–4000 cm^{-1} region. The SH-SY5Y cells, seeded at a density of 1.5×10^6 cells/well, were treated with SiO$_2$-CGA, 5 wt% and SiO$_2$-CGA, 15 wt% (2.0 mg). After 72 h exposure, cells were harvested by centrifugation at $200 \times g$ and 4 °C and washed ($2\times$) with an aqueous solution of sterile 0.9% NaCl. The pellet was well suspended in NaCl solution (0.9%, 300 μL) and then placed on a glass slide. After 12 h of drying, a cell smear was analyzed by ATR (attenuated total reflectance, AIM-8800, Shimadzu, Tokyo, Japan) by setting the parameters of observation to 65 scans and 8 cm^{-1} resolutions. The spectra were then processed with the software IRsolution (1.30 Version, Shimadzu, Tokyo, Japan).

4. Conclusions

The possibility of combining the versatility of sol–gel technology with the important biological properties of natural phenols, well known for their antioxidant efficacy, represents a new challenge. Herein, chlorogenic acid, broadly distributed in the plant kingdom and mainly abundant in coffee beans and in some organs of plants belonging to Asteraceae family, was embedded into a silica matrix providing new bioactive and antioxidant materials. The biocompatibility of the synthetized materials seemed to strongly depend on the amount of entrapped phenol. Cell-specific effects were observed. In particular, when hybrids with the highest dose of chlorogenic acid were tested towards tumorigenic cells, cytotoxicity was more pronounced. Data from ATR analysis suggested the occurrence of apoptosis. These preliminary acquired data lay the foundation to further analysis aimed at deeply investigating the suitable target of use of these new organic-inorganic hybrids, mainly based on the amount of the phenol therein, for a their safe and effective future application.

Author Contributions: M. Catauro and S. Pacifico conceived and designed the experiments. Moreover, M. Catauro performed the synthesis of the materials, the FTIR, the bioactivity test and the ATR-FTIR; S. Pacifico performed the UV-Visible, tested the DPPH and ABTS$^{\bullet+}$ Scavenging Capacity, and performed MTT assasy. Both Authors contribute to analyze the data and to write the report.

Conflicts of Interest: The authors declare no conflict of interest.

References

1. Piccolella, S.; Pacifico, S. Chapter five—Plant-derived polyphenols: A chemopreventive and chemoprotectant worth-exploring resource in toxicology. In *Advances in Molecular Toxicology*; James, C.F., Jacqueline, M.H., Eds.; Elsevier: Amsterdam, The Netherlands, 2015; Volume 9, pp. 161–214.

2. Galasso, S.; Pacifico, S.; Kretschmer, N.; Pan, S.-P.; Marciano, S.; Piccolella, S.; Monaco, P.; Bauer, R. Influence of seasonal variation on Thymus longicaulis C. Presl chemical composition and its antioxidant and anti-inflammatory properties. *Phytochemistry* **2014**, *107*, 80–90. [CrossRef] [PubMed]

3. Russo, G.L.; Tedesco, I.; Spagnuolo, C.; Russo, M. Antioxidant polyphenols in cancer treatment: Friend, foe or foil? *Semin. Cancer Biol.* **2017**. [CrossRef] [PubMed]

4. Li, W.; Guo, Y.; Zhang, C.; Wu, R.; Yang, A.Y.; Gaspar, J.; Kong, A.-N.T. Dietary phytochemicals and cancer chemoprevention: A perspective on oxidative stress, inflammation, and epigenetics. *Chem. Res. Toxicol.* **2016**, *29*, 2071–2095. [CrossRef] [PubMed]

5. Tao, L.; Park, J.-Y.; Lambert, J.D. Differential prooxidative effects of the green tea polyphenol, (−)-epigallocatechin-3-gallate, in normal and oral cancer cells are related to differences in sirtuin 3 signaling. *Mol. Nutr. Food Res.* **2015**, *59*, 203–211. [CrossRef] [PubMed]

6. Santana-Gálvez, J.; Cisneros-Zevallos, L.; Jacobo-Velázquez, D. Chlorogenic acid: Recent advances on its dual role as a food additive and a nutraceutical against metabolic syndrome. *Molecules* **2017**, *22*, 358. [CrossRef] [PubMed]

7. Ali, N.; Rashid, S.; Nafees, S.; Hasan, S.K.; Shahid, A.; Majed, F.; Sultana, S. Protective effect of chlorogenic acid against methotrexate induced oxidative stress, inflammation and apoptosis in rat liver: An experimental approach. *Chem.-Biol. Interact.* **2017**, *272*, 80–91. [CrossRef] [PubMed]

8. Nallamuthu, I.; Devi, A.; Khanum, F. Chlorogenic acid loaded chitosan nanoparticles with sustained release property, retained antioxidant activity and enhanced bioavailability. *Asian J. Pharm. Sci.* **2015**, *10*, 203–211. [CrossRef]

9. Fu, S.; Wu, C.; Wu, T.; Yu, H.; Yang, S.; Hu, Y. Preparation and characterisation of chlorogenic acid-gelatin: A type of biologically active film for coating preservation. *Food Chem.* **2017**, *221*, 657–663. [CrossRef] [PubMed]

10. Salimi, A.; Hallaj, R. Adsorption and reactivity of chlorogenic acid at a hydrophobic carbon ceramic composite electrode: Application for the amperometric detection of hydrazine. *Electroanalysis* **2004**, *16*, 1964–1971. [CrossRef]

11. Catauro, M.; Bollino, F.; Mozzati, M.C.; Ferrara, C.; Mustarelli, P. Structure and magnetic properties of SiO_2/PCL novel sol-gel organic-inorganic hybrid materials. *J. Solid State Chem.* **2013**, *203*, 92–99. [CrossRef]

12. Catauro, M.; Bollino, F.; Papale, F.; Leonelli, C. Influence of the drying treatment on the performance of V-Nb mixed oxides catalysts synthesised via sol-gel. *J. Non-Cryst. Solids* **2013**, *380*, 1–5. [CrossRef]

13. Catauro, M.; Papale, F.; Bollino, F. Characterization and biological properties of TiO_2/PCL hybrid layers prepared via sol-gel dip coating for surface modification of titanium implants. *J. Non-Cryst. Solids* **2015**, *415*, 9–15. [CrossRef]

14. Catauro, M.; Bollino, F.; Nocera, P.; Piccolella, S.; Pacifico, S. Entrapping quercetin in silica/polyethylene glycol hybrid materials: Chemical characterization and biocompatibility. *Mater. Sci. Eng. C* **2016**, *68*, 205–212. [CrossRef] [PubMed]

15. Catauro, M.; Bollino, F.; Papale, F.; Piccolella, S.; Pacifico, S. Sol-gel synthesis and characterization of SiO_2/PCL hybrid materials containing quercetin as new materials for antioxidant implants. *Mater. Sci. Eng. C* **2016**, *58*, 945–952. [CrossRef] [PubMed]

16. Catauro, M.; Papale, F.; Bollino, F.; Piccolella, S.; Marciano, S.; Nocera, P.; Pacifico, S. Silica/quercetin sol-gel hybrids as antioxidant dental implant materials. *Sci. Technol. Adv. Mater.* **2015**, *16*, 035001. [CrossRef] [PubMed]

17. Boukamp, P.; Petrussevska, R.T.; Breitkreutz, D.; Hornung, J.; Markham, A.; Fusenig, N.E. Normal keratinization in a spontaneously immortalized aneuploid human keratinocyte cell line. *J. Cell Biol.* **1988**, *106*, 761–771. [CrossRef] [PubMed]

18. Adeogun, M.J.; Hay, J.N. Structure control in sol-gel silica synthesis using ionene polymers. 2: Evidence from spectroscopic analysis. *J. Sol-Gel Sci. Technol.* **2001**, *20*, 119–128. [CrossRef]

19. Innocenzi, P. Infrared spectroscopy of sol-gel derived silica-based films: A spectra-microstructure overview. *J. Non-Cryst. Solids* **2003**, *316*, 309–319. [CrossRef]

20. Catauro, M.; Bollino, F.; Papale, F.; Ferrara, C.; Mustarelli, P. Silica-polyethylene glycol hybrids synthesized by sol-gel: Biocompatibility improvement of titanium implants by coating. *Mater. Sci. Eng. C* **2015**, *55*, 118–125. [CrossRef] [PubMed]

21. Han, S.; Hou, W.; Zhang, C.; Sun, D.; Huang, X.; Gouting Wang, A. Structure and the point of zero charge of magnesium aluminium hydroxide. *J. Chem. Soc. Faraday Trans.* **1998**, *94*, 915–918. [CrossRef]

22. Nedelec, J.M.; Hench, L.L. Ab initio molecular orbital calculations on silica rings. *J. Non-Cryst. Solids* **1999**, *255*, 163–170. [CrossRef]

23. Mishra, S.; Tandon, P.; Eravuchira, P.J.; El-Abassy, R.M.; Materny, A. Vibrational spectroscopy and density functional theory analysis of 3-O-caffeoylquinic acid. *Spectrochim. Acta Part A Mol. Biomol. Spectrosc.* **2013**, *104*, 358–367. [CrossRef] [PubMed]

24. Bajko, E.; Kalinowska, M.; Borowski, P.; Siergiejczyk, L.; Lewandowski, W. 5-O-caffeoylquinic acid: A spectroscopic study and biological screening for antimicrobial activity. *LWT Food Sci. Technol.* **2016**, *65*, 471–479. [CrossRef]

25. Angélica Alvarez Lemus, M.; Castañeda, O.J.O.; Hernández Pérez, A.D.; González, R.L. An alcohol-free SiO_2 sol-gel matrix functionalized with acetic acid as drug reservoir for the controlled release of pentoxifylline. *J. Nanomater.* **2014**, *2014*, 108. [CrossRef]

26. Catauro, M.; Bollino, F.; Renella, R.A.; Papale, F. Sol-gel synthesis of SiO_2–CaO–P_2O_5 glasses: Influence of the heat treatment on their bioactivity and biocompatibility. *Ceram. Int.* **2015**, *41*, 12578–12588. [CrossRef]

27. Catauro, M.; Renella, R.A.; Papale, F.; Vecchio Ciprioti, S. Investigation of bioactivity, biocompatibility and thermal behavior of sol-gel silica glass containing a high peg percentage. *Mater. Sci. Eng. C* **2016**, *61*, 51–55. [CrossRef] [PubMed]

28. Kokubo, T.; Takadama, H. How useful is sbf in predicting in vivo bone bioactivity? *Biomaterials* **2006**, *27*, 2907–2915. [CrossRef] [PubMed]

29. Jiang, X.-W.; Bai, J.-P.; Zhang, Q.; Hu, X.-L.; Tian, X.; Zhu, J.; Liu, J.; Meng, W.-H.; Zhao, Q.-C. Caffeoylquinic acid derivatives from the roots of *Arctium lappa* L. (burdock) and their structure–activity relationships (sars) of free radical scavenging activities. *Phytochem. Lett.* **2016**, *15*, 159–163. [CrossRef]

30. Messer, R.L.W.; Lockwood, P.E.; Wataha, J.C.; Lewis, J.B.; Norris, S.; Bouillaguet, S. In vitro cytotoxicity of traditional versus contemporary dental ceramics. *J. Prosthet. Dent.* **2003**, *90*, 452–458. [CrossRef]

31. Xue, N.; Zhou, Q.; Ji, M.; Jin, J.; Lai, F.; Chen, J.; Zhang, M.; Jia, J.; Yang, H.; Zhang, J.; et al. Chlorogenic acid inhibits glioblastoma growth through repolarizating macrophage from M2 to M1 phenotype. *Sci. Rep.* **2017**, *7*, 39011. [CrossRef] [PubMed]

32. Martin, K.R.; Appel, C.L. Polyphenols as dietary supplements: A double-edged sword. *Nutr. Diet. Suppl.* **2010**, *2*, 1–12. [CrossRef]

33. Gasparri, F.; Muzio, M. Monitoring of apoptosis of hl60 cells by fourier-transform infrared spectroscopy. *Biochem. J.* **2003**, *369*, 239–248. [CrossRef] [PubMed]

34. Goormaghtigh, E.; Cabiaux, V.; Ruysschaert, J.-M. Determination of soluble and membrane protein structure by fourier transform infrared spectroscopy. In *Physicochemical Methods in the Study of Biomembranes*; Hilderson, H.J., Ralston, G.B., Eds.; Springer: Boston, MA, USA, 1994; pp. 405–450.

Article

The Influence of the Polymer Amount on the Biological Properties of PCL/ZrO$_2$ Hybrid Materials Synthesized via Sol-Gel Technique

Michelina Catauro [1,*], Elisabetta Tranquillo [1,2], Michela Illiano [2], Luigi Sapio [2], Annamaria Spina [2] and Silvio Naviglio [2]

[1] Department of Industrial and Information Engineering, University of Campania "Luigi Vanvitelli", Via Roma 29, 81031 Aversa, Italy; elisabetta.tranquillo@unicampania.it

[2] Department of Biochemistry, Biophysics and General Pathology, Medical School, University of Campania "Luigi Vanvitelli", Via L. De Crecchio 7, 80138 Naples, Italy; Michela.Illiano@unicampania.it (M.I.); luigi.sapio@unicampania.it (L.S.); annamaria.spina@unicampania.it (A.S.); silvio.naviglio@unicampania.it (S.N.)

*   Correspondence: michelina.catauro@unicampania.it; Tel.: +39-081-501-0360

Received: 21 August 2017; Accepted: 13 October 2017; Published: 17 October 2017

Abstract: Organic/inorganic hybrid materials are attracting considerable attention in the biomedical area. The sol-gel process provides a convenient way to produce many bioactive organic–inorganic hybrids. Among those, poly(e-caprolactone)/zirconia (PCL/ZrO$_2$) hybrids have proved to be bioactive with no toxic materials. The aim of this study was to investigate the effects of these materials on the cellular response as a function of the PCL content, in order to evaluate their potential use in the biomedical field. For this purpose, PCL/ZrO$_2$ hybrids containing 6, 12, 24, and 50 wt % of PCL were synthesized by the sol-gel method. The effects of their presence on the NIH-3T3 fibroblast cell line carrying out direct cell number counting, MTT, cell damage assays, flow cytometry-based analysis of cell-cycle progression, and immunoblotting experiments. The results confirm and extend the findings that PCL/ZrO$_2$ hybrids are free from toxicity. The hybrids containing 12 and 24 wt % PCL, (more than 6 and 50 wt % ones) enhance cell proliferation when compared to pure ZrO$_2$ by affecting cell cycle progression. The finding that the content of PCL in PCL/ZrO$_2$ hybrids differently supports cell proliferation suggests that PCL/ZrO$_2$ hybrids could be useful tools with different potential clinical applications.

Keywords: sol-gel technique; biomaterials; cell proliferation; cell cycle

---

1. Introduction

Organic/inorganic hybrid materials are attracting considerable attention today. They are biphasic systems in which the organic and inorganic components are connected on a nanometer scale. A classification of the hybrid materials based on interactions between the phases was proposed by Judenstein and Sanchez [1]. They defined these materials as first and second class hybrids. In particular, Class I consists of organic and inorganic compounds bonded through hydrogen, van der Waals, or ionic bonds, whereas in Class II, the organic and inorganic phases are linked through strong chemical bonds (covalent or polar covalent bonds). Many organic/inorganic hybrid materials can be developed using the sol-gel method. The sol-gel chemistry is based on the hydrolysis and polycondensation reactions of the precursor metal alkoxides M(OR)x, where M represents a network-forming element (such as Si, Sn, Zr, Ti, Al), and R is usually an alkyl group. The sol-gel reactions are dependent on many parameters, such as structure and concentration of the reactants, solvents, and catalysts, as well as reaction temperature and rate of removal of by-products and solvents [2,3]. The sol-gel process provides a convenient way to produce porous [4], biocompatible, and bioactive materials [5]. The last property is due to the presence of –OH groups on the material surface, which can promote apatite layer formation on the biomaterial surface [6]. Therefore, sol-gel materials may be easily osseointegrated when implanted in vivo [7,8]. This is a property required for a material to be used in the dental and orthopedic fields. Moreover, the sol-gel method shows several advantages over most traditional syntheses. It gives homogeneous materials and leads to a fine control of the chemical composition. The low processing temperature allows one to add thermolabile molecules, such as polymers, drugs, and/or biomolecules, to the glassy matrix [9], obtaining organic/inorganic hybrids. The main idea, when a hybrid material is developed, is to overcome the disadvantages of each of the components and to retain their advantages. Among the synthetic polymers, poly(e-caprolactone) (PCL), which is a biodegradable aliphatic polyester [10–12], has already been proposed for many biomedical applications, such as drug delivery and tissue engineering [12–21]. However, PCL is too flexible and weak to satisfy the mechanical requirements for this specific application, while polymer-based nanocomposite materials provide an alternative choice to overcome these problems [22–25]. On the other hand, it is known that glass matrices, such as zirconia, are bioactive and biocompatible [26,27], but they cannot be used in some applications due to their poor mechanical properties.

Elsewhere, Catauro et al. proposed the use of organic/inorganic hybrid materials based on zirconia (ZrO_2), with different percentages of PCL embedded, for biomedical applications [28]. These materials were extensively characterized [29]. The presence of hydrogen bonds between the carboxylic groups of the polymer and the hydroxyl groups of the inorganic matrix was proved by Fourier transform infrared (FTIR) analysis, and is strongly supported by solid-state NMR [29]. SEM analysis confirmed that they can be considered hybrid materials and no appreciable difference between their morphologies was observed [29]. Furthermore, the materials are nanocomposites, as shown by AFM analysis [30]. XRD diffractograms showed that ZrO_2/PCL gel exhibits broad humps that are characteristic of amorphous materials, even though the PCL diffractogram showed the sharp peaks typical of a crystalline material [31].

PCL/ZrO_2 hybrids were proposed as matrices for the controlled release of drugs [30], or as coatings able to improve the biological properties of bio-inert titanium implants or to release drugs in a controlled manner [28,32]. The obtained films, indeed, appeared to be more bioactive and biocompatible than uncoated titanium. The coatings induced the formation of hydroxyapatite when soaked in SBF, did not show cytotoxicity and were also supportive of cell proliferation at all compositions. In order to evaluate the potential use of such materials as biomaterials, further knowledge about their biological properties is needed. Therefore, the aim of the present study has been to investigate, more in depth, the effects of the presence of PCL/ZrO_2 materials as a function of the PCL content on the NIH-3T3 fibroblast cellular response and the underlying mechanisms. Therefore, five PCL/ZrO_2 hybrid samples containing different amount of PCL (6, 12, 24, and 50 wt %) have been synthesized via the sol-gel route as reported elsewhere [29] and the effects of their presence on cell viability and cell cycle progression, as well as membrane integrity have been evaluated. All of the results are reported as a function of PCL content.

2. Results and Discussion

2.1. PCL/ZrO₂ Hybrid Materials Do Not Have Cytotoxic Effects

The mouse NIH-3T3 cell line is a well-established model system of fibroblasts, largely used to study cellular responses, including those induced by biomaterials [33–38].

In order to evaluate the cytotoxicity of the bioactive PCL/ZrO_2 hybrid materials (synthesized as shown in the Materials and Methods section) as a function of PCL content, NIH-3T3 fibroblasts were grown in the absence and presence of the pure ZrO_2 and PCL/ZrO_2 hybrid materials, each containing different percentages (6, 12, 24, 50 wt %) of PCL, for up to 72 h. Over the time course, the plates were observed daily under phase-contrast microscopy and relative pictures were taken (Figure 1). Notably, in Figure 1 it is shown that the cells grown in the presence of our preparations appeared not to suffer and shared a morphology similar to that of control untreated cells. In addition, it is also observed that the 12% and 24% PCL/ZrO$_2$ hybrids treated plates contain more cells as compared to plates treated with PCL-free ZrO$_2$, 6% and 50% PCL/ZrO$_2$ hybrids, with a cell density similar to that of control untreated plate.

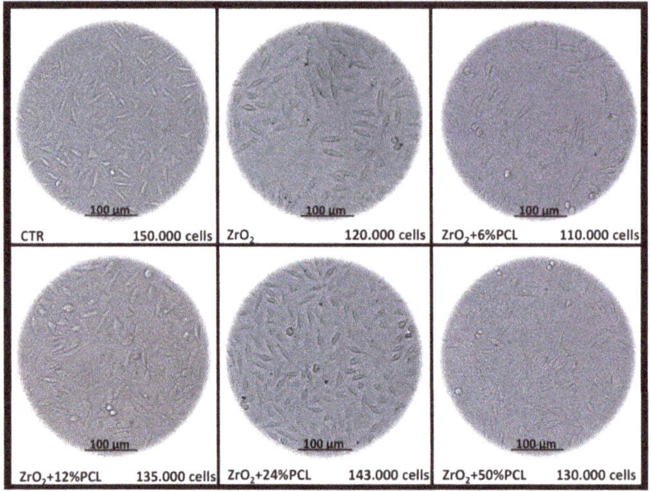

Figure 1. Typical phase-contrast images of NIH 3T3 cells treated with and without PCL-free ZrO$_2$ material and poly(e-caprolactone) (PCL)/ZrO$_2$ hybrids with different percentages of PCL (6 wt %, 12 wt %, 24 wt %, 50 wt %) for 48 h.

Then, cells were detached from monolayers by brief exposure to a solution of trypsin and a Trypan Blue dye exclusion assay was performed (Figure 2). Interestingly, Figure 2 shows that the number of Trypan Blue positive dead cells increased in the presence of pure ZrO$_2$ ($\approx$15%) as compared to control untreated cells, with less evident variations in the presence of PCL/ZrO$_2$ hybrid materials.

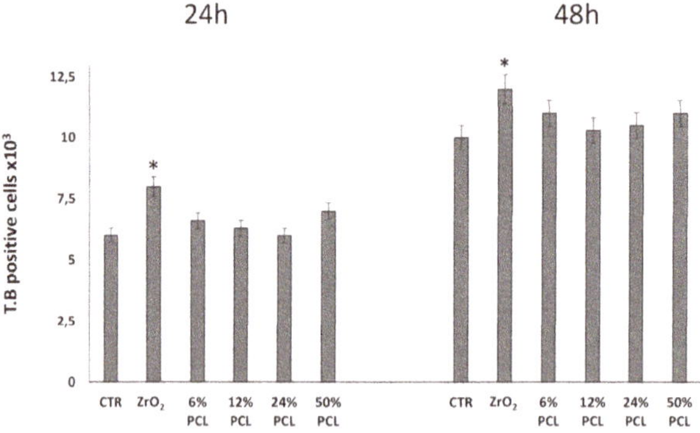

Figure 2. Trypan Blue Exclusion Dye test. NIH-3T3 cells were cultured in 10% serum containing medium in the absence (control, CTR) or presence of PCL-free ZrO$_2$ and PCL/ZrO$_2$ hybrid materials with different percentages (6 wt %, 12 wt %, 24 wt %, 50 wt %) for 24 and 48 h. Then, the cell number was recorded as stated in Materials and Methods. Data represent the average of three independent experiments. The means and S.D. are shown. *, $p < 0.05$ vs. control cells.

Finally, we recovered the culturing medium in which NIH-3T3 fibroblasts were grown, and tested it, by using the Abbott Lab Chemistry Analyzer ci 8200 (usually used to test blood samples), looking for the presence of any known intracellular enzymatic activities, such as lactate dehydrogenase, (LDH), aspartate aminotransferase (AST), and alanine aminotransferase (ALT) [39,40], to evaluate their release upon possible cell damage in response to the materials (Table 1). The results, reported in Table 1, show a clear increase of both LDH and AST activities in media from NIH-3T3 fibroblasts treated with the cytotoxic agent doxorubicin (DOXO) (1 µM) (used as positive control) [41] as compared to the untreated ones. Moreover, media from plates without cells were tested as a negative control. In addition, total proteins and ALT transaminase were measured as further internal controls [40]. As expected, according to the absence of ALT activity in fibroblasts [42,43], no increase of such activity is seen in any samples.

Table 1. Effects of pure ZrO_2 material, PCL/ZrO_2 hybrid materials with different percentages (6 wt %, 12 wt %, 24 wt %, 50 wt %) of PCL, and doxorubicin on the release of AST, ALT, LDH intracellular enzymatic activities. NIH-3T3 cells were cultured in the absence (control) or presence of the indicated substances for 72 h. Then, the medium from cultured cells plates and "blank" plates was collected and tested for the indicated parameters. Values are means ± S.D. of triplicate samples of a typical experiment. *, $p < 0.05$ hybrids vs. PCL-free ZrO_2 treated-cells and Doxo vs. control.

Sample Labels	Total Proteins g/dL	AST U/L	ALT U/L	LDH U/L
Blank	1.0 ± 0.05	3 ± 0.1	6 ± 0.2	54 ± 1
Control	1.1 ± 0.04	5 ± 0.1	6 ± 0.1	58 ± 2
ZrO_2	1.1 ± 0.03	6 ± 0.3	6 ± 0.2	61 ± 3
6 wt % PCL	1.0 ± 0.03	5 ± 0.2	6 ± 0.1	64 ± 2 *
12 wt % PCL	1.1 ± 0.02	5 ± 0.3	6 ± 0.3	63 ± 3
24 wt % PCL	1.1 ± 0.05	6 ± 0.2	6 ± 0.4	62 ± 3
50 wt % PCL	1.0 ± 0.02	5 ± 0.1	6 ± 0.1	64 ± 2 *
DOXO	1.1 ± 0.05	23 ± 1 *	6 ± 0.2	130 ± 5 *

Remarkably, Table 1 shows that no increase of AST transaminase activity is detected in medium from NIH-3T3 fibroblasts grown for 72 h in the presence of PCL/ZrO_2 hybrid materials, as compared to the PCL-free ZrO_2, whereas only a very slight increase of LDH activity (from 61 to 64 U/L) is recorded. This results suggests that minimal, if any, cell membrane injury occurs in response to PCL/ZrO_2 hybrid materials.

Overall, the above data clearly indicate that the content of PCL in the PCL/ZrO_2 hybrid materials does not cause cell damage on NIH-3T3 fibroblasts.

2.2. PCL Content in the PCL/ZrO₂ Hybrids Positively Affects the Cell Proliferation

To test the influence of PCL content on cell proliferation, NIH-3T3 fibroblasts were grown for 24 and 48 h in 10% serum containing medium in the absence (control cells) or presence of the following preparations: PCL-free ZrO_2 and PCL/ZrO_2 hybrid materials with different percentages (6 wt %, 12 wt %, 24 wt %, 50 wt %) of PCL. Thereafter, cell proliferation was determined by direct cell number counting and by conventional MTT assay (Figure 3).

Figure 3 shows that the cell population is significantly higher, at both time points, in plates containing 12 wt % and 24 wt % PCL/ZrO_2 hybrid materials, as compared to cell plates cultured in the presence of pure ZrO_2, whereas no significant variations in cell number are evident in the presence of 6 wt % and 50 wt % PCL/ZrO_2 hybrid materials. In agreement with previous findings [34,44], a decrease of cell number and cell viability in response to PCL-free ZrO_2 is evident as compared to control fibroblasts (grown in the absence of materials), according to a growth impairment induced by such material. Recently, indeed, Meesser et al. [44] proved that most ceramics cause a mild suppression of cell functions in vitro. However, the levels of the suppression observed would be acceptable on the basis of standards used to evaluate alloys and composites (<25% suppression of dehydrogenases activity).

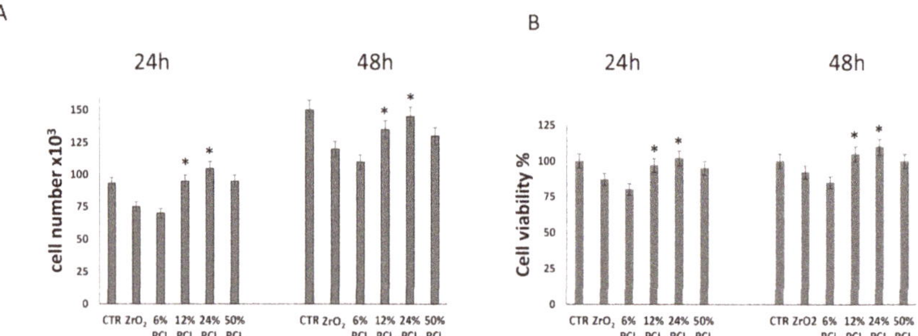

Figure 3. Effects of PCL-free ZrO_2 material, PCL/ZrO_2 hybrid materials with different percentages (6 wt %, 12 wt %, 24 wt %, 50 wt %) of PCL on cell proliferation. NIH-3T3 cells were cultured in 10% serum containing medium in the absence (control, CTR) or presence of the indicated substances for 24 and 48 h. Then, the cell number was recorded (**A**); and cell viability was measured by MTT assay (**B**). Data represent the average of three independent experiments. The means and S.D. are shown. * $p < 0.05$ vs. PCL-free ZrO_2 treated-cells.

Altogether, the above results suggest that the PCL/ZrO_2 hybrids, mainly those containing 12 wt % and 24 wt % PCL, positively affect cell proliferation when compared to the inorganic material in the non-hybrid form. Our data are in agreement with previous similar findings based mainly on cell viability MTT assay [28,34], supporting the evidence that the presence of different PCL amounts favors the preparation of biomaterials with improved biocompatibility. In fact, it is known that its slow degradation kinetics, biocompatibility, and semicrystalline nanofibers can result in stimulation of the extracellular matrix providing a good scaffold for cell proliferation and engineering [25,45].

The slightly lower positive effect observed in the plate containing the system with 6 wt % PCL as compared to those ones containing 12 wt % and 24 wt % suggests that when the polymer is added at a low level, the effect of the ZrO_2 matrix predominates. Probably, the inorganic network masks the polymer that is embedded and linked within it by the formation of –OH bonds. On the other hand, the decrease of the positive effect recorded when the PCL content is higher than 24%, is in agreement with other works in literature [46,47], which ascribed the effect of slight inhibition to the hydrophobic nature of PCL inhibitings cells adhesion. This, in turn, hinders cell proliferation, and thus causes a decrease of cells vitality. However, it has been proven [48] that the addition of hydrophilic SiO_2 to PCL coatings leads to an increase of coating hydrophilicity. Therefore, up to 24 wt % of PCL the presence of zirconia matrix make the synthesized hybrid materials hydrophilic and the negative effect of PCL is observed only for higher polymer amounts. The dependence between cell proliferation and the PCL content can be due to a different degradation rate of the polymer in the different hybrid materials. It is reported in literature that a high PCL degradation rate negatively affects cell proliferation [22–25] and when 50 wt % of polymer is added in the zirconia matrix, part of the PCL cannot forms H-bonds because all hydroxyl groups of the matrix are already involved in H-bonds. Therefore, the polymer can be highly subjected to degradation process. This is in accordance with the results of Vecchio Ciprioti et al. [49], who showed the protective role of the zirconia matrix in ZrO2/PEG hybrid materials against the thermal degradation of the polymer. To increase the specificity and reliability of the above data, we tested the effects of the five formulations on NIH-3T3 cell proliferation, also in a different experimental growth condition, i.e., by culturing fibroblasts in low serum (1%) containing medium for 24 and 48 h. As expected, in all 1% serum samples we found less cells when compared to the 10% serum samples (Figure 4) [50]. Interestingly, Figure 4 also shows that the cell population again is higher in plates containing 12 wt % and 24 wt % PCL/ZrO$_2$ hybrid materials, as compared to cell plates cultured in presence of the PCL-free ZrO$_2$.

Overall, these data confirm that even in low serum conditions, i.e., in adverse growth conditions, no deleterious cytotoxic effect due to PCL content in PCL/ZrO$_2$ hybrid materials could be seen and also that the hybrids with PCL percentages of 12 wt % and 24 wt % result in a growth advantage to NIH-3T3 fibroblasts.

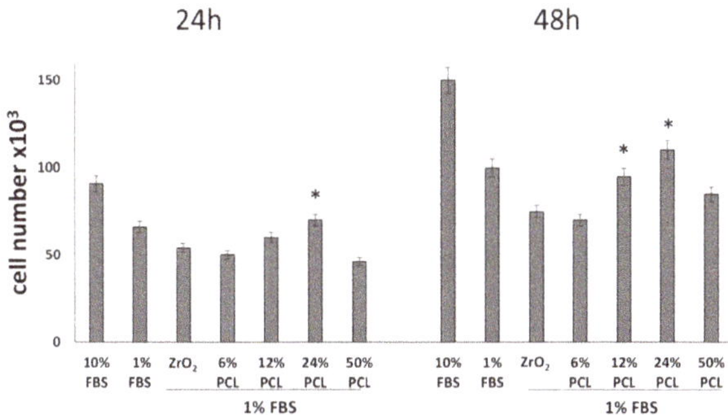

Figure 4. Effects of pure ZrO$_2$, PCL/ZrO$_2$ hybrid materials with different PCL percentages (6 wt %, 12 wt %, 24 wt %, 50 wt %) in low serum (1%) containing medium on cell proliferation. NIH-3T3 cells were cultured in low serum (1%) containing medium in the absence (control) or presence of the indicated substances for 24 and 48 h. Then, the cell number was recorded. Data represent the average of three independent experiments. The means and S.D. are shown. * $p < 0.05$ vs. PCL-free ZrO$_2$ treated-cells.

2.3. PCL Content in the PCL/ZrO$_2$ Hybrids Affects the Cell Cycle Progression

To further explore the influence of PCL content on NIH-3T3 cell proliferation, the distribution of cells in the G0/G1, S, G2/M cell cycle phases was evaluated by flow cytometric analysis of propidium iodide-stained cells. (Figure 5). The sub-G1 population (i.e., the proportion of cells with hypoploid DNA content), characteristic of cells having undergone DNA fragmentation (the biochemical hallmark of apoptotic cell death) was also looked at. Firstly, Figure 5 shows that the percentages of asynchronously growing control NIH-3T3 fibroblasts in the G0/G1, S, and G2/M phases are 49%, 42%, 9%, respectively, in full agreement with previous result [51]. Notably, in agreement with the described weak cytotoxic effect of PCL-free ZrO$_2$ [34,44] (Figure 2), in Figure 5 the appearance of a sub-G1 cell death population (7%) was observed in ZrO$_2$ pure treated cells, as compared to control untreated cells, whereas such a sub-G1 population is either completely absent or only present to a low extent in PCL/ZrO$_2$ hybrid materials, suggesting a protective/growth supporting effect of the PCL component. In addition, interestingly, fibroblasts grown in the presence of 12 wt % PCL/ZrO$_2$ hybrid materials accumulate more in the S phase (62%), whereas fibroblasts grown in the presence of 24 wt % PCL/ZrO$_2$ hybrid materials have a more evident G0/G1 phase (53%) compared to those cultured in the presence of PCL-free ZrO$_2$, suggesting that the PCL content can differently affect cell cycle progression. These results are in accordance with the data obtained via cell counting and the MTT assay. The different modification of the cell cycle progression, indeed, may be attributed to the different concentration of the PCL degradation products in the medium, due to the different PCL degradation rate in the hybrid system [52,53]. This, in turn, can differently affect the pathway(s) involved in the progression of the cell cycle [22–25].

To obtain further insights into the possible effects of PCL/ZrO$_2$ hybrid materials on the cell division cycle of NIH-3T3 cells, we studied the expression of some relevant cell cycle regulating proteins in response to the following preparations: ZrO$_2$ and PCL/ZrO$_2$ hybrid materials with different percentages (6 wt %, 12 wt %, 24 wt %, 50 wt %) of PCL (Figure 5B).

To this aim, NIH-3T3 cells were exposed to the materials for 48 h. The cell extracts were analyzed by Western blotting to examine the levels of cyclin-dependent kinase inhibitor p27 and of cyclin A proteins.

Remarkably, it was noted that cyclin A and p27 protein levels, as well as those of other relevant cell cycle proteins, oscillate depending on cell cycle phases and are strongly involved in cell cycle progression [50,54,55]. In Figure 5B it is shown that the amounts of the positive cell cycle regulator cyclin A and of cell cycle inhibitor p27 increased and decreased, respectively, in response to PCL/ZrO_2 hybrid materials. Moreover, there is evidence that p27 (that is a cell cycle inhibitor) is more abundant in ZrO_2 pure treated cells when compared to control untreated cells and decreases in response to hybrids, whereas cyclin A (that is a positive regulator of cell cycle) increases (Figure 5B). Interestingly, this is in agreement with the growth promoting effect of the hybrid materials, as compared to PCL-free ZrO_2 which, on the contrary, has a growth inhibitory effect. Overall, the above cell cycle and Western blot data suggest that 12 wt % and 24 wt % PCL/ZrO_2 hybrid materials affect the cell cycle progression of NIH-3T3 cells more than the others compositions.

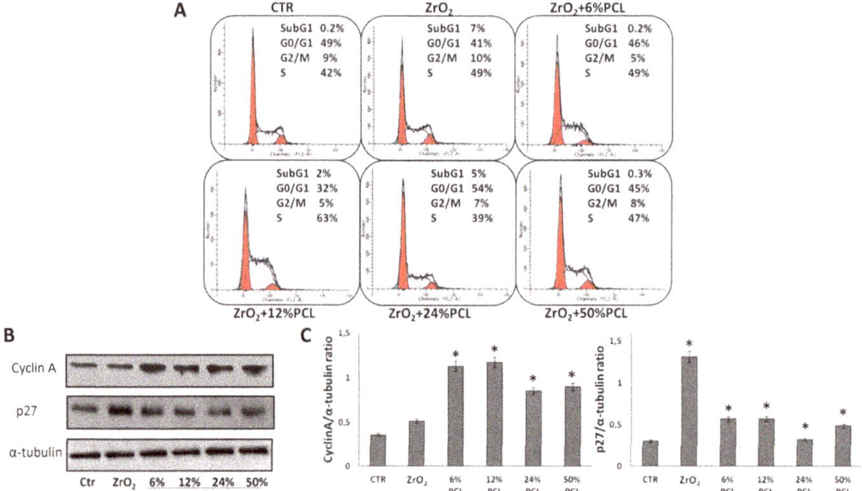

Figure 5. Effects of ZrO_2 material alone and PCL/ZrO_2 hybrid materials with different PCL percentages (6 wt %, 12 wt %, 24 wt %, 50 wt %) on the distribution of NIH-3T3 cells in cell cycle phases and on the levels of some relevant cell cycle regulating proteins. NIH-3T3 cells were cultured in the absence (control, CTR) or presence of the indicated substances for 48 h. (**A**) FACS analysis of propidium iodide stained cells. Profiles and quantitative data indicating the percentage of G0/G1, S, and G2/M cells from a typical representative experiment; (**B**) 30 µg of cell extracts were subjected to SDS-PAGE and blotted with antibodies against the indicated proteins (α-tubulin was used as a standard for the equal loading of protein in the lanes). The image is representative of two immunoblotting analyseis from two different cellular preparations with similar results; (**C**) Graphs showing the densitometric intensity of Cyc A and p27/tubulin bands ratio. The intensities of signals were expressed as arbitrary units. * $p < 0.05$ vs. control untreated cells.

3. Materials and Methods

3.1. Sol-Gel Synthesis

The organic/inorganic PCL/ZrO_2 hybrids materials, containing 0, 6, 12, 24, and 50 wt % of the organic component, were synthesized by the sol-gel process as elsewhere [29]. A zirconium(IV) propoxide solution ($Zr(OC_3H_7)_4$ 70 wt % in n-propanol, Sigma Aldrich) and PCL (Mn = 10,000) were used as precursors of the inorganic and organic phases, respectively. A solution of PCL in chloroform was added to the solution of $Zr(OC_3H_7)_4$ in an ethanol–acetylacetone–water mixture. Acetylacetone was used to control the hydrolytic activity of zirconium alkoxide. The solution was stirred by a magnetic stirrer until the resulting sols were uniform and homogeneous. The molar ratios between the reagent achieved were: $CH_3CH_2OH/Zr(OCH_2CH_2CH_3)_4 = 5.7$; $Zr(OCH_2CH_2CH_3)_4/AcAc = 4.5$.

After gelation, all of the wet gels were air dried at 45 °C for 48 h to remove residual solvents without any polymer degradation. As indicated in the text, Figure 6 shows the flow chart of hybrid synthesis by the sol-gel method.

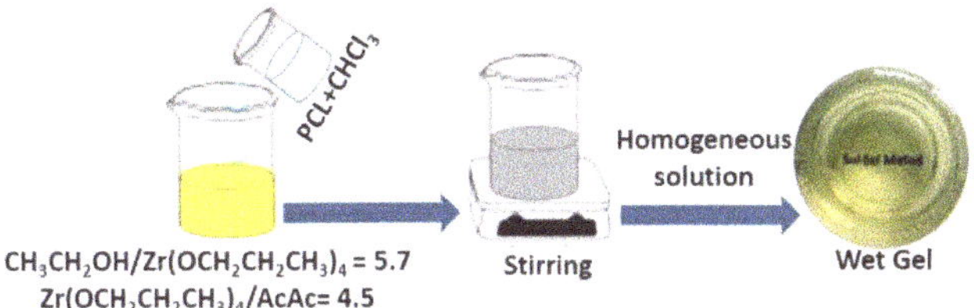

Figure 6. Flow chart of hybrids synthesis by the sol-gel method and the molar ratios between the reagents achieved in the sol.

3.2. Cell Culture and Treatments

The mouse fibroblast NIH-3T3 cell line was obtained from the American Type Culture Collection (Rockville, MD, USA). NIH-3T3 cells were grown in Dulbecco's Modified Eagle's Medium (DMEM) supplemented with 2 mM glutamine, 100 U/mL penicillin, 100 mg/mL streptomycin, and 10% fetal bovine serum (FBS), and cultured at 37 °C in a 5% CO_2 humidified atmosphere. Typically, cells were split (5×10^5/10 cm plate, 6×10^4/6-well plate) and grown in 10% serum containing medium. After 24 h, the medium was removed, the cells were washed with PBS and incubated with 10% (in some experiments 1%) FBS fresh medium (Time 0), in the absence (control cells) or in the presence of the organic/inorganic PCL/ZrO_2 hybrid materials, at 0.3 mg/mL concentration, containing 0, 6, 12, 24, and 50 wt % of the organic component PCL and grown for the times indicated in the text and figures. After treatment, both floating cells and adherent cells were recovered and counted (by centrifugation from culture medium and harvested by trypsinization, respectively). In the case of experiments designed for fluorescence-activated cell sorting (FACS) analysis and the evaluation of the level of proteins involved in cell cycle progression, adherent cells were collected, washed twice with ice-cold PBS, and divided into two aliquots (see below).

3.3. Trypan Blue Exclusion Test

The Trypan Blue (dye) exclusion test is used to determine the number of dead cells present in a cell suspension. Briefly, cells were seeded in 10% PBS-containing medium in a 6-well plate at a density of 6×10^4 cell/well for 24 and 48 h at 37 °C and cultured, as above. At the indicated times, after cell collection, cell counting was performed by mixing 10 μL of cell suspension with an equal volume of Trypan Blue (TB; 0.4%, *v/v*). The number of blue staining cells (dead cells) and of viable cells (that exclude the Trypan Blue) was recorded. TB experiments were performed three times (in replicates of six wells for each data point in each experiment). Data are presented as means ± standard deviation for a representative experiment.

3.4. Cell Viability Assay

Cell viability assay was performed, as described previously [41]. Briefly, cells were seeded in 10% PBS-containing medium in a 96-well plate at a density of 3×10^3 cells/well for 24 and 48 h at 37 °C and cultured, as above. Viable cells were determined by the 3-[4,5-dimethylthiazol-2-yl]-2,5-

diphenyltetrazolium bromide (MTT) assay and cell viability was assessed by adding MTT solution in PBS to a final concentration of 0.5 mg/mL. The plates were then incubated at 37 °C for 4 h and the MTT-formazan crystals were solubilized in a solution of 4% 1N isopropanol/hydrochloric acid at 37 °C on a shaking table for 20 min. The absorbance values of the solution in each well were measured at 570 nm using a Bio-Rad 550 microplate reader (Bio-Rad Laboratories, Milan, Italy). Cell viability was expressed as a percentage of absorbance values in treated samples with respect to that of control (100%). MTT experiments were performed three times (in replicates of 6 wells for each data point in each experiment). Data are presented as means ± standard deviation for a representative experiment.

3.5. Evaluation of Cell Cycle Phases by Flow Cytometry

After treatment, the cells were recovered as previously described in "Cell culture and treatments", fixed by resuspension in 70% ice-cold methanol/PBS and incubated overnight at 4 °C. The cells were then centrifuged at 1200 RPM ((revolutions per minute) for 5 min, and the pellets obtained were washed with ice-cold PBS, centrifuged for 5 min, resuspended in 0.5 mL DNA staining solution (50 µg/mL PI and 100 µg RNase A in PBS), and incubated overnight at 37 °C in the dark. The samples were then transferred to 5-mL Falcon tubes and stored oin ice until assayed. Flow cytometry analysis was performed using a FACSCalibur flow cytometer (Becton Dickinson, San Jose, CA, USA) interfaced with a Hewlett-Packard computer (model 310) for data analysis. For the evaluation of intracellular DNA content, at least 20,000 events for each point were analyzed, and regions were set up to acquire quantitative data of cells that fell into the normal G1, S, and G2/M regions [50]. Results were analyzed using ModiFIT Cell Cycle Analysis software (Version 3.0, verity software house, Inc., Topsham, ME, USA).

3.6. Preparation of Cell Lysates

Cell extracts were prepared as previously described [41]. Briefly, after treatments, the cells were recovered as previously described in "Cell culture and treatments"; 3–5 volumes of RIPA buffer (PBS, 1% NP-40, 0.5% sodium deoxycholate, 0.1% SDS) containing 10 µg/mL aprotinin, leupeptin, and 1 mM phenylmethylsulfonyl fluoride (PMSF) were added to the recovered cells. After incubation on ice for 1 h, samples were centrifuged at 18,000 g in an Eppendorf microcentrifuge for 15 min at 4 °C and the supernatant (SDS total extract) was recovered. Some aliquots were taken for protein quantification according to the Bradford method; others were diluted in 4× (to be used at 1:4 dilution) Laemmli buffer, boiled and stored as samples for immunoblotting analysis.

3.7. Western Blot Analysis

Typically, we employed 20–40 µg of total extracts for immunoblotting [56]. Proteins from cell preparations were separated by sodium dodecyl sulphate-polyacrylamide gel electrophoresis (SDS-PAGE) and transferred onto nitrocellulose sheets (Schleicher & Schuell, Dassel, Germany) by a Mini Trans-Blot apparatus (BioRad, Hercules, CA, USA). Membranes were washed in TBST (10 mM Tris, pH 8.0, 150 mM NaCl, 0.05% Tween 20), and were blocked with TBST supplemented with 5% nonfat dry milk. Then the membranes were incubated with different primary antibodies in TBST and 5% nonfat dry milk, washed, and incubated with horseradish peroxide-conjugated secondary goat anti-rabbit or anti-mouse antibodies, conjugated with horseradish peroxidase (BioRad). Enhanced chemiluminescence detection reagents were used as a detection system (ECL), according to the manufacturer's instructions (Amersham Biosciences, Little Chalfont, UK).

3.8. Statistical Analysis

Most of the experiments were performed at least three times with replicate samples, except where otherwise indicated. Data are plotted as mean ± SD (standard deviation). The means were compared using analysis of variance (ANOVA) plus Bonferroni's *t*-test. p values of less than 0.05 were considered significant.

4. Conclusions

Organic/inorganic hybrid materials represent a very attractive field in biomedical area. The sol-gel method provides a versatile way to produce bioactive hybrid materials and many organic/inorganic hybrids, including poly(e-caprolactone)/Zirconia (PCL/ZrO$_2$), have been developed by the sol-gel process.

In this study, we describe by multiple assays (including those aimed to test the release of intracellular enzymatic activities such as transaminases and lactate dehydrogenase upon cell damage), that PCL/ZrO$_2$ hybrid materials are free from toxicity. We also provide interesting evidence that the hybrids containing 12 wt % and 24 wt % PCL, more than 6 wt % and 50 wt % ones, enhance cell proliferation when compared to PCL-free ZrO$_2$, and impacted cell cycle progression.

The underlying molecular mechanisms by which PCL/ZrO$_2$ hybrid materials support cell cycle progression and cell proliferation, remain unclear and need to be investigated. It is to be noted that we are planning this for future studies. Overall, whatever the exact mechanism(s), our results enrich the evidence of favourable biological properties of PCL/ZrO$_2$ hybrid materials and suggest the further development of such hybrids for dental and orthopedic applications.

Author Contributions: M.C. and S.N. conceived and designed the experiments and wrote the paper; M.C. assisted by E.T. performed the material synthesis; MI., L.S. and E.T. performed the biological experiments; A.S. analyzed the experimental data.

Conflicts of Interest: The authors declare no conflict of interest.

References

1. Judeinstein, P.; Sanchez, C. Hybrid organic-inorganic materials: A land of multidisciplinarity. *J. Mater. Chem.* **1996**, *6*, 511–525. [CrossRef]
2. Brinker, C.J.; Scherer, G.W. *Sol-Gel Science: The Physics and Chemistry of Sol-Gel Processing*; Academic Press: Cambridge, MA, USA, 2013.
3. Klein, L.C. *Sol-Gel Technology for Thin Films, Fibers, Preforms, Electronics, and Specialty Shapes*; William Andrew Publishing: Norwich, NY, USA, 1988.
4. Lu, B.; Lin, Y. Sol-gel synthesis and characterization of mesoporous yttria-stabilized zirconia membranes with graded pore structure. *J. Mater. Sci.* **2011**, *46*, 7056–7066. [CrossRef]
5. Catauro, M.; Papale, F.; Sapio, L.; Naviglio, S. Biological influence of Ca/P ratio on calcium phosphate coatings by sol-gel processing. *Mater. Sci. Eng. C* **2016**, *65*, 188–193. [CrossRef] [PubMed]
6. Kokubo, T.; Kim, H.-M.; Kawashita, M. Novel bioactive materials with different mechanical properties. *Biomaterials* **2003**, *24*, 2161–2175. [CrossRef]
7. Gupta, R.; Kumar, A. Bioactive materials for biomedical applications using sol-gel technology. *Biomed. Mater.* **2008**, *3*, 034005. [CrossRef] [PubMed]
8. Martin, R.A.; Yue, S.; Hanna, J.V.; Lee, P.; Newport, R.J.; Smith, M.E.; Jones, J.R. Characterizing the hierarchical structures of bioactive sol–gel silicate glass and hybrid scaffolds for bone regeneration. *Phil. Trans. R. Soc. A* **2012**, *370*, 1422–1443. [CrossRef] [PubMed]
9. Sanchez, C.; Ribot, F. Design of hybrid organic-inorganic materials synthesized via sol-gel chemistry. *New J. Chem.* **1994**, *18*, 1007–1047.
10. Eldsäter, C.; Erlandsson, B.; Renstad, R.; Albertsson, A.-C.; Karlsson, S. The biodegradation of amorphous and crystalline regions in film-blown Poly (ε-caprolactone). *Polymer* **2000**, *41*, 1297–1304. [CrossRef]
11. Choi, E.J.; Kim, C.H.; Park, J.K. Synthesis and characterization of starch-g-polycaprolactone copolymer. *Macromolecules* **1999**, *32*, 7402–7408. [CrossRef]
12. Causa, F.; Battista, E.; Della Moglie, R.; Guarnieri, D.; Iannone, M.; Netti, P.A. Surface investigation on biomimetic materials to control cell adhesion: The case of rgd conjugation on PCL. *Langmuir* **2010**, *26*, 9875–9884. [CrossRef] [PubMed]
13. Zhong, Z.; Sun, X.S. Properties of soy protein isolate/polycaprolactone blends compatibilized by methylene diphenyl diisocyanate. *Polymer* **2001**, *42*, 6961–6969. [CrossRef]

14. De Santis, R.; Gloria, A.; Russo, T.; D'Amora, U.; Zeppetelli, S.; Dionigi, C.; Sytcheva, A.; Herrmannsdörfer, T.; Dediu, V.; Ambrosio, L. A basic approach toward the development of nanocomposite magnetic scaffolds for advanced bone tissue engineering. *J. Appl. Polym. Sci.* **2011**, *122*, 3599–3605. [CrossRef]

15. Hutmacher, D.W.; Schantz, T.; Zein, I.; Ng, K.W.; Teoh, S.H.; Tan, K.C. Mechanical properties and cell cultural response of polycaprolactone scaffolds designed and fabricated via fused deposition modeling. *J. Biomed. Mater. Res. A* **2001**, *55*, 203–216. [CrossRef]

16. Kyriakidou, K.; Lucarini, G.; Zizzi, A.; Salvolini, E.; Mattioli Belmonte, M.; Mollica, F.; Gloria, A.; Ambrosio, L. Dynamic co-seeding of osteoblast and endothelial cells on 3d polycaprolactone scaffolds for enhanced bone tissue engineering. *J. Bioact. Compat. Polym.* **2008**, *23*, 227–243. [CrossRef]

17. Bañobre-López, M.; Pineiro-Redondo, Y.; De Santis, R.; Gloria, A.; Ambrosio, L.; Tampieri, A.; Dediu, V.; Rivas, J. Poly (caprolactone) based magnetic scaffolds for bone tissue engineering. *J. Appl. Phys.* **2011**, *109*, 07B313. [CrossRef]

18. Bartolo, P.; Domingos, M.; Gloria, A.; Ciurana, J. Biocell printing: Integrated automated assembly system for tissue engineering constructs. *CIRP Ann.-Manuf. Technol.* **2011**, *60*, 271–274. [CrossRef]

19. Catauro, M.; Bollino, F.; Papale, F.; Piccolella, S.; Pacifico, S. Sol-gel synthesis and characterization of SiO_2/PCL hybrid materials containing quercetin as new materials for antioxidant implants. *Mater. Sci. Eng. C* **2016**, *58*, 945–952. [CrossRef] [PubMed]

20. Domingos, M.; Chiellini, F.; Gloria, A.; Ambrosio, L.; Bartolo, P.; Chiellini, E. Effect of process parameters on the morphological and mechanical properties of 3d bioextruded poly (ε-caprolactone) scaffolds. *Rapid Prototyp. J.* **2012**, *18*, 56–67. [CrossRef]

21. Catauro, M.; Bollino, F. Anti-inflammatory entrapment in polycaprolactone/silica hybrid material prepared by sol-gel route, characterization, bioactivity and in vitro release behavior. *J. Appl. Biomater. Funct. Mater.* **2013**, *11*, 172–179. [CrossRef] [PubMed]

22. Ghorbani, F.; Moradi, L.; Shadmehr, M.B.; Bonakdar, S.; Droodinia, A.; Safshekan, F. In-vivo characterization of a 3d hybrid scaffold based on PCL/decellularized aorta for tracheal tissue engineering. *Mater. Sci. Eng. C* **2017**, *81*, 74–83. [CrossRef] [PubMed]

23. Perrin, D.E.; English, J.P. Polycaprolactone. In *Handbook of Biodegradable Polymers*; Domb, A.J., Kost, J.K., Wiseman, D.M., Eds.; CRC Press: Amsterdam, The Netherlands, 1997; Volume 7, pp. 63–77.

24. Lee, S.J.; Liu, J.; Oh, S.H.; Soker, S.; Atala, A.; Yoo, J.J. Development of a composite vascular scaffolding system that withstands physiological vascular conditions. *Biomaterials* **2008**, *29*, 2891–2898. [CrossRef] [PubMed]

25. Sung, H.-J.; Meredith, C.; Johnson, C.; Galis, Z.S. The effect of scaffold degradation rate on three-dimensional cell growth and angiogenesis. *Biomaterials* **2004**, *25*, 5735–5742. [CrossRef] [PubMed]

26. Catauro, M.; Bollino, F.; Papale, F. Biocompatibility improvement of titanium implants by coating with hybrid materials synthesized by sol-gel technique. *J. Biomed. Mater. Res. A* **2014**, *102*, 4473–4479. [CrossRef] [PubMed]

27. Catauro, M.; Bollino, F.; Papale, F.; Mozzati, M.C.; Ferrara, C.; Mustarelli, P. ZrO_2/PEG hybrid nanocomposites synthesized via sol-gel: Characterization and evaluation of the magnetic properties. *J. Non-Cryst. Solids* **2015**, *413*, 1–7. [CrossRef]

28. Catauro, M.; Bollino, F.; Papale, F.; Mozetic, P.; Rainer, A.; Trombetta, M. Biological response of human mesenchymal stromal cells to titanium grade 4 implants coated with PCL/ZrO_2 hybrid materials synthesized by sol-gel route: In vitro evaluation. *Mater. Sci. Eng. C* **2014**, *45*, 395–401. [CrossRef] [PubMed]

29. Catauro, M.; Mozzati, M.C.; Bollino, F. Sol–gel hybrid materials for aerospace applications: Chemical characterization and comparative investigation of the magnetic properties. *Acta Astronaut.* **2015**, *117*, 153–162. [CrossRef]

30. Catauro, M.; Raucci, M.; Ausanio, G. Sol-gel processing of drug delivery zirconia/polycaprolactone hybrid materials. *J. Mater. Sci.-Mater. M* **2008**, *19*, 531–540. [CrossRef] [PubMed]

31. Catauro, M.; Bollino, F.; Papale, F. Preparation, characterization, and biological properties of organic-inorganic nanocomposite coatings on titanium substrates prepared by sol-gel. *J. Biomed. Mater. Res. A* **2014**, *102*, 392–399. [CrossRef] [PubMed]

32. Catauro, M.; Bollino, F.; Papale, F.; Pacifico, S.; Galasso, S.; Ferrara, C.; Mustarelli, P. Synthesis of zirconia/polyethylene glycol hybrid materials by sol-gel processing and connections between structure and release kinetic of indomethacin. *Drug Deliv.* **2014**, *21*, 595–604. [CrossRef] [PubMed]

33. Naviglio, S.; Spina, A.; Chiosi, E.; Fusco, A.; Illiano, F.; Pagano, M.; Romano, M.; Senatore, G.; Sorrentino, A.; Sorvillo, L. Inorganic phosphate inhibits growth of human osteosarcoma U2OS cells via adenylate cyclase/camp pathway. *J. Cell. Biochem.* **2006**, *98*, 1584–1596. [CrossRef] [PubMed]

34. Catauro, M.; Papale, F.; Bollino, F.; Gallicchio, M.; Pacifico, S. Biological evaluation of zirconia/peg hybrid materials synthesized via sol-gel technique. *Mater. Sci. Eng. C* **2014**, *40*, 253–259. [CrossRef] [PubMed]

35. Ramanathan, G.; Singaravelu, S.; Muthukumar, T.; Thyagarajan, S.; Perumal, P.T.; Sivagnanam, U.T. Design and characterization of 3d hybrid collagen matrixes as a dermal substitute in skin tissue engineering. *Mater. Sci. Eng. C* **2017**, *72*, 359–370. [CrossRef] [PubMed]

36. Park, J.; Lee, S.J.; Chung, S.; Lee, J.H.; Kim, W.D.; Lee, J.Y.; Park, S.A. Cell-laden 3d bioprinting hydrogel matrix depending on different compositions for soft tissue engineering: Characterization and evaluation. *Mater. Sci. Eng. C* **2017**, *71*, 678–684. [CrossRef] [PubMed]

37. Ivanova, S.I.; Chakarov, S.; Momchilova, A.; Pankov, R. Live-cell biosensor for assessment of adhesion qualities of biomaterials. *Mater. Sci. Eng. C* **2017**, *78*, 230–238. [CrossRef] [PubMed]

38. Navarro, L.; Mogosanu, D.-E.; Ceaglio, N.; Luna, J.; Dubruel, P.; Rintoul, I. Novel poly (diol sebacate) s as additives to modify paclitaxel release from poly (lactic-co-glycolic acid) thin films. *J. Pharm. Sci.* **2017**, *106*, 2106–2114. [CrossRef] [PubMed]

39. Morgenstern, S.; Flor, R.; Kessler, G.; Klein, B. Automated determination of nad-coupled enzymes. Determination of lactic dehydrogenase. *Anal. Biochem.* **1965**, *13*, 149–161. [CrossRef]

40. Ferri, F.F. *Ferri's Best Test E-Book: A Practical Guide to Laboratory Medicine and Diagnostic Imaging*, 3rd ed.; Elsevier Health Sciences: Philadelphia, PA, USA, 2014.

41. Spina, A.; Sorvillo, L.; Di Maiolo, F.; Esposito, A.; D'Auria, R.; Di Gesto, D.; Chiosi, E.; Naviglio, S. Inorganic phosphate enhances sensitivity of human osteosarcoma U2OS cells to doxorubicin via a p53-dependent pathway. *J. Cell. Physiol.* **2013**, *228*, 198–206. [CrossRef] [PubMed]

42. McKeehan, W.L. Glutaminolysis in animal cells. In *Carbohydrate Metabolism in Cultured Cells*; Morgan, M.J., Ed.; Springer: Boston, MA, USA, 1986; pp. 111–150.

43. Sekiya, S.; Suzuki, A. Direct conversion of mouse fibroblasts to hepatocyte-like cells by defined factors. *Nature* **2011**, *475*, 390–393. [CrossRef] [PubMed]

44. Messer, R.L.; Lockwood, P.E.; Wataha, J.C.; Lewis, J.B.; Norris, S.; Bouillaguet, S. In vitro cytotoxicity of traditional versus contemporary dental ceramics. *J. Prosthet. Dent.* **2003**, *90*, 452–458. [CrossRef]

45. Sims-Mourtada, J.; Niamat, R.A.; Samuel, S.; Eskridge, C.; Kmiec, E.B. Enrichment of breast cancer stem-like cells by growth on electrospun polycaprolactone-chitosan nanofiber scaffolds. *Int. J. Nanomed.* **2014**, *9*, 995–1003. [CrossRef] [PubMed]

46. Allo, B.A.; Rizkalla, A.S.; Mequanint, K. Hydroxyapatite formation on sol-gel derived poly (ε-caprolactone)/bioactive glass hybrid biomaterials. *ACS Appl. Mater. Interfaces* **2012**, *4*, 3148–3156. [CrossRef] [PubMed]

47. Catauro, M.; Bollino, F.; Papale, F. Surface modifications of titanium implants by coating with bioactive and biocompatible poly (ε-caprolactone)/sio 2 hybrids synthesized via sol-gel. *Arab. J. Chem.* **2015**. [CrossRef]

48. Teng, S.-H.; Wang, P.; Dong, J.-Q. Bioactive hybrid coatings of poly (ε-caprolactone)–silica xerogel on titanium for biomedical applications. *Mater. Lett.* **2014**, *129*, 209–212. [CrossRef]

49. Ciprioti, S.V.; Bollino, F.; Tranquillo, E.; Catauro, M. Synthesis, thermal behavior and physicochemical characterization of ZrO_2/PEG inorganic/organic hybrid materials via sol-gel technique. *J. Therm. Anal. Calorim.* **2017**, *130*, 535–540. [CrossRef]

50. Naviglio, S.; Matteucci, C.; Matoskova, B.; Nagase, T.; Nomura, N.; Di Fiore, P.P.; Draetta, G.F. Ubpy: A growth-regulated human ubiquitin isopeptidase. *EMBO J.* **1998**, *17*, 3241–3250. [CrossRef] [PubMed]

51. Moraleva, A.; Magoulas, C.; Polzikov, M.; Hacot, S.; Mertani, H.C.; Diaz, J.J.; Zatsepina, O. Involvement of the specific nucleolar protein surf6 in regulation of proliferation and ribosome biogenesis in mouse nih/3t3 fibroblasts. *Cell Cycle* **2017**, *16*, 1979–1991. [CrossRef] [PubMed]

52. Ogawa, Y.; Yamamoto, M.; Okada, H.; Yashiki, T.; Shimamoto, T. A new technique to efficiently entrap leuprolide acetate into microcapsules of polylactic acid or copoly (lactic/glycolic) acid. *Chem. Pharm. Bull.* **1988**, *36*, 1095–1103. [CrossRef] [PubMed]

53. Tokiwa, Y.; Suzuki, T. Hydrolysis of polyesters by lipases. *Nature* **1977**, *270*, 76–78. [CrossRef] [PubMed]

54. O'Brien, V.; Campo, M.S. Bpv-4 E8 transforms NIH 3T3 cells, up-regulates cyclin a and cyclin a-associated kinase activity and de-regulates expression of the cdk inhibitor p27 Kip1. *Oncogene* **1998**, *17*, 293–301. [CrossRef] [PubMed]

55. Guardavaccaro, D.; Pagano, M. Stabilizers and destabilizers controlling cell cycle oscillators. *Mol. Cell* **2006**, *22*, 1–4. [CrossRef] [PubMed]

56. Sapio, L.; Sorvillo, L.; Illiano, M.; Chiosi, E.; Spina, A.; Naviglio, S. Inorganic phosphate prevents Erk1/2 and Stat3 activation and improves sensitivity to doxorubicin of mda-mb-231 breast cancer cells. *Molecules* **2015**, *20*, 15910–15928. [CrossRef] [PubMed]

Review

Synthesis of Hollow Sphere and 1D Structural Materials by Sol-Gel Process

Fa-Liang Li [1,2,*] **and Hai-Jun Zhang** [1]

[1]   The State Key Laboratory of Refractories and Metallurgy, Wuhan University of Science and Technology, Wuhan 430081, China; zhanghaijun@wust.edu.cn
[2]   Jiangxi Engineering Research Center of Industrial Ceramics, Pingxiang 337022, China
*   Correspondence: lfliang@wust.edu.cn; Tel.: +86-027-6886-2258 (ext. 2829)

Received: 31 July 2017; Accepted: 21 August 2017; Published: 25 August 2017

Abstract: The sol-gel method is a simple and facile wet chemical process for fabricating advanced materials with high homogeneity, high purity, and excellent chemical reactivity at a relatively low temperature. By adjusting the processing parameters, the sol-gel technique can be used to prepare hollow sphere and 1D structural materials that exhibit a wide application in the fields of catalyst, drug or gene carriers, photoactive, sensors and Li-ion batteries. This feature article reviewed the development of the preparation of hollow sphere and 1D structural materials using the sol-gel method. The effects of calcination temperature, soaking time, pH value, surfactant, etc., on the preparation of hollow sphere and 1D structural materials were summarized, and their formation mechanisms were generalized. Finally, possible future research directions of the sol-gel technique were outlined.

Keywords: sol-gel; hollow sphere; 1D structure

---

1. Introduction

The sol-gel method has attracted much attention as it can be used to fabricate high purity products with a fine particle size and good chemical homogeneity at low temperatures. With outstanding advantages of accuracy, stability, low reaction temperature, and a high purity of targeted products, the sol-gel process is considered as one of synthetic material methods with the most potential.

In the past few decades, the sol-gel method (combined with other techniques including microwave heating, ultrasonication, spin-coating, dip coating, laminar flow coating, and strong field induction) have been used to prepare various kinds of advanced materials with different morphologies including magnetic [1–10], optical [11–13], electronic [14,15], and structural [16–25] particles, and hollow spheres [26], fibers [27], nanowires [28] and films [29–34]. Many reviews have been published on the preparation of particles and coatings using the sol-gel process; for instance, Guo et al. [35] summarized the sol-gel synthesis and application of monodisperse nanoparticles and granules including Li_2TiO_3, $ZnO-B_2O_3-SiO_2$ additive, $La_{1-x}Sr_xCoO_{3-\delta}$ (x = 0.1 − 0.7) and $Y_3Al_5O_{12}$. Zhang et al. [36] outlined the sol-gel process synthesis of high-temperature non-oxide ultrafine powders including nitride, carbide and boride. This demonstrated that the sol-gel process offered many advantages in the preparation of powders, such as lower formation temperatures, shorter soaking times, and the ability to synthesize submicron and nano-crystalline ultrafine powders. Moreover, the resultant powders possessed a narrow and uniform distribution, higher homogeneity and purity. Meanwhile, the corresponding electronic ceramics prepared by the as-prepared nanopowders could be sintered at a lower temperature, and showed good temperature stability and high electrical properties. This also indicated that the process suffered several drawbacks including the requirement of expensive precursors and additives, and it was difficult to control the structure of the targeted powders. On the other hand, Guo et al. [35] summarized the sol-gel preparation of many types of functional coatings, including anti-reflection, anticorrosion, wearable, and anti-soiling coatings. Livage et al. [37] reviewed sol-gel electrochromic coatings including WO_3, MoO_3, V_2O_5, Nb_2O_5, TiO_2, CeO_2, IrO_2, Fe_2O_3, and NiO. Yoldas [38] pointed out that sol-gel coatings could be used on optical fibers and photovoltaic cells, as well as demonstrated that sol-gel coatings offered many excellent properties such as easy doping, low cost, no shape limitation, good adhesion, and crack-free.

Recently, studies have been carried out to prepare hollow spheres and 1D structural materials as it is well-known that hollow spheres can be widely used in the fields of therapeutics, energy storage, electronics, environmental remediation, medical ultrasounds, biosensors, non-destructive testing, electronic devices, and low-density transducer arrays given their excellent properties of low density, high specific surface area, high energy conversion efficiency, high adsorption capacity, and large light-harvesting efficiencies [39–51]. Furthermore, 1D structural materials such as fibers, nanorods, and nanowires have attracted a lot of attention in the areas of catalysis, reinforcement, sensors, solar cells, super-capacitors, optics, electronics, etc., due to their unique properties such as light volume-weight, high strength and modulus, high thermo-mechanical stability, and enhanced photocatalytic activity [52–63].

In this review, we provide an overview of recent developments in the fabrication of hollow sphere and 1D structural materials using the sol-gel technique. Finally, probable improvements and future outlooks of the sol-gel method are outlined.

2. Synthesis of Hollow Sphere Materials by the Sol-Gel Method

The template method (including hard template and soft template methods) is considered as the most effective route for preparing hollow structures. Various kinds of hollow spheres have been prepared by the sol-gel technique based on the template method.

Binary metal borides exhibit excellent mechanical properties. Among them, zirconium diboride (ZrB_2) is seen as one of the most promising ultra-high temperature ceramics owing to its high melting point (3040 °C), high hardness (22 GPa), good chemical stability, good corrosion resistance, and good thermal shock resistance [64]. Our group successfully synthesized ZrB_2 ultrafine hollow spherical powders using sol-gel combined with a boro/carbothermal reduction method using zirconium oxychloride, boric acid, and glucose as the main raw materials [65]. The effects of reaction temperature, amount of B, Zr, and C on the formation of ZrB_2 were studied and indicated that ZrB_2 could be synthesized at 1200 °C, and that ZrB_2 hollow spheres were successfully prepared at 1500 °C for 2 h when the molar ratios of B/Zr = 2.5 and C/Zr = 6.5. The formation temperature was about 300 °C lower than that demanded by the conventional solid-state mixing method [32]. Scanning electron microscopy SEM results showed that the diameters of ZrB_2 ultrafine hollow spheres ranged from 100 to 500 nm with an average shell thickness of about 30 nm (Figure 1a). The possible mechanism for the formation of ZrB_2 ultrafine hollow spheres is illustrated in Figure 1b. First, large carbonaceous spheres with sizes ranging from 100 to 500 nm were formed during the carbonization of glucose and citric acid. Second, the as-formed carbonaceous spheres attached to ZrO_2/B_2O_3 precursors on their surfaces. Finally, when the gel was heated at high temperature, the oxide precursors reacted with the carbonaceous sphere templates to form in situ ZrB_2 hollow spheres. The results also indicated a low yield of hollow spheres via this method.

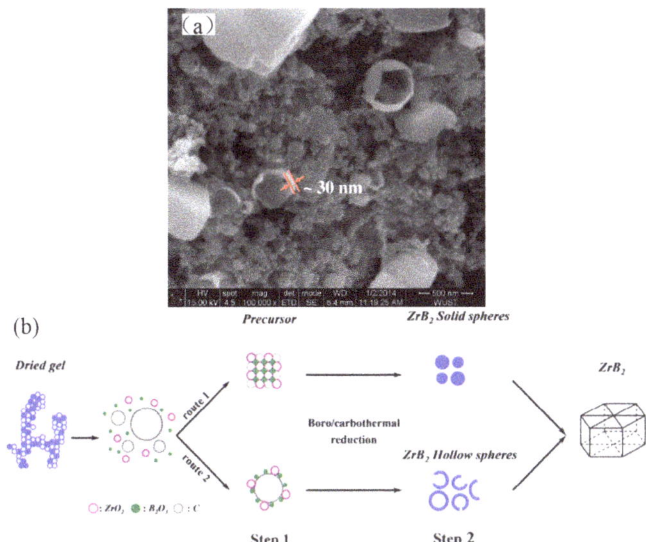

Figure 1. (**a**) SEM image of ZrB$_2$ prepared at 1500 °C for 2 h; (**b**) Schematic illustration of the grown process of hollow ZrB$_2$ spheres crystals [65].

Given the outstanding properties of high strength, high thermal conductivity, high chemical stability, wide band gap energy (~2.4 eV), low negative conduction band potential (−1.40 V), and good environmental friendliness, silicon carbide (SiC) has been widely used from metallurgy to aerospace. Wang et al. [66] successfully fabricated openmouthed β-SiC hollow spheres through the environmentally friendly sol-gel method using glucose as a carbon resource and poly(ethylene oxide)–poly(propylene oxide)–poly(ethylene oxide) (PEO–PPO–PEO) as a silicon source, and the effects of carbothermal reduction temperature and time on the synthesis of β-SiC were investigated. They indicated that the preparation of β-SiC hollow spheres was significantly affected by the calcination temperature and time. When the precursors were fired at 1450 °C for 8 h, the as-prepared SiC samples exhibited openmouthed hollow microsphere morphology with sizes of about 10 μm. When the calcination temperature was increased to 1500 °C, no whole hollow microspheres were obtained. The fabrication of β-SiC hollow spheres was suggested through the following five stages: (1) The PEO-PPO-PEO and glucose were wrapped together to form a microsphere template, and the SiO$_2$ particles derived from the hydrolysis of tetraethyl orthosilicate were self-assembled onto the template spherical surface; (2) Heating the PEO-PPO-PEO and glucose microspheres at 550 °C under an N$_2$ atmosphere to form carbon spheres; (3) SiC sphere shells were produced via the reaction between the outer SiO$_2$ and carbon sphere core by increasing the heating temperature to 1450 °C; (4) The unreacted carbon core was burned out when the samples were calcined at 600 °C in an O$_2$ atmosphere, which resulted in an open mouth on the shell of SiC sphere; and (5) Unreacted SiO$_2$ was removed using an NaOH solution, and finally openmouthed β-SiC hollow spheres with rough surface were obtained (Figure 2). The UV-vis spectra revealed that the as-prepared β-SiC hollow spheres displayed a blue shift of absorption edges, which was caused by the little change in crystal structure of SiC with calcination temperature and time. Furthermore, the prepared β-SiC products could photocatalytically reduce CO$_2$ to CH$_4$ with better efficiency than the standard photocatalyst P25 TiO$_2$. It is believed that one reason was the very negative conduction band potential of −1.40 V of the resulting β-SiC products, and the other reason was their special morphology. The special hollow sphere structure brought a higher BET surface area and allowed multi-reflections of illumination light within its interior cavities, which was an efficient use of the light source and finally enhanced photocatalytic activity.

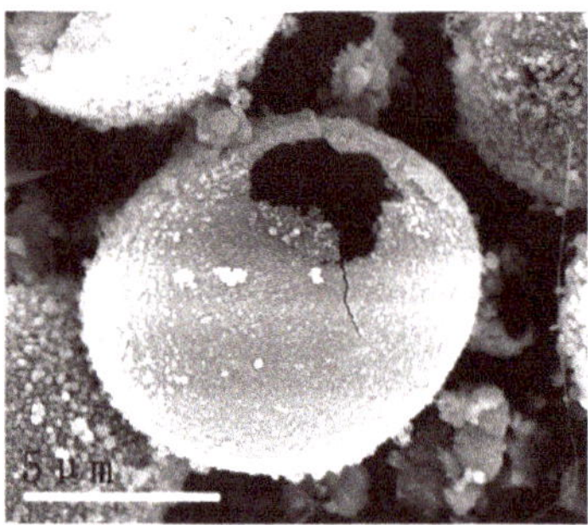

Figure 2. SEM image of as-prepared β-SiC [66].

Hollow mesoporous silica spheres which have been widely used in the fields of delivery, separation and catalysis can be also synthesized through the sol-gel method. Wang et al. [67] synthesized hollow mesoporous silica spheres using the sol-gel/emulsion method with cetyltrimethylammonium bromide (CTAB) as the surfactant to stabilize the tetraethoxysilane (TEOS) droplet, and the effects of ethanol on the formation of silica hollow spheres were studied. Transmission electron microscopy (TEM) images revealed that hollow silica spheres could form at an ethanol to water ratio from 0.62 to 0.47, and that the wall thickness of the hollow silica spheres decreased with decreasing ethanol content. Additionally, the HR-TEM images showed that mesopores radiated throughout the spheres and that CTAB concentration played an important role on the synthesis of hollow mesoporous silica spheres. Interfacial energy was reduced by introducing more CTAB into the system that resulted in smaller TEOS droplets, which were enclosed by CTAB micelles that served as templates for the formation of silica hollow spheres. By adding a solution of aqueous ammonia, TEOS on the interface was hydrolyzed into silica and then deposited onto the interface. Along with a decrease of the concentration of TEOS on the interface, the TEOS would also diffuse from the inside to the interface for concentration equilibrium. Finally, hollow silica spheres were obtained after calcination. The proposed schematic of the formation of silica spheres is shown in Figure 3. It also indicates that ethanol content is crucial for the formation of SiO_2 hollow spheres. If too little ethanol was introduced into the system, the hydrolysis speed of TEOS was faster than their diffusion speed in the droplets. As a result, the synthesized silica spheres were all solid, and the resultant hollow mesoporous silica spheres could be used as catalyst support.

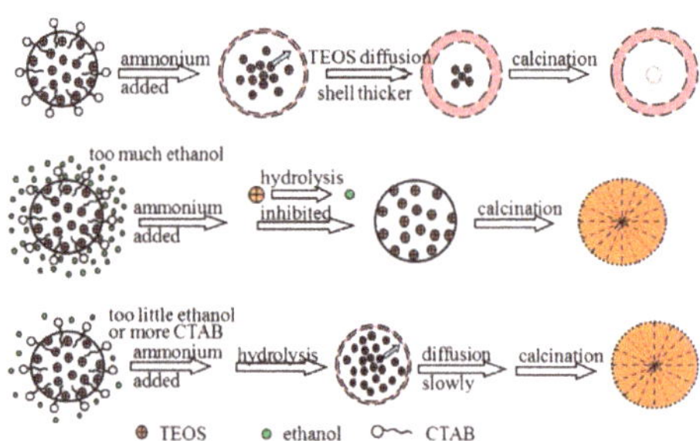

Figure 3. Schematic illustrations for the controllable synthesis of silica hollow sphere [67].

Rare earth oxide hollow spheres can also be prepared with the sol-gel method. Europium sesquioxide (Eu_2O_3) is widely used in scintillators, catalysis, electrochemical energy-storage devices, and luminescent materials. Zhang et al. [68] successfully synthesized Eu_2O_3 hollow spheres via a sol-gel method using polystyrene/polyelectrolyte microspheres as the templates. The X-ray diffraction XRD results indicated that phase pure cubic structured Eu_2O_3 was obtained when the precursors were fired at 700 $^\circ$C for 3 h.

TEM and FFSEM images showed that the shell thickness and outer diameter of the as-prepared Eu_2O_3 hollow spheres was about 75 and 690 nm, respectively (Figure 4). The mechanism of the formation of hollow sphere Eu_2O_3 was also proposed as follows: (1) when the polystyrene/polyelectrolyte templates were dipped into the mixed solution (containing europium nitrate and urea), the pores of the templates were easier to fill with the solution owing to the lower viscosity; (2) in the next heating treatment, urea was hydrolyzed to form OH^- ions that increased the pH value of the solution, and the OH^- ions reacted with europium to form a $Eu[(OH)_3](H_2O)_y$ sol, which took place simultaneously within the pores and the solution; and (3) during the calcination process, the sol is transformed to gel by the condensation reactions within the polyelectrolyte microreactor. Meanwhile, the polystyrene/polyelectrolyte template was removed and finally formed a hollow sphere structure. Luminescence spectra demonstrated that the excitation wavelength of the resultant Eu_2O_3 hollow sphere was 514.5 nm at room temperature. Compared with bulk Eu_2O_3, the peak in luminescence spectra of the obtained sample was obviously broadened.

Figure 4. Transmission electron microscopy (TEM) image of Eu_2O_3 hollow sphere [68].

The sol-gel method can also be used to prepare srilankite-type zirconium titanate ($ZrTiO_4$), which has been widely used in microwave telecommunications, the manufacture of high temperature pigments, catalysis, and photocatalysis. Syoufian et al. [69] successfully synthesized $ZrTiO_4$ hollow spheres of a submicrometer size via the sol-gel method using sulfonated polystyrene latex particles as the template. TEM images indicated the hollow structure of the resultant $ZrTiO_4$ spheres and revealed that the spherical shell consisted of a dense arrangement of $ZrTiO_4$ nano-crystals with relatively smooth surfaces. The average outer and void diameter of $ZrTiO_4$ hollow spheres was approximately 190 nm and 160 nm, respectively. The homogeneous spherical shell was determined by the amphiphilic nature of the sulfonated polystyrene latex particles which allowed them to be easily distributed in the solvent, and provided suitable sites for the attachment of tetrabutyl titanate and tetrabutyl zirconate. UV-vis absorption spectra revealed that the band gap energy E_g of $ZrTiO_4$ hollow spheres was higher than that of the TiO_2 powders (Degussa P25), which illustrated that their potential redox in the photocatalytic system also increased. This may be attributed to the following two factors: (1) The smaller and denser building blocks of the hollow sphere shell wall, whose special structure led to the blue-shift in absorption spectra because of the quantum size effect; and (2) the existence of Zr (as ZrO_2) within the TiO_2 framework, which would enhance the UV-responsiveness as ZrO_2 is a direct band gap semiconductor. Therefore, the prepared $ZrTiO_4$ hollow spheres are a promising photocatalyst candidate with higher redox potential.

As one of the most promising piezoelectric materials, lead zirconate titanate (PZT) has excellent properties such as large electromechanical coupling coefficients, high resistance to depolarization, and high temperature stability. Yang et al. [70] successfully fabricated PZT hollow spheres by using the sol-gel method where polyacrylamide latex solid microspheres and PZT sol were prepared before the polyacrylamide spheres were poured into the PZT sol. The PZT sol was gelled inside the solid polyacrylamide spheres. After that, the gel precursors were fired at high temperatures to obtain hollow sphere PZT. SEM images revealed that the outer diameter and wall thickness of the resultant PZT hollow sphere was respectively about 1–2 mm and 100 μm. According to the SEM images of the outer surface, small cracks and pores were observed on the surface of the sphere and the grain size on the surface was about 0.8 μm. In contrast, the SEM images of the inner surface showed that the prepared PZT hollow sphere exhibited a rough inner surface owing to non-uniform penetration of the PZT sol into the polyacrylamide sphere. Additionally, rib-like structures were also found on the inner surface, which provided strength to the hollow PZT spheres. It was noted that the density of resultant PZT hollow spheres and the hollow spheres wall was 1.123 and 3.10 $g \cdot cm^{-3}$, respectively, suggesting that not only were the prepared PZT spheres hollow, but that the hollow spheres wall also exhibited a porous structure considering the theoretical density of PZT (about 8.0 $g \cdot cm^{-3}$). Moreover, the planar coupling factor of resultant PZT hollow spheres was lower than that of the dense PZT discs. Therefore, the PZT hollow spheres can be used as light-weight transducers in medical ultrasonics and underwater applications.

The sol-gel method has also been used to prepare bioactive glasses, which have been widely used for bone tissue regeneration due to their good bioactive, resorbable, and osteo-productive properties. Hu et al. [71] successfully fabricated hollow mesoporous bioactive glass sub-micron spheres (HMBGS) via the sol-gel technique using CTAB as a template agent. SEM (combined with the TEM images) revealed that the microstructure of HMBGS was significantly affected by the concentration of CTAB. When introducing 3.3 mM CTAB, the average particle diameter and shell thickness of as-prepared HMBGS was 294 and 32 nm, respectively. With an increase in CTAB concentration from 3.3 to 5.9 mM, the spheres became solid with an average particle diameter of 87 nm. The N_2 absorption–desorption isotherm results showed that the specific surface area of the prepared HMBGS was larger than 444.0 $m^2 \cdot g^{-1}$ and the average pore size of the prepared HMBGS was larger than 4.6 nm. This indicated that the formation of hollow mesoporous structures was dominated by a surfactant-template mechanism. As shown in Figure 5, CTAB self-assembled into spherical vesicles in the ethanol-water solution after stirring first. With increasing CTAB concentration in the system, the hydrolysis of TEOS was accelerated. As a result, the nucleation rate was faster than the CTAB self-assembling rate, and a large number of bioactive glass sol particles were formed. Second, the CTAB molecules adsorbed to the surface of the sol particles by hydrogen bonding interactions. Finally, CTAB was removed by calcination at high temperature and solid mesoporous bioactive glass spheres were obtained. In contrast, if the CTAB self-assembling rate was faster than the nucleation rate, the CTAB molecules self-assembled into spherical vesicles. Next, the bioactive glass sols were adsorbed onto the surface of the vesicle by hydrogen bonding interactions. Finally, HMBGS were obtained after the removal of CTAB through high temperature treatment. The prepared HMBGS can be used as good candidates for drug or gene carriers in bone tissue regeneration.

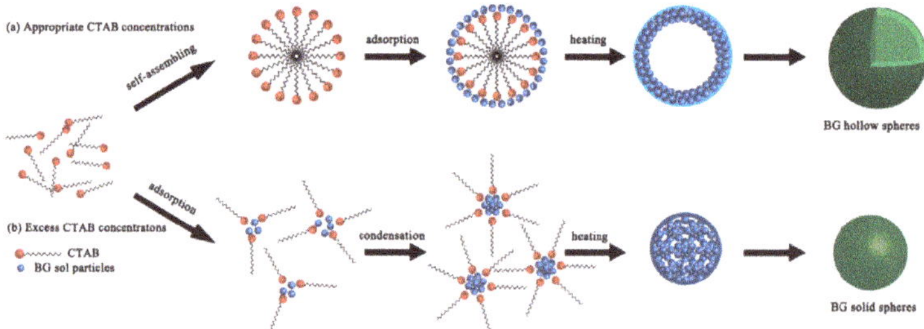

Figure 5. Schematic illustration of the formation processes of hollow mesoporous bioactive glass sub-micron spheres (HMBGS) [71].

Not only single phase hollow spheres, but also composite hollow sphere materials can be prepared by the sol-gel method. Toyama et al. [72] prepared hollow silica-alumina (SiO_2-Al_2O_3) composite spheres via the sol-gel method using polystyrene particles as a template. Methanol, ethanol, and 2-propanol were separately used as solvents to investigate the effect of alcohol solvents on the morphology of hollow SiO_2-Al_2O_3 composite spheres. When methanol was used as a solvent, only aggregate particles formed after 6 h. By increasing the coating time to 17 h, half hollow spheres and hollow spheres with diameters of approximately 260 nm were observed in the samples. If the coating time was allowed to continue to 36 h, hollow spherical particles with diameters of about 260 nm were formed. The formation of hollow spheres prepared by the polystyrene template method is schematically illustrated in Figure 6. First, the positively charged polystyrene particles were prepared via emulsion polymerization using azo diisobutyl amidine hydrochloride as the initiator, and the silica-alumina composite primary particles with a negative charge were then prepared in a basic solution. Next, the silica-alumina composite primary particles were attracted to the polystyrene templates through electrostatic interaction, and the silica-alumina composite primary particles were sparsely coated on polystyrene template particles to form composite shells. Finally, the polystyrene template was burned out at high temperature and hollow SiO_2-Al_2O_3 composite spheres were obtained. This showed that the amount and rate of hydrogen evolution of the prepared hollow spheres were dependent on the solvent. When methanol or ethanol was used as the solvent, the as-prepared hollow SiO_2-Al_2O_3 composite spheres showed a higher hydrolytic dehydrogenation of NH_3BH_3.

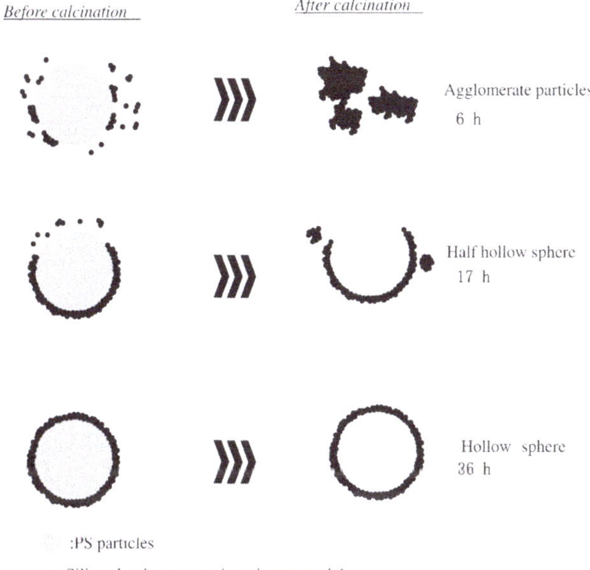

Figure 6. Schematic illustration of formation process of the hollow spheres with different coating time (6, 17, 36 h) [72].

Aside from the above-mentioned hollow spheres, other hollow sphere materials such as Pt-doped TiO_2 hollow spheres [73], organosilica spheres [74], Fe/CeO_2 hollow spheres [75], $SrTiO_3$ hollow spheres [76], etc. have also been fabricated by the sol-gel technique. According to the work that has been reported, it can be seen that both oxide and non-oxide hollow sphere materials can be fabricated at relatively low temperatures and short duration times by using the sol-gel technique based on template method. The size, shell thickness, and porosity of the hollow spheres can be tailored by controlling the processing parameters of the method. Additionally, the prepared hollow sphere products always showed a relatively narrow particle size distribution and good dispersibility. Therefore, both single phase and composite hollow sphere materials can be fabricated by the sol-gel method. However, it should be noted that there are still several drawbacks to overcome to prepare hollow spheres using the present sol-gel process. For example, cracks or pores were always found on the surface of the hollow spheres due to the escaping of organic materials, the yield of hollow spheres was low and a long gelling time was usually required, and the solvent used to remove the template was often harmful. Therefore, a new process should be developed for improving product yield. On the other hand, the contents of organic matter should be decreased to reduce defects in the targeted products.

3. Synthesis of 1D Structural Materials by the Sol-Gel Method

1D structural materials can be prepared by various techniques such as template, electrolysis, sol-gel synthesis, hydrothermal growth, and viscous solution spinning [77]. Up to now, different kinds of 1D structural materials have been prepared by the sol-gel based method.

Magnesium boride (MgB_2) is a well-known superconductor material with a relative transition temperature of approximately $-234\,^\circ$C [78]. Nath et al. [79] prepared MgB_2 nanowires by a simple sol-gel method with magnesium bromide and sodium borohydride as the main raw materials. First, a precursor gel was synthesized by mixing magnesium bromide and sodium borohydride reagents in the presence of CTAB. Then, the resulted gel was pyrolized under a diborane-N_2 atmosphere for forming MgB_2. SEM images showed that the as-prepared MgB_2 nanowires with very smooth surfaces were at least 20 μm in length and about 50–100 nm in diameter (Figure 7). TEM images showed that the synthesized MgB_2 nanowires were solid, straight, and very uniform in diameter along their lengths. Some nanowires had a rounded tip, while others exhibited a flat rectangular or polygonal tip. The selected area electron diffraction (SAED) patterns of some individual nanowires demonstrated that the nanowires were crystalline. The formation of nanowires was mainly determined by the gel in the initial step, and the prearrangement of the precursor particles could template the formation of the one-dimensional morphology of the nanowires with the participation of a CTAB surfactant. Additionally, B_2H_6 gas played an important role in the synthesis process. First, B_2H_6 gas can be used to maintain a B-rich atmosphere. Second, B_2H_6 reacts with even minute amounts of O_2 to form solid B-oxide phase and prevent oxygen from making contact with the reactants. The magnetic susceptibility data revealed that the nanowires morphology did not affect the transition temperature, and the synthesized MgB_2 nanowires exhibited a superconducting temperature of about $-234.4\,^\circ$C, which was close to that of bulk MgB_2 ($-234\,^\circ$C). Magnetization vs. field measurement demonstrated that MgB_2 was a type II superconductor.

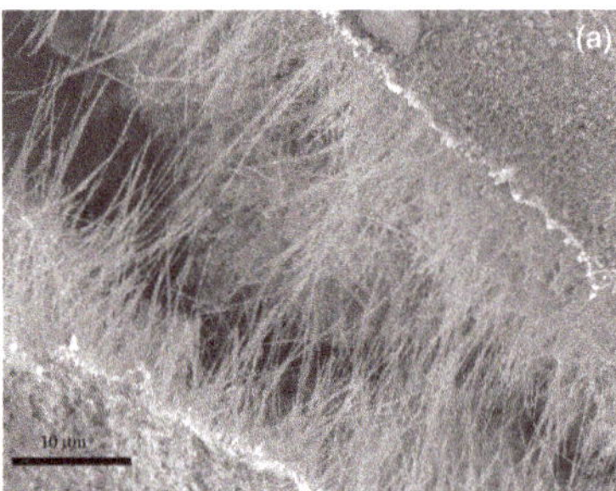
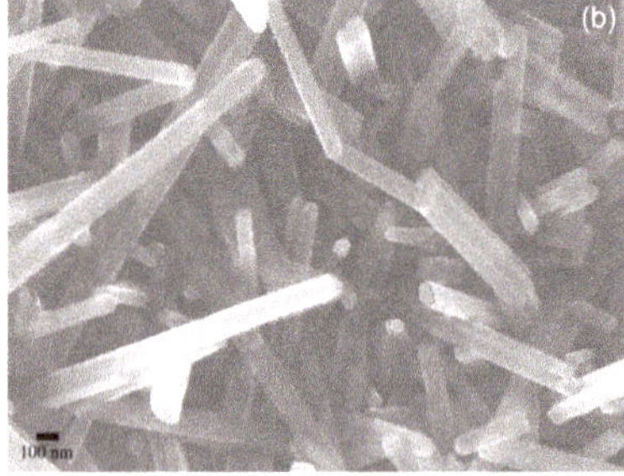

Figure 7. SEM images of the MgB_2 nanowires. (**a**) A thick mesh of nanowires; (**b**) Higher-magnification image of the vertically oriented nanowires [79].

The sol-gel technique has also been used to prepare SiBON fibers, which can be used as spacecraft material with good wavetransparent and mechanical properties. Li et al. [80] successfully synthesized SiBON fibers through the sol-gel method with boric acid, melamine, and TEOS as raw materials. The SiBON fiber precursor was first prepared via the sol-gel technique and then nitrided at 1400–1800 $^\circ$C in N_2 to obtain SiBON fibers, and the effects of reaction temperature and pH value on the fabrication of SiBON fibers were investigated. XRD patterns showed that the resultant SiBON fiber was almost amorphous when the precursor was nitrided at 1400 °C and the crystallization of SiBON fiber was improved with increasing heating temperature. The morphologies of the SiBON fiber were affected significantly by pH value; when pH = 8, the SiBON fiber had a shorter and more uniform distribution when compared with the precursor fiber. After being nitrided, the resultant SiBON fiber became looser and bifurcated at both ends with even thickness when pH = 6. With decreasing pH value to 4, the aspect ratio of fibers (after being nitride) decreased, while the surface turned rough. Fourier transform infrared spectroscopy (FTIR) results revealed that the resultant SiBON fiber had B-N-Si and B-O-Si bonds. It was suggested that the SiBON fiber precursor growth underwent different reactions. First, Si-O and B-O groups were generated in a mixture of TEOS and boric acid. When melamine was added, the O of Si-O and B-O groups was generated by the reaction in solution. As the N of -NH$_2$ groups of melamine are strong negative centers, it is easy to form B-N-Si chains. Second, the etherification and dehydration of Si-OH groups derived from the hydrolysis of TEOS formed Si-O-Si chain segments. Si-O-Si chains grew along the SiBON radial direction and eventually formed SiBON fiber precursors with a certain aspect ratio.

Manganese oxides have been used in catalysts, absorbents, and Li$^+$ related batteries as they have outstanding structural flexibility and a multitude of oxidation states (Mn^{2+}, Mn^{3+}, Mn^{4+}). Tang et al. [81] prepared ultrafine MnO$_2$ nanowires and nanorods via the sol-gel method with different surfactants in an ethanol solvent. Four different surfactants including CTAB, polyvinyl pyrrolidone (PVP, K30), pluronic P123 triblock copolymer (EO$_{20}$PO$_{70}$EO$_{20}$), and sodium dodecyl sulfate were used in the process, and the effects of the surfactants on the fabrication of MnO$_2$ nanowires and nanorods were studied. SEM and TEM images showed that products with lengths up to several micrometers consisted of a large amount of highly dispersed ultrafine wire-like structures, which had a diameter of 7 nm (Figure 8). SEM images also revealed that the P123-derived products possessed irregular particle structures, which were aggregated by a large number of nanowires. Based on higher magnification images, nanowires were found with diameters of about 10 nm and with lengths of about 200 nm. A HRTEM image revealed that the interplanar spacing of the lattice planes was about 0.24 nm, which can be ascribed to the (211) crystal planes of the tetragonal MnO$_2$. By adding PVP as a surfactant, the obtained XRD patterns were similar to those of the P123-derived products. When sodium dodecyl sulfate was used as the surfactant, SEM images showed that the as-prepared particles were aggregated with a free-standing sheet structure. TEM images revealed that the sheet-like structures consisted of several nanorods with diameters of about 10 nm and lengths of approximately 50 nm. Therefore, the structure of the products was sharply determined by the surfactants.

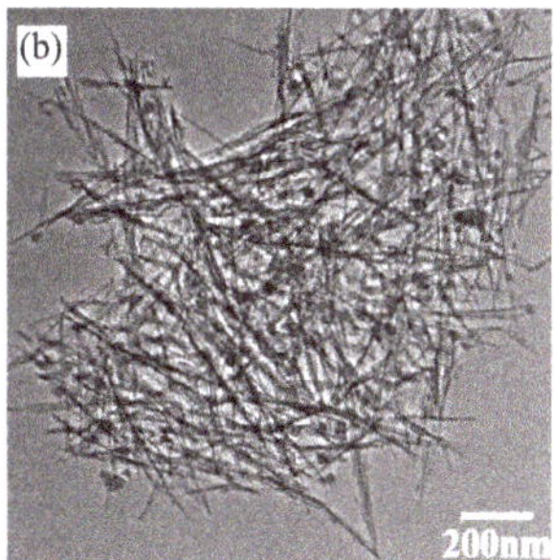

Figure 8. (**a**) SEM image and (**b**) TEM image of the cetyltrimethylammonium bromide (CTAB)-derived MnO$_2$ nanowires [81].

Manganese titanate (MnTiO$_3$) has been regarded as a promising material in solar energy systems with strong absorption in the visible region [82]. Nakhowong et al. [83] successfully fabricated MnTiO$_3$ nanofibers using the sol-gel assisted electrospinning method with polyvinylacetate, manganese acetate, and titanium (IV) isopropoxide as the main raw materials, and the effect of temperature on the synthesis of $MnTiO_3$ nanofibers was also investigated. SEM images showed that the microstructure of the prepared $MnTiO_3$ was significantly affected by the calcination temperature. Before heat treatment, the composite fiber precursors exhibited a smooth surface with an average diameter of about 850 nm. When the precursors were heated at 800 °C, the resultant fibers shrunk to an average diameter of about 328 nm with a rough surface (Figure 9). This phenomenon was caused by the decomposition of polyvinylacetate and subsequent crystallization. Increasing the calcination temperature to 900 °C, the obtained $MnTiO_3$ nanofibers became discrete in length and the average diameter increased to 415 nm. By increasing the calcination temperature to 1000 °C, only $MnTiO_3$ particles were observed in the resultant sample. The FTIR spectrum of prepared $MnTiO_3$ nanofibers revealed that the absorption peaks were found at 452 cm^{-1} and 532 cm^{-1}, which was associated with Ti-O and Mn-O bands, respectively, and demonstrated the formation of $MnTiO_3$ crystalline.

Figure 9. SEM image of $MnTiO_3$ fibers calcined at 800 °C [83].

Spinel structure materials also can be prepared by the sol-gel method. $ZnMn_2O_4$ has been widely used in Li-ion batteries, supercapacitors, sensors, and thermistors. Shamitha et al. successfully fabricated $ZnMn_2O_4$ nanofibers via the sol-gel assisted electrospinning method using poly(styrene-*co*-acrylonitrile) as a sacrificial polymeric binder, and the influence of calcination temperature on the synthesis of $ZnMn_2O_4$ nanofibers was investigated [84]. Before heat treatment, the average diameter of the composite fiber precursors was about 281 nm. When the precursors were fired at high temperature, the surface of the resultant fibers became rough and the average diameter of the resultant fibers obviously decreased, which was caused by the elimination of the organic phases. When the calcination temperature increased from 500 to 700 °C, the average diameter of the resultant fibers decreased from 243 to 181 nm due to crystallite growth and densification. The SAED patterns confirmed the formation of $ZnMn_2O_4$ nanofibers and demonstrated that the crystallinity of $ZnMn_2O_4$ nanofibers were enhanced with increasing calcination temperature. The nitrogen adsorption-desorption isotherms revealed that the prepared $ZnMn_2O_4$ nanofibers were mesoporous, which was caused by the elimination of a styrene-acrylonitrile random copolymer and the decomposition of metal acetates during heat treatment. The highest surface area of the prepared $ZnMn_2O_4$ nanofibers was about 79.51 $m^2 \cdot g^{-1}$, which was higher than that reported elsewhere. The reactions that may have occurred during the whole process were as follows:

$$MnCO_3 \rightarrow MnO + CO_2 \tag{1}$$

$$Mn_3O_4 + CO \rightarrow 3MnO + CO_2 \tag{2}$$

$$Mn_2O_3 + CO \rightarrow 2MnO + CO_2 \tag{3}$$

$$MnO_2 + CO_2 \rightarrow MnO + CO_2 \tag{4}$$

$$2ZnO + 2MnO + O_2 \rightarrow 2ZnMnO_3 \tag{5}$$

$$ZnMnO_3 + MnO \rightarrow ZnMn_2O_4 \tag{6}$$

It was noteworthy that the prepared $ZnMn_2O_4$ nanofibers could be a better electrode material for lithium ion batteries with a superior surface area.

The sol-gel method can be also applied to fabricate mullite fibers, which have exhibited good chemical and thermal stability, excellent high temperature mechanical strength, low thermal expansion coefficient, and thermal conductivity. Wei et al. [85] prepared flexible mullite nanofibers via electrospinning based on a nonhydrolytic sol-gel method using anhydrous aluminum chloride ($AlCl_3$), TEOS, PVP, and dichloromethane as the main starting materials. The nonhydrolytic sol was prepared first and then fired at 1000 °C for 1 h to obtain flexible mullite nanofibers, and the effect of calcination temperature on the synthesis of mullite fibers was studied. The study indicated that mullite fibers could be synthesized at relatively low heating temperatures. SEM images revealed that when mullite precursors were heated at 800 °C for 1 h, the average fiber diameter decreased from about 395 nm to about 250 nm (Figure 10). This phenomenon may have been induced by the burning-out of PVP after calcination. Meanwhile, the removal of PVP would have also left smooth surfaces on the nanofibers. By increasing the calcination temperature to 1000 °C, the average diameter of fibers was ulteriorly decreased to about 213 nm, which can be ascribed to the complete burn-up of PVP and further shrinkage of the fibers. TEM images of the product fired at 1000 °C for 1 h indicated that the size of the obtained mullite fibers was about 140 nm and the size of mullite grain in fibers was about 20 nm. On the other hand, the digital images of the mullite nanofibers fabricated at 1000 °C for 1 h revealed that the prepared fibers could be easily folded many times without breakage, which illustrated that the as-prepared mullite nanofibers were very soft and flexible. The resultant mullite fiber products that exhibit good high temperature properties could be used in high-temperature industrial and aerospace fields.

Figure 10. SEM image of the electrospun mullite fibers after calcination at 800 °C [85].

As the thermodynamically stable yttrium aluminum oxide ceramic, yttrium aluminum garnet ($Al_5Y_3O_{12}$ or YAG) has been widely used as high temperature structural materials. Ma et al. [86] prepared chromia-yttrium aluminum garnet (Cr-YAG) long fibers through the sol-gel method using aluminum chloride, aluminum powder, yttrium oxide, chromium trioxide, and acetic acid as the raw materials. The gel fibers were prepared by pulling out the thin glass rod immersed in the spinning sol slowly at room temperature, and the effect of heating temperature on the fabrication of Cr-YAG fibers was investigated. The study indicated that the YAG crystallized directly from the amorphous precursor without forming any intermediate phase. Due to the complete dissolution of Cr_2O_3 in solid solution, the Cr_2O_3 phase was not detected in the fiber products. SEM images revealed that the grain diameter of YAG and Cr-YAG fibers was about 1.45 and 1.38 nm, respectively (Figure 11), which illustrated that the solid solution ion may affect grain growth by grain growth pinning. The grain growth exponent (n) of Cr-YAG fibers (2.88) was slower than that of YAG ($n = 3$), indicating that the grain growth rate was reduced by adding Cr in YAG.

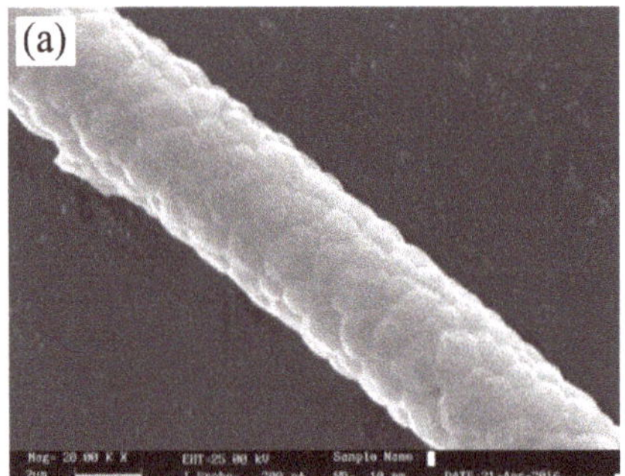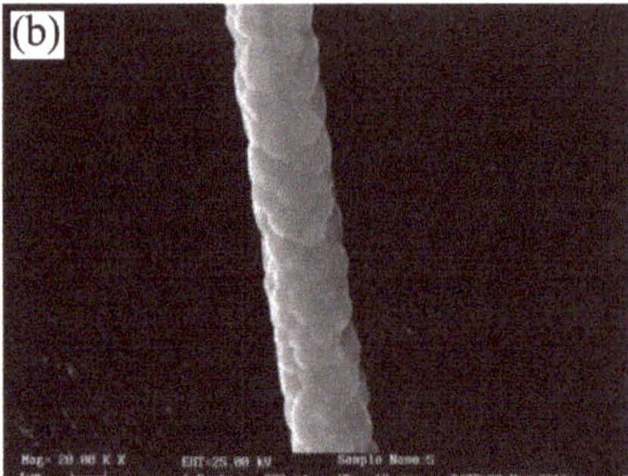

Figure 11. SEM microstructures of (**a**) YAG and (**b**) Cr-YAG precursor gel fibers heated at 1600 °C for 2 h and 6 h, respectively [86].

Composite fibers can also be fabricated by the sol-gel method. Lead magnesium niobate-lead titanate (PMN-PT) single crystals had ultrahigh piezoelectric properties when compared with traditional piezoelectric ceramics. Lam et al. [87] successfully prepared PMN-PT ($0.65Pb(Mg_{1/3}Nb_{2/3})O_3$-$0.35PbTiO_3$) ceramic fibers with the sol-gel method with lead (II) acetate trihydrate, magnesium nitrate salt, niobium(V) ethoxide, and titanium(IV) n-butoxide as the main raw materials. The effect of sintering temperature on the fabrication of PMN-PT ceramic fibers was studied, and XRD results revealed that samples sintered at different temperatures (1150, 1200, 1250 °C) had similar perovskite phases. SEM images showed that the microstructure of the PMN-PT fibers was dependent on the sintering temperature, and small cracks were observed in the resultant PMN-PT fibers due to the escaption of the organics during sintering. Interestingly, the relative permittivity and electromechanical coupling coefficient of the prepared PMN-PT ceramic fibers were larger than that of the ceramic disc, thus PMN-PT fibers can be used as reinforcements in 1–3 composites for high-frequency ultrasonic transducer applications.

Aside from the above-mentioned 1D structural materials, other 1D structural materials, including $CaZrO_3$ fibers [88], Al_2O_3-YAG nanostructured fibers [89], NiO nanofibers [90], Mg_2Si/CNT thermoelectric nanofibers [91], NbN fibers [92], etc. have also been prepared by the sol-gel technique. With the introduction of the sol-gel technique, 1D structural materials with high quality stoichiometric, uniform diameter, high purity, high homogeneity, and good continuity can be fabricated in large quantity under relatively low temperatures. Furthermore, the morphology (fiber, nanorod, nanowire, and nanotube), shape (hexagonal, round), aspect ratio, density and orientation can be tailored in the sol-gel process. However, cracks and rough surfaces were always observed in the final products. Furthermore, serious shrinkage and loosening structures were usually induced by the elimination of organic impurities. To overcome the existing drawbacks, assisted techniques should be used to decrease the gelling time, and inorganic binders are encouraged for reducing the influence induced by the removal of the organic components.

4. Conclusions

As a facile synthesis method, the sol-gel method exhibits some outstanding advantages for fabricating hollow sphere and 1D structural materials such as low reaction temperature, short soaking time, fine particle size, high purity products, and good chemical homogeneity. The as-prepared materials can be used in the fields of high temperature, semiconductor, photoelectric, magnetic, and so on. To extend the use of the sol-gel method, other technologies were introduced such as surface modification, sol coating, organic-inorganic hybridization, templating, etc. More encouragingly, the grain size, size distribution, surface, morphology and homogeneity of the targeted products could be tailored in the process of sol-gel.

However, long gelling time and various kinds of organic materials were always required to achieve a high quality gel precursor. In the process of removing organic matter, cracks, pores, rough surface, loosened structure and great volume shrinkage may be induced. To improve the quality of the targeted products, the used organics should be removed by calcination, dissolution, or etching. When organic components were burned out, cracks, pores, rough surface and great volume shrinkage may be induced. On the other hand, some of the solvents used to remove the organics were harmful. To enhance the green and environmentally-friendly ability of the sol-gel technique, the amount of volatile solvents and organic additives used in the sol-gel process should be decreased. Furthermore, new processing techniques should be developed to decrease the gelling time and improve the dispersibility of the targeted products. Additionally, the application of the sol-gel technique should be extended from single phase materials to complex systems (binary and ternary system) to fulfill increased demands across various fields.

We have tried to present the abilities and advantages of the sol-gel method used for preparing hollow sphere and 1D structural materials; however, it was impossible to review all the works carried out on this field. Therefore, only representative investigations were presented in detail with other works listed as references, and we extend our apologies to any overlooked contributions. Although some drawbacks were found in the sol-gel process, there was no doubt that more materials will be prepared by various sol-gel based routes.

Acknowledgments: This work was financially supported by the National Natural Science Foundation of China (General program, 51502211, 51472184, 51672194), the China Postdoctoral Science Foundation (2016M590721), and the Program for Innovative Teams of Outstanding Young and Middle-aged Researchers in the Higher Education Institutions of Hubei Province (T201602).

Conflicts of Interest: The authors declare no conflict of interest.

References

1. Zeynali, H.; Akbari, H. Magnetic Properties of $L1_0(FePt)_{100-x}Ag_x$ Nanoparticles Synthesized by the Sol-Gel Method. *J. Supercond. Nov. Magn.* **2016**, *29*, 1865–1869. [CrossRef]
2. Jang, M.S.; Roh, I.J.; Park, J.M.; Kang, C.Y.; Choi, W.J.; Baek, S.H.; Park, S.S.; Yoo, J.W.; Lee, K.S. Dramatic enhancement of the saturation magnetization of a sol-gel synthesized $Y_3Fe_5O_{12}$ by a mechanical pressing process. *J. Alloys Compd.* **2017**, *711*, 693–697. [CrossRef]
3. Zhang, H.; Liu, Z.; Ma, C.; Yao, X.; Zhang, L.; Wu, M. Preparation and microwave properties of Co- and Ti-doped barium ferrite by citrate sol-gel process. *Mater. Chem. Phys.* **2003**, *80*, 129–134. [CrossRef]
4. Zhang, H.; Liu, Z.; Yao, X.; Zhang, L.; Wu, M. Dielectric and magnetic properties of ZnCo-substituted X hexaferrites prepared by citrate sol-gel process. *Mater. Res. Bull.* **2003**, *38*, 363–372. [CrossRef]
5. Zhang, H.; Liu, Z.; Ma, C.; Yao, X.; Zhang, L.; Wu, M. Complex permittivity, permeability, and microwave absorption of Zn- and Ti-substituted barium ferrite by citrate sol-gel process. *Mater. Sci. Eng. B* **2002**, *96*, 289–295. [CrossRef]
6. Zhang, H.; Liu, Z.; Yao, X.; Zhang, L.; Wu, M. The Synthesis, Characterization and Microwave Properties of ZnCo-Substituted W-Type Barium Hexaferrite, from a Sol-Gel Precursor. *J. Sol-Gel Sci. Technol.* **2003**, *27*, 277–285. [CrossRef]
7. Zhang, H.; Yao, X.; Zhang, L. The preparation and microwave properties of $BaZn_{2-z}Co_zFe_{16}O_{27}$ ferrite obtained by a sol-gel process. *Ceram. Int.* **2002**, *28*, 171–175. [CrossRef]
8. Zhang, H.; Yao, X.; Zhang, L. The preparation and microwave properties of $Ba_2Zn_zCo_{2-z}Fe_{12}O_{22}$ hexaferrites. *J. Eur. Ceram. Soc.* **2002**, *22*, 835–840. [CrossRef]
9. Zhang, H.; Yao, X.; Zhang, L. The preparation and microwave properties of $Ba_2Zn_xCo_{2-x}Fe_{28}O_{46}$ hexaferrites. *J. Magn. Magn. Mater.* **2002**, *241*, 441–446. [CrossRef]

10. Fernández, C.P.; Zabotto, F.L.; Garcia, D.; Kiminami, R.H.G.A. In Situ sol-gel co-synthesis at as low hydrolysis rate and microwave sintering of PZT/Fe_2CoO_4 magnetoelectric composite ceramics. *Ceram. Int.* **2017**, *43*, 5925–5933. [CrossRef]

11. Kahouadji, B.; Guerbous, L.; Boukerika, A.; Dolić, S.D.; Jovanović, D.J.; Dramićanin, M.D. Sol gel synthesis and pH effect on the luminescent and structural properties of YPO_4: Pr^{3+} nanophosphors. *Opt. Mater.* **2017**, *70*, 138–143. [CrossRef]

12. Nouri, M.S.; Kompany, A.; Khorsand Zak, A.; Khorrami, Gh.H. Characterization of $Ce_{(1-x)}Zr_xO_2$ yellow nanopigments synthesized by a green sol-gel method. *Ceram. Int.* **2017**, *43*, 8482–8487. [CrossRef]

13. Almamoun, O.; Ma, S. Effect of Mn doping on the structural, morphological and optical properties of SnO_2 nanoparticles prepared by Sol-gel method. *Mater. Lett.* **2017**, *199*, 172–175. [CrossRef]

14. Hu, R.; Zhao, J.; Zheng, J. Synthesis of SnO_2/rGO hybrid materials by sol-gel/thermal reduction method and its application in electrochemical capacitors. *Mater. Lett.* **2017**, *197*, 59–62. [CrossRef]

15. Cao, E.; Wang, H.; Wang, X.; Yang, Y.; Hao, W.; Sun, L.; Zhang, Y. Enhanced ethanol sensing performance for chlorine doped nanocrystalline $LaFeO_{3-\delta}$ powders by citric sol-gel method. *Sens. Actuators B Chem.* **2017**, *251*, 885–893. [CrossRef]

16. Li, F.; Zhang, H.; Zhang, S.; Liang, F.; Liu, J.; Cao, Y. Low-temperature preparation of ZrC powders using a combined sol-gel and microwave carbothermal reduction method. *J. Ceram. Soc. Jpn.* **2016**, *124*, 1171–1174. [CrossRef]

17. Zhang, H.; Li, F.; Lu, L.; Zhang, S.; Cao, Y. Preparation and characterization of ultrafine ZrB_2-SiC composite powders by a combined sol-gel and microwave boro/carbothermal reduction method. *Ceram. Int.* **2015**, *41*, 7823–7829. [CrossRef]

18. Li, F.; Fu, F.; Lu, L.; Zhang, H.; Zhang, S. Preparation and artificial neural networks analysis of ultrafine β-Sialon powders by microwave-assisted carbothermal reduction nitridation of sol-gel derived powder precursors. *Adv. Powder Technol.* **2015**, *26*, 1417–1422. [CrossRef]

19. Zhang, H.; Li, F.; Jia, Q.; Ye, G. Preparation of titanium carbide powders by sol-gel and microwave carbothermal reduction methods at low temperature. *J. Sol-Gel Sci. Technol.* **2008**, *46*, 217–222. [CrossRef]

20. Zhang, H.; Li, F. Preparation and microstructure evolution of diboride ultrafine powder by sol-gel and microwave carbothermal reduction method. *J. Sol-Gel Sci. Technol.* **2008**, *45*, 205–211. [CrossRef]

21. Zhang, H.; Wang, Z.; Zhang, H. Synthesis of O′-SiAlON ultrafine powder. *Am. Ceram. Soc. Bull.* **2007**, *86*, 9401–9408.

22. Zhang, H.; Zhang, H.; Miao, J.; Wang, Z.; Jia, Q.; Jia, X. Preparation of Ultrafine β-Sialon Powder by Citrate Sol-Gel and Carbothermal Reduction Nitridation. *Key Eng. Mater.* **2007**, *336–338*, 927–929. [CrossRef]

23. Zhang, H.; Yan, Y.; Liu, Z. Effect of seeds on the synthesis of mullite powder by the citrate sol-gel method. *Interceram* **2005**, *54*, 328–331.

24. Zhang, H.; Jia, X.; Yan, Y.; Liu, Z.; Yang, D.; Li, Z. The effect of the concentration of citric acid and pH values on the preparation of $MgAl_2O_4$ ultrafine powder by citrate sol-gel process. *Mater. Res. Bull.* **2004**, *39*, 839–850. [CrossRef]

25. Zhang, H.; Jia, X.; Liu, Z.; Li, Z. The low temperature preparation of nanocrystalline $MgAl_2O_4$ spinel by citrate sol-gel process. *Mater. Lett.* **2004**, *58*, 1625–1628. [CrossRef]

26. Xiao, L.; Zhao, Y.; Yin, J.; Zhang, L. Clewlike ZnV_2O_4 hollow spheres: Nonaqueous sol-gel synthesis, formation mechanism, and lithium storage properties. *Chem. Eur. J.* **2009**, *15*, 9442–9450. [CrossRef] [PubMed]

27. Li, J.; Jiao, X.; Chen, D. Preparation of Y-TZP ceramic fibers by electrolysis-sol-gel method. *J. Mater. Sci.* **2007**, *42*, 5562–5569. [CrossRef]

28. Yang, Q.; Sha, J.; Ma, X.; Yang, D. Synthesis of NiO nanowires by a sol-gel process. *Mater. Lett.* **2005**, *59*, 1967–1970. [CrossRef]

29. Chelouche, A.; Touam, T.; Tazerout, M.; Djouadi, D.; Boudjouan, F. Effect of Li codoping on highly oriented sol-gel Ce-doped ZnO thin films properties. *J. Lumin.* **2017**, *188*, 331–336. [CrossRef]

30. Chen, R.; Zhang, Y.; Liu, T.; Xu, B.; Shen, Y.; Lin, Y.; Nan, C. Improvement of the conductivity of sol-gel derived Li-La-Zr-O thin films by the addition of surfactant. *Ceram. Int.* **2017**, in press. [CrossRef]

31. Ivanova, T.; Harizanova, A.; Koutzarova, T.; Vertruyen, B. Optical characterization of sol-gel ZnO:Al thin films. *Superlattice Microstruct.* **2015**, *85*, 101–111. [CrossRef]

32. Kayani, Z.N.; Riaz, S.; Naseem, S. Study of Nickel Nitride Thin Films Deposited by Sol-Gel Route. *Trans. Indian Inst. Met.* **2017**, *70*, 1097–1101. [CrossRef]

33. Predoana, L.; Stanciu, I.; Anastasescu, M.; Calderon-Moreno, J.M.; Stoica, M.; Preda, S.; Gartner, M.; Zaharescu, M. Structure and properties of the V-doped TiO_2 thin films obtained by sol-gel and microwave-assisted sol-gel method. *J. Sol-Gel Sci. Technol.* **2016**, *78*, 589–599. [CrossRef]

34. Ren, Q.; Zhang, Y.; Chen, Y.; Wang, G.; Dong, X.; Tang, X. Structure and magnetic properties of $La_{0.67}Sr_{0.33}MnO_3$ thin films prepared by sol-gel method. *J. Sol-Gel Sci. Technol.* **2013**, *67*, 170–174. [CrossRef]

35. Guo, X.; Zhang, Q.; Ding, X.; Shen, Q.; Wu, C.; Zhang, L.; Yang, H. Synthesis and application of several sol-gel-derived materials via sol-gel process combining with other technologies: A review. *J. Sol-Gel Sci. Technol.* **2016**, *79*, 328–358. [CrossRef]

36. Zhang, H.; Fu, F.; Cao, Y.; Du, S.; Lu, L.; Zhang, S. Sol-Gel Process Synthesis of High-Temperature Non-oxide Ultrafine Powders. *Interceram* **2013**, *62*, 282–286.

37. Livage, J.; Ganguli, D. Sol-gel electrochromic coatings and devices:A review. *Sol. Energy Mater. Sol. Cells* **2001**, *68*, 365–381. [CrossRef]

38. Yoldas, B.E. Technological significance of Sol-Gel process and process-induced variations in Sol-Gel materials and coatings. *J. Sol-Gel Sci. Technol.* **1993**, *1*, 65–77. [CrossRef]

39. Du, X.; He, J. Facile preparation of titania hollow spheres by combination of the mixed solvent method and the sol-gel process and post-calcination. *Mater. Res. Bull.* **2009**, *44*, 1238–1243. [CrossRef]

40. Dobó, D.G.; Berkesi, D.; Kukovecz, Á. Morphology conserving aminopropyl functionalization of hollow silica nanospheres in toluene. *J. Mol. Struct.* **2017**, *1140*, 83–88. [CrossRef]

41. Dai, Z.; Meiser, F.; Möhwald, H. Nanoengineering of iron oxide and iron oxide/silica hollow spheres by sequential layering combined with a sol-gel process. *J. Colloid Interf. Sci.* **2005**, *288*, 298–300. [CrossRef] [PubMed]

42. Qiao, M.; Wu, S.; Chen, Q.; Shen, J. Novel triethanolamine assisted sol-gel synthesis of N-doped TiO_2 hollow spheres. *Mater. Lett.* **2010**, *64*, 1398–1400. [CrossRef]

43. Chen, Z.; Wang, F.; Zhang, H.; Yang, T.; Cao, S.; Xu, Y.; Jiang, X. Synthesis of uniform hollow TiO_2 and SiO_2 microspheres via a freezing assisted reverse microemulsion-templated sol-gel method. *Mater. Lett.* **2015**, *151*, 16–19. [CrossRef]

44. Teng, Z.; Han, Y.; Li, J.; Yan, F.; Yang, W. Preparation of hollow mesoporous silica spheres by a sol-gel/emulsion approach. *Microporous Mesoporous Mater.* **2010**, *127*, 67–72. [CrossRef]

45. Yin, H.; Wang, X.; Wang, L.; Yuan, Q.; Zhao, H. Self-doped TiO_2 hierarchical hollow spheres with enhanced visible-light photocatalytic activity. *J. Alloys Compd.* **2015**, *640*, 68–74. [CrossRef]

46. Li, X.; Zhang, D.; Chen, Y. Silicone rubber/hollow silica spheres composites with enhanced mechanical and electrical insulating performances. *Mater. Lett.* **2017**, *205*, 240–244. [CrossRef]

47. Zhang, Y.; Li, G.; Wu, Y.; Xie, T. Sol-gel synthesis of titania hollow spheres. *Mater. Res. Bull.* **2005**, *40*, 1993–1999. [CrossRef]

48. Fan, H.; Lei, Z.; Jia, H.; Zhao, X. Sol-gel synthesis, microstructure and adsorption properties of hollow silica spheres. *Mater. Lett.* **2011**, *65*, 1811–1814. [CrossRef]

49. Ashuri, M.; He, Q.; Zhang, K.; Emani, S.; Shaw, L.L. Synthesis of hollow silicon nanospheres encapsulated with a carbon shell through sol-gel coating of polystyrene nanoparticles. *J. Sol-Gel Sci. Technol.* **2017**, *82*, 201–213. [CrossRef]

50. Deng, W.; Chen, D.; Chen, L. Synthesis of monodisperse CeO_2 hollow spheres with enhanced photocatalytic activity. *Ceram. Int.* **2015**, *41*, 11570–11575. [CrossRef]

51. Lin, X.; Rong, F.; Ji, X.; Fu, D. Visible light photocatalytic activity and Photoelectrochemical property of Fe-doped TiO_2 hollow spheres by sol-gel method. *J. Sol-Gel Sci. Technol.* **2011**, *59*, 283–289. [CrossRef]

52. Pullar, R.C.; Taylor, M.D.; Bhattacharya, A.K. Blow spun strontium zirconate fibers produced from a sol-gel precursor. *J. Mater. Sci.* **1998**, *33*, 3229–3232. [CrossRef]

53. Venkatesh, R.; Ramanan, S.R. Effect of organic additives on the properties of sol-gel spun alumina fibers. *J. Eur. Ceram. Soc.* **2000**, *20*, 2543–2549. [CrossRef]

54. Chandradass, J.; Balasubramanian, M. Extrusion of alumina fiber using sol-gel precursor. *J. Mater. Sci.* **2006**, *41*, 6026–6030. [CrossRef]

55. Lee, J.H.; Kim, Y.J. Hydroxyapatite nanofibers fabricated through electrospinning and sol-gel process. *Ceram. Int.* **2014**, *40*, 3361–3369. [CrossRef]

56. Tan, H.; Ding, Y.; Yang, J. Mullite fibers preparation by aqueous sol-gel process and activation energy of mullitization. *J. Alloys Compd.* **2010**, *492*, 396–401. [CrossRef]

57. Granger, G.; Restoin, C.; Roy, P.; Jamier, R.; Roungier, S.; Lecomte, A.; Blondy, J.M. Nanostructured optical fibers in the SiO_2/SnO_2 system by the sol-gel method. *Mater. Lett.* **2014**, *120*, 292–294. [CrossRef]

58. Tan, H.; Ma, X.; Fu, M. Preparation of continuous alumina gel fibers by aqueous sol-gel process. *Bull. Mater. Sci.* **2013**, *36*, 153–156. [CrossRef]

59. You, Y.; Zhang, S.; Wan, L.; Xu, D. Preparation of continuous TiO_2 fibers by sol-gel method and its photocatalytic degradation on formaldehyde. *Appl. Surf. Sci.* **2012**, *258*, 3469–3474. [CrossRef]

60. Liu, X.; Wang, J.; Zhang, J.; Yang, S. Sol-gel template synthesis of LiV_3O_8 nanowires. *J. Mater. Sci.* **2007**, *42*, 867–871. [CrossRef]

61. Senthil, T.; Anandhan, S. Structure-property relationship of sol-gel electrospun ZnO nanofibers developed for ammonia gas sensing. *J. Colloid Interface Sci.* **2014**, *432*, 285–296. [CrossRef] [PubMed]

62. Admaiai, L.F.; Daza, L.; Grange, P.; Delmon, B. Synthesis of $YBa_2Cu_3O_{7-x}$ superconductor with fiber structure by the sol-gel method. *J. Mater. Sci. Lett.* **1994**, *13*, 668–670. [CrossRef]

63. Boulton, J.M.; Jones, K.; Emblem, H.G. The preparation of spinel fiber by a sol-gel route. *J. Mater. Sci. Lett.* **1990**, *9*, 914–915. [CrossRef]

64. Ji, G.; Ji, H.; Li, M.; Li, X.; Sun, X. Synthesis of zirconium diboride nano-powders by novel complex sol-gel technology at low temperature. *J. Sol-Gel Sci. Technol.* **2014**, *69*, 114–119. [CrossRef]

65. Cao, Y.; Du, S.; Wang, J.; Zhang, H.; Li, F.; Lu, L.; Zhang, S.; Deng, X. Preparation of zirconium diboride ultrafine hollow spheres by a combined sol-gel and boro/carbothermal reduction technique. *J. Sol-Gel Sci. Technol.* **2014**, *72*, 130–136. [CrossRef]

66. Wang, Y.; Zhang, L.; Zhang, X.; Zhang, Z.; Tong, Y.; Li, F.; Wu, J.C.S.; Wang, X. Openmouthed β-SiC hollow-sphere with highly photocatalytic activity for reduction of CO_2 with H_2O. *Appl. Catal. B Environ.* **2017**, *206*, 158–167. [CrossRef]

67. Wang, T.; Ma, W.; Shangguan, J.; Jiang, W.; Zhong, Q. Controllable synthesis of hollow mesoporous silica spheres and application as support of nano-gold. *J. Solid State Chem.* **2014**, *215*, 67–73. [CrossRef]

68. Zhang, L.; Luo, J.; Wu, M.; Jiu, H.; Chen, Q. Synthesis of Eu_2O_3 hollow submicrometer spheres through a sol-gel template approach. *Mater. Lett.* **2007**, *61*, 4452–4455. [CrossRef]

69. Syoufian, A.; Manako, Y.; Nakashima, K. Sol-gel preparation of photoactive srilankite-type zirconium titanate hollow spheres by templating sulfonated polystyrene latex particles. *Powder Technol.* **2015**, *280*, 207–210. [CrossRef]

70. Yang, X.; Chaki, T.K. Millimetre-sized hollow spheres of lead zirconate titanate by a sol-gel method. *J. Mater. Sci.* **1996**, *31*, 2563–2567. [CrossRef]

71. Hu, Q.; Li, Y.; Zhao, N.; Ning, C.; Chen, X. Facile synthesis of hollow mesoporous bioactive glass sub-micron spheres with a tunable cavity size. *Mater. Lett.* **2014**, *134*, 130–133. [CrossRef]

72. Toyama, N.; Ohki, S.; Tansho, S.; Shimizu, T.; Umegaki, T.; Kojima, Y. Influence of alcohol solvents on morphology of hollow silica–alumina composite spheres and their activity for hydrolytic dehydrogenation of ammonia borane. *J. Sol-Gel Sci. Technol.* **2017**, *82*, 92–100. [CrossRef]

73. Zhu, Z.; Kao, C.T.; Tang, B.; Chang, W.; Wu, R. Efficient hydrogen production by photocatalytic water-splitting using Pt-doped TiO_2 hollow spheres under visible light. *Ceram. Int.* **2016**, *42*, 6749–6754. [CrossRef]

74. Lu, H.T.; Tseng, I.H. Fabrication of organosilica hollow spheres using organosiloxane-templated sol-gel process. *J. Sol-Gel Sci. Technol.* **2015**, *76*, 465–468. [CrossRef]

75. Mkhalid, I.A.; Abdulsalam, A.A. Photocatalytic reduction of Hg using core-shell Fe/CeO_2 hollow sphere nanocomposites. *Ceram. Int.* **2015**, *41*, 5614–5620. [CrossRef]

76. Katagiri, K.; Kamiya, J.; Koumoto, K.; Inumaru, K. Preparation of hollow titania and strontium titanate spheres using sol-gel derived silica gel particles as templates. *J. Sol-Gel Sci. Technol.* **2012**, *63*, 366–372. [CrossRef]

77. Chronakis, I.S. Novel nanocomposites and nanoceramics based on polymer nanofibers using electrospinning process-A review. *J. Mater. Process. Technol.* **2005**, *167*, 283–293. [CrossRef]

78. Nagamatsu, J.; Nakagawa, N.; Muranaka, T.; Zenitani, Y.; Akimitsu, J. Superconductivity at 39 K in magnesium diboride. *Nature* **2001**, *410*, 63–64. [CrossRef] [PubMed]

79. Nath, M.; Parkinson, B.A. A Simple Sol-Gel Synthesis of Superconducting MgB_2 Nanowires. *Adv. Mater.* **2006**, *18*, 1865–1868. [CrossRef]
80. Li, J.; Zhang, Y.; Li, G. Preparation and characterization of SiBON fiber. *Mater. Lett.* **2012**, *89*, 266–268. [CrossRef]
81. Tang, W.; Shan, X.; Li, S.; Liu, H.; Wu, X.; Chen, Y. Sol-gel process for the synthesis of ultrafine MnO_2 nanowires and nanorods. *Mater. Lett.* **2014**, *132*, 317–321. [CrossRef]
82. Zhou, G.; Kang, Y.S. Synthesis and structural properties of manganese titanate $MnTiO_3$ nanoparticle. *Mater. Sci. Eng. C* **2004**, *24*, 71–74. [CrossRef]
83. Nakhowong, R. Fabrication and characterization of $MnTiO_3$ nanofibers by sol-gel assisted electrospinning. *Mater. Lett.* **2015**, *161*, 468–470. [CrossRef]
84. Shamitha, C.; Senthil, T.; Wu, L.; Kumar, B.S.; Anandhan, S. Sol-gel electrospun mesoporous $ZnMn_2O_4$ nanofibers with superior specific surface area. *J. Mater. Sci. Mater. Electron.* **2017**, *20*, 1–15. [CrossRef]
85. Wei, H.; Li, H.; Cui, Y.; Sang, R.; Wang, H.; Wang, P.; Bu, J.; Dong, G. Synthesis of flexible mullite nanofibers by electrospinning based on nonhydrolytic sol-gel method. *J. Sol-Gel Sci. Technol.* **2017**, *82*, 718–727. [CrossRef]
86. Ma, X.; Lv, Z.; Tan, H.; Nan, J.; Wang, C.; Wang, X. Preparation and grain-growth of chromia-yttrium aluminum garnet composites fibers by sol-gel method. *J. Sol-Gel Sci. Technol.* **2017**, *83*, 275–280. [CrossRef]
87. Lam, K.H.; Li, K.; Chan, H.L.W. Lead magnesium niobate-lead titanate fibers by a modified sol-gel method. *Mater. Res. Bull.* **2005**, *40*, 1955–1967. [CrossRef]
88. Liu, B.; Lin, X.; Zhu, L.; Wang, X.; Xu, D. Fabrication of calcium zirconate fibers by the sol-gel method. *Ceram. Int.* **2014**, *40*, 12525–12531. [CrossRef]
89. Shojaie-Bahaabad, M.; Taheri-Nassaj, E.; Naghizadeh, R. An alumina-YAG nanostructured fiber prepared from an aqueous sol-gel precursor: Preparation, rheological behavior and spinnability. *Ceram. Int.* **2008**, *34*, 1893–1902. [CrossRef]
90. George, G.; Anandhan, S. Comparison of structural, spectral and magnetic properties of NiO nanofibers obtained by sol-gel electrospinning from two different polymeric binders. *Mater. Sci. Semicond. Process.* **2015**, *32*, 40–48. [CrossRef]
91. Kikuchi, K.; Yamamoto, K.; Nomura, N.; Kawasaki, A. Synthesis of n-type Mg_2Si/CNT Thermoelectric Nanofibers. *Nanoscale Res. Lett.* **2017**, *12*, 343. [CrossRef] [PubMed]
92. Nomura, K.; Takasuka, Y.; Kamiya, K.; Nasu, H. Preparation of NbN fibers by nitridation of sol-gel derived Nb_2O_5 fibers. *J. Mater. Sci. Mater. Electron.* **1994**, *5*, 53–58. [CrossRef]

 *materials* 

Article

Zirconia/Hydroxyapatite Composites Synthesized Via Sol-Gel: Influence of Hydroxyapatite Content and Heating on Their Biological Properties

Flavia Bollino [1,*], Emilia Armenia [2] and Elisabetta Tranquillo [1]

[1] Department of Industrial and Information Engineering, University of Campania "Luigi Vanvitelli", 81031 Aversa, Italy; elisabetta.tranquillo@unicampania.it

[2] Department of Cardiothoracic and Respiratory Sciences, University of Campania "Luigi Vanvitelli", 80131 Naples, Italy; emiliaarmenia@hotmail.it

* Correspondence: flavia.bollino@unina2.it; Tel.: +39-081-501-0483

Academic Editor: Michelina Catauro
Received: 17 May 2017; Accepted: 30 June 2017; Published: 5 July 2017

Abstract: Zirconia (ZrO_2) and zirconia-based glasses and ceramics are materials proposed for use in the dental and orthopedic fields. In this work, ZrO_2 glass was modified by adding different amounts of bioactive and biocompatible hydroxyapatite (HAp). ZrO_2/HAp composites were synthesized via the sol-gel method and heated to different temperatures to induce modifications of their chemical structure, as ascertained by Fourier transform infrared spectroscopy (FTIR) analysis. The aim was to investigate the effect of both HAp content and heating on the biological performances of ZrO_2. The materials' bioactivity was studied by soaking samples in a simulated body fluid (SBF). FTIR and scanning electron microscopy (SEM)) analyses carried out after exposure to SBF showed that all materials are bioactive, i.e., they are able to form a hydroxyapatite layer on their surface. Moreover, the samples were soaked in a solution containing bovine serum albumin (BSA). FTIR analysis proved that the synthesized materials are able to adsorb the blood protein, the first step of cell adhesion. WST-8 ([2-(2-methoxy-4-nitrophenyl)-3-(4-nitrophenyl)-5-(2,4-disulfophenyl)-2H-tetrazolium, monosodium salt]) assay showed that no cytotoxicity effects were induced by the materials' extract. However, the results proved that bioactivity increases with both the HAp content and the temperature used for the thermal treatment, whereas biocompatibility increases with heating but is not affected by the HAp content.

Keywords: sol-gel method; Fourier transform infrared spectroscopy (FTIR) analysis; bioactivity; biocompatibility

---

1. Introduction

Zirconia and zirconia-based glasses and ceramics have attracted considerable interest as materials to be used in the biomedical field [1–5]. In vivo and in vitro studies showed that zirconia and zirconia-based glasses and ceramics do not induce any cytotoxicity effects in either soft or hard tissue [3,5]. When a zirconia prosthesis was implanted in vivo, it was encapsulated by connective tissue, and thus any local or systemic toxic effect was not recorded after its insertion [3,5]. However, although zirconia is well tolerated, it is a bioinert material, as it does not show either the ability of direct bone bonding or osteoconduction behavior. Moreover, release of residues or degradation phenomena of zirconia implants were not detected and a low bacterial growth was observed [3,5]. In addition to these biological properties, the good mechanical behavior of zirconia has stimulated the interest of researchers in the biomaterials field. ZrO_2 is an oxide which presents three types of crystalline structures at ambient pressure: (i) the monoclinic phase (m-ZrO_2), which is stable from room temperature up to 1170 °C and exhibits poor mechanical properties; (ii) the tetragonal phase (t-ZrO$_2$), which is stable in the temperature range 1170–2370 °C and has good mechanical properties; and (iii) the cubic phase (c-ZrO$_2$), which is stable above 2370 °C and has moderate mechanical properties [6–8]. In order to stabilize the tetragonal phase at room temperature, zirconia can be mixed with other metallic oxides (e.g., MgO, La$_2$O$_3$, CaO, Y$_2$O$_3$) to obtain strong ceramics. Among such materials, yttrium-stabilized zirconia (known as tetragonal zirconia polycrystal (TZP)) is one of the most studied [9]. As their mechanical properties, such as resistance to traction and compression resistance, are similar to those of metals (e.g., stainless steel) [10], and they have good biocompatibility, zirconia and zirconia-based ceramics were first used in orthopedics as a substitute for titanium and alumina in hip head prostheses [11]. Afterwards, they were successfully used in dentistry, where they are still considered to be a material of choice for root canal posts, fixed partial dentures, and dental implants [12]. Moreover, some studies [13–16] report the use of TZP as fillers to reinforce the mechanical properties of synthetic hydroxyapatite (HAp) (a biodegradable and biocompatible calcium phosphate ceramic with a bone-like structure, capable of forming strong chemical bonds with natural bone tissue [17]). Other studies investigated the use of zirconia-based glasses, synthesized via the sol-gel method, as drug delivery systems [2,18] or as coatings capable of improving the biological performance of titanium implants [19–22]. Zirconia-based glasses, synthesized by means of the sol-gel method, showed the ability to slightly improve cell viability of the human osteosarcoma cell line (SAOS-2) [20] and human mesenchymal stromal cells (hMSCs) [22]. This is ascribable mainly to the preparation method. The sol-gel technique, indeed, is a versatile process used to make glass and ceramic materials at low temperature, and has been extensively used to prepare a wide variety of materials with different applications, including the bioglass [23–27]. The process starts from a solution of metal alkoxide or metal salt precursors in water-alcohol and involves their hydrolysis and condensation reactions, which lead to the formation of a 3D rigid gel [28]. By drying the obtained wet gel, it is possible to prepare xerogels (by exposure to low temperatures), aerogels (by solvent extraction under supercritical conditions), or dense ceramics and glasses by means of a further heat treatment at higher temperatures. Glasses and ceramics synthesized via the sol–gel method exhibit higher bioactivity and biocompatibility than materials with the same composition but prepared by melt–quenching [29–31]. Sol-gel-derived glasses have an inherent mesoporosity that gives them a larger surface area and potentially more rapid degradation rates than melt-derived glasses of a similar composition. Moreover, the presence of –OH groups on their surface stimulates hydroxyapatite nucleation, promoting easier osseointegration.

The aim of the present study has been to synthesize zirconia-based composites containing different amounts of HAp (xZrO$_2$·(1 − x)HAp, with x = no stabilized ZrO$_2$ mole fraction) via the sol-gel method in order to investigate the influence of the biocompatible and bioactive HAp on the biological response of the synthesized composites. Moreover, it is known [32,33] that both sol-gel zirconia and HAp glasses can crystallize by heating, leading to different crystalline phases. Therefore, the gel materials were heated at 120 °C, 600 °C, and 1000 °C to induce phase transformation and, in turn, to study the effect of thermal treatment on the biological response of the obtained materials. The choice of the temperature has been based on published studies [32,33]. Fourier transform infrared spectroscopy (FTIR) analysis was carried out to follow the materials' structural modification, induced by both HAp addition and heating. Moreover, bioactivity and biocompatibility of all samples were studied by in vitro preliminary tests after the different thermal treatments.

2. Results and Discussion

2.1. Chemical Characterization

The heat treatments of ZrO$_2$ and xZrO$_2$·(1 − x)HAp gels led to transformations visible to the naked eye. The samples heated to 120 °C are yellow. After heating to 600 °C they became black, whereas after heating to 1000 °C the powders became white, according to literature [34,35]. The change in the sample color is ascribable to the transformation of the ZrO_2 structure induced by heating and studied by [35,36] X-ray diffraction (XRD) and thermal analyses of zirconia, as reported in the literature [32,35–37]. It was shown that Acetil Acetone (AcAc)-containing zirconia, synthesized via the sol-gel method, is an amorphous and yellow-brown material, because a complex between the zirconium and the AcAc is formed in the sol phase, which exhibits strong absorption in the UV region and low-intensity absorption in the visible spectral region [38].

The increase in temperature leads to the condensation of surface –OH groups with the formation of H_2O. The sites involved in this reaction are also the sites for the promotion of the nucleus formation of the tetragonal phase. At about 400 °C, the tetragonal phase begins to crystallize [32,37] and the material gradually darkens [34,35]. At 600 °C the material is entirely converted into the tetragonal phase (t-ZrO_2) and it appears completely black [34,35]. This phenomenon is correlated to the formation of defects in the material, according to Wachsman et al. [35]. The dihydroxylation process, which takes place on the material's surface above 100 °C, leads to oxygen surface desorption and, thus, to the formation of oxygen vacancies [32]. The gradual transformation of t-ZrO_2 into the monoclinic phase (m-ZrO_2) occurs [32,37] when the temperature is further increased. Therefore, when the material was heat treated at 1000 °C, it was transformed completely into the monoclinic structure and appears white, indicating a lack of oxygen vacancies, as reported by Wachsman et al. [35].

Despite the fact that HAp appears as a white powder regardless of the temperature used for heat treatment, the $xZrO_2 \cdot (1 - x)HAp$ composites retain the zirconia color variations. To follow the material transformations induced by heating, FTIR analysis of all samples was carried out as a function of the temperature (Figures 1–3).

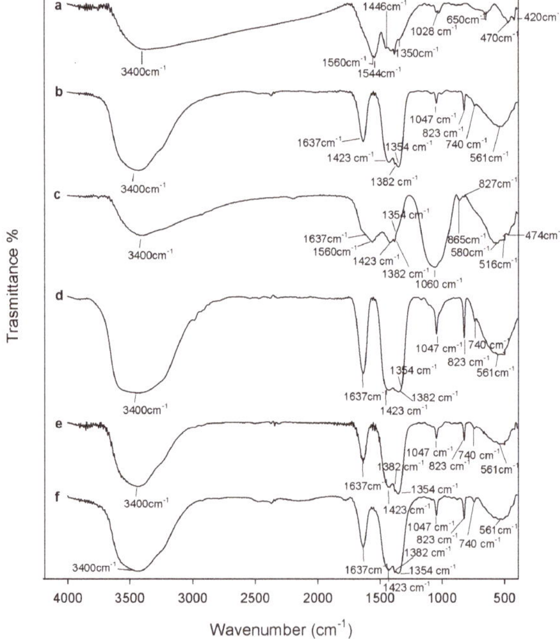

Figure 1. Fourier transform infrared spectroscopy (FTIR) of (**a**) ZrO_2; (**b**) hydroxyapatite (Hap); (**c**) $9ZrO_2 \cdot 1Hap$; (**d**) $7ZrO_2 \cdot 3Hap$; and (**e**) $5ZrO_2 \cdot 5HAp$ gels, heated to 120 °C and (**f**) $Ca(NO_3)_2$ $4H_2O$.

Figure 1 shows the FTIR spectra of all samples after drying at 120 °C. The FTIR spectrum of ZrO_2 after 120 °C heating (Figure 2a) shows all peaks typical of amorphous acetyl acetone (AcAc) containing zirconia sol-gel materials [36,39]. The broad intense band at 3400 cm^{-1} is due to the vibrations of –OH groups in both physically bonded water and Zr-OH groups in the matrix. Moreover, the bands observed at 1560 cm^{-1} and 1446 cm^{-1} are assigned to C=O vibrations of AcAc bidentate binding. The bands at 1544 cm^{-1} and 1350 cm^{-1} are due to C–C vibrations. The peaks at 1028 cm^{-1} are assigned to C–C–H bending, mixed with stretching C–C vibrations of AcAc [39]. The bands at 650 cm^{-1} and 470 cm^{-1} are due to Zr–OH and Zr–O–Zr stretching, respectively [10,11], whereas the peak at 420 cm^{-1} is attributed to Zr–OAcAc vibrations [39].

The spectrum of HAp after 120 °C heating (Figure 1b) is very similar to the calcium nitrate spectrum (Figure 1f). Only the bands of the –OH stretching and banding (at 3400 cm^{-1} and 1637 cm^{-1} respectively), in the adsorbed water and of nitrate vibrations are visible, such as the signals related to asymmetric and symmetric stretching of nitrate ions at 1423, 1382, and 1354 cm^{-1} [42], the sharp peaks at 1047, 823 cm^{-1}, and the weak peak at 738 cm^{-1}, ascribable to the bending modes of the nitrate ions [42]. Calcium nitrate, indeed, was soluble in the sol and it remained in solution as the particles formed and coalesced. During drying at 120 °C, calcium nitrate coats the formed particles, therefore only the signals of the salt are visible, while those of the phosphate are masked. Only after the thermal degradation of nitrate ions (at temperatures over 500 °C [30]) does the calcium enter the network by diffusion, and the nitrate by-products are driven off. Therefore, a cleavage of the bridging oxygen bonds and the formation of non-oxygen bonds (and thus, ionic-crosslinkage with Ca^{2+} ions) occurs in the glass network [30,43,44]. Also, Catauro et al. [45] observed a similar phenomenon in SiO$_2$-CaO-P$_2$O$_5$ ternary systems synthesized via sol-gel, where calcium nitrate was used as precursor of the CaO phase. The FTIR spectra of the materials dried at 120 °C were dominated by nitrate signals. However, the authors proved the presence of the silica and phosphate phases by FTIR analysis of the materials after soaking in a water solution. Only the signals of the silica and phosphate phases, and no signals of nitrates, were visible in the spectra. Moreover, IC analysis of the water solution showed that nitrate ions were released.

For the same reason the spectra of 0.7ZrO$_2$·0.3HAp and 0.5ZrO$_2$·0.5HAp samples (Figure 1d,e) are also very similar to the Ca(NO$_3$)$_2$·4H$_2$O spectrum, whereas the spectrum of the 0.9ZrO$_2$·0.1HAp (Figure 1c) sample also shows signals, due to ZrO$_2$ and the phosphate phase. In particular, two intense bands are still visible in the region 1700–1300 cm^{-1}, but with some differences in the shape and position compared to the ZrO$_2$ spectrum. This can be due to the influence of: (i) nitrate vibrations in the region 1500–1300 cm^{-1}, which also cause the appearance of the weak peak at 827 cm^{-1} [42]; and (ii) vibrations of carbonate ions not incorporated in the apatitic structure, in the region 1500–1400 cm^{-1} [46]. The presence of carbonate ions, due to the solubilization of atmospheric CO$_2$ in the sol, is also proved by the presence of the weak peak at 865 cm^{-1} [14,15,46]. Moreover, (iii) it can be hypothesized that a new complex between calcium ions and AcAc in the sol phase is formed, and originates IR bands in the same region. It has been reported in literature [47], indeed, that calcium bis(acetylacetonate) is synthesized by adding calcium nitrate tetrahydrate and AcAc to an aqueous solution of ammonium hydroxide, which are reagents present in the sol of the xZrO$_2$·(1 − x)HAp composites. Moreover, the intense band at 1060 cm^{-1} (ascribable to PO$_4^{3-}$ vibration [13,45,48,49]) and the broad band in the region 750–400 cm^{-1} suggest that an amorphous calcium phosphate was formed [46,50]. The presence of the PO$_4^{3-}$ vibration suggests that in this material the calcium nitrate does not coat the glass network. It can be explained by the formation of the complex between calcium ions and AcAc, which subtracts the Ca^{2+} ions by the solution, reducing the interaction between nitrate ions and the glass network. As in the 0.9ZrO$_2$·0.1HAp sample, a lower ratio between the amount of HAp and ZrO$_2$ is present; a higher ratio AcAc/Ca(NO$_3$)$_2$·4H$_2$O was present in the sol of this sample, compared to the other samples. Therefore, a higher amount of the Ca-AcAc complex was formed.

Figure 2 shows FTIR spectra of all samples after heating at 600 °C. In the FTIR spectrum of ZrO$_2$ (Figure 2a) the peaks related to AcAc disappear due to its degradation [51], while sharp peaks in the region (generally assigned to the vibrations of Zr-O-Zr and Zr-OH bonds in a crystalline structure) appear [52,53], such as those at 780, 578, 499, and 430 cm^{-1}.

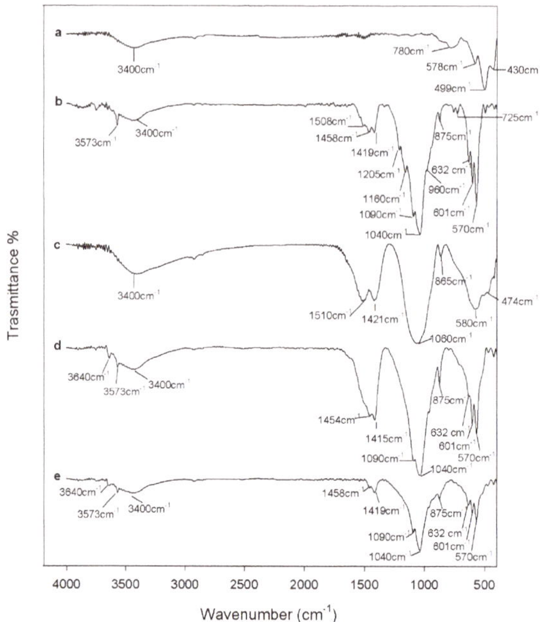

Figure 2. FTIR of (**a**) ZrO_2; (**b**) Hap; (**c**) $9ZrO_2 \cdot 1Hap$; (**d**) $7ZrO_2 \cdot 3Hap$; and (**e**) $5ZrO_2 \cdot 5HAp$ gels, heated to 600 °C.

After 600 °C heating, degradation of the nitrate ions occurs and calcium ions enter into the network by diffusion [30]. Therefore, the spectra of HAp, of $0.7ZrO_2 \cdot 0.3Hap$, and of $0.5ZrO_2 \cdot 0.5HAp$ composites (Figure 2b,d,e) completely change shape. In particular, FTIR analysis confirms that pure HAp was obtained (Figure 2b). All hydroxyapatite typical peaks, indeed, are visible. The intense bands at 1090 and 1040 cm^{-1}, as well as the shoulder at 960 cm^{-1}, are due to asymmetric and symmetric stretching in PO_4^{3-} ions, respectively. The doublet at 601 and 570 cm^{-1} is due to phosphate bending modes and the sharp peaks at 3573 and 632 cm^{-1} are assigned to stretching and wagging vibrations of –OH groups in the crystalline apatite structure [14,54]. However, a calcium-deficient hydroxyapatite was obtained as proven by the signals of CO_3^{2-} and HPO_4^{2-} ions. In particular, the broad band in the range 1500–1400 cm^{-1} and the weak peak at 875 cm^{-1} are assigned to stretching and bending in the CO_3^{2-} ions, which can substitute OH^- (type A substitution) or PO_4^{3-} ions (more commune type B substitution) [54]. The signal at 875 cm^{-1} can also be ascribed to the presence of HPO_4^{2-} that characterizes the non-stoichiometric HAp and indicates the formation of anhydrous dicalcium phosphate (DCPA, Ca_2HPO_4) [54,55]. As the two signals overlap, it is generally difficult to distinguish between CO_3^{2-} and HPO_4^{2-} groups. However, in the recorded HAp spectrum, the sharp peaks at 1205 cm^{-1}, 1160 cm^{-1}, and 725 cm^{-1} confirm the formation of DCPA. Those peaks, indeed, are due to the P-O vibrations in pyrophosphate ions, which originate by the condensation of two HPO_4^{2-} to $P_2O_7^{4-}$ [55,56]. The spectra of $0.7ZrO_2 \cdot 0.3HAp$ and $0.5ZrO_2 \cdot 0.5HAp$ composites (Figure 2d,e) show the typical signals of carbonated HAp, but the carbonate bands have a different shape and intensity. The shape of this band mainly depends on the presence of "non-apatitic" or "apatitic" CO_3^{2-} ions, and in the latter case, on the substitution type in the HAp lattice (A or B type) [46]. The complexity of this band in the HAp spectrum (Figure 2b) suggests that different types of CO_3^{2-} ions are present. Generally, a doublet shape in the region 1450–1410 cm^{-1}, coupled to a weak peak at 870–875 cm^{-1}, is due to the vibrations of type B carbonate, whereas in the region 1450–1550 cm^{-1} vibrations of type A carbonate are visible, which generally are coupled with a band at 880 cm^{-1} [57,58]. However, at

1500 and 1420 cm^{-1}, vibrations of non-apatitic CO$_3{}^{2-}$ ions are also present, coupled with a weak peak at about 866 cm^{-1} [58]. Therefore, in HAp samples, where peaks at 1530 cm^{-1}, 1455 cm^{-1}, and 1415 cm^{-1} are distinguishable, apatitic type A and B carbonate ions are also present [59]. In the spectra of 0.7ZrO$_2$·0.3HAp and 0.5ZrO$_2$·0.5HAp composites (Figure 2d,e), the typical doublet of apatitic CO$_3{}^{2}$ type B is visible at 1455 cm^{-1} and 1415 cm^{-1} [57]. It cannot be excluded that the vibrations of CaCO$_3$, produced by the thermal decomposition of calcium acetylacetonate [47], contribute to the increase of the doublet intensity, visible by comparing the HAp spectrum to the composite spectra and the composite spectra to each other. Its production and, thus, the band intensity, grows with zirconia content in the materials, due to the higher amount of AcAc used in the synthesis. The presence of the sharp peak at 3640 cm^{-1}, ascribable to –OH stretching in calcium hydroxide, suggests that the formation of a CaO phase [50] also occurred in the 0.7ZrO$_2$·0.3HAp samples. The spectrum of the 0.9ZrO$_2$·0.1HAp sample (Figure 2c) shows that the formation of the HAp structure still had not occurred. This can be due to the high amount of zirconia. Other works [60,61] in literature reported the reduction of the HAp crystallinity extent, due to the incorporation of silica or zirconia in the HAp lattice. Therefore, not many differences are evident by comparing this spectrum to the spectra of the 120 °C heated samples (Figure 1c), except for the position and intensity of the doublet in the region 1500–1300 cm^{-1}. The signals are shifted to 1510 cm^{-1} and 1420 cm^{-1}, and the weak peak at 865 cm^{-1} is still present in the spectrum. This suggests the prevalence of non-apatitic carbonate ions. Moreover, its higher intensity compared to pure Hap, 0.7ZrO$_2$·0.3HAp and 0.5ZrO$_2$·0.5HAp composites can be ascribed to the degradation of the higher calcium acetylacetonate content, due to the use of the higher amount of AcAc in the synthesis.

Figure 3 shows the FTIR spectra of all samples after 1000 °C heating. All peaks in the spectrum of the ZrO$_2$ sample (Figure 3a) appear more intense and sharp compared to the spectrum of ZrO$_2$ after 600 °C heating (Figure 2a). This observation suggests that an increase of sample crystallinity occurred. Moreover, the broad weak band at 718 cm^{-1} was replaced by an intense peak at 748 cm^{-1}, typical of the m-ZrO$_2$ spectrum [52].

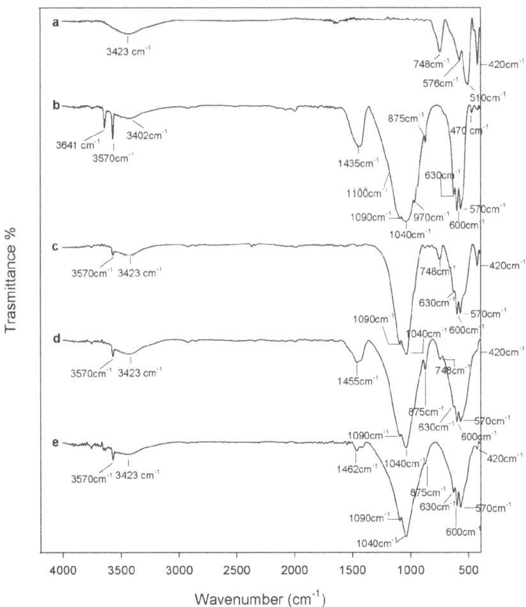

Figure 3. FTIR of (**a**) ZrO$_2$; (**b**) Hap; (**c**) 9ZrO$_2$·1Hap; (**d**) 7ZrO$_2$·3Hap; and (**e**) 5ZrO$_2$·5HAp gels, heated to 1000 °C.

After 1000° heating of the pure HAp sample the degradation of DCPA occurred, as proved by the disappearance of $P_2O_7^{4-}$ peaks from the spectrum (Figure 3b) [54–56]. It is known [62] that calcium-deficient hydroxyapatite contains hydrogen phosphate ions which condense to pyrophosphate in the temperature range 600–700 °C. The pyrophosphate formed, in turn, degrades at higher temperatures producing HAp and tricalcium phosphate polymorphs β (β-TCP). The formation of β-TCP as a secondary phase is confirmed by the shoulders at 970 and 1100 cm^{-1} [54]. Moreover, the carbonate peaks are still present after 1000 °C heating, but with different shapes. Part of the $CaCO_3$, indeed, decomposes into CO_2 and CaO. The first is released as a volatile gas; the second is retained in the material, as proved by the sharp peak at 3641 cm^{-1} [50,63]. Therefore, the single band at 1435 cm^{-1} and the weak peak at 875 cm^{-1} are ascribable to the residual apatitic B type carbonate. After 1000 °C heating, the formation of HAp also occurred in the 0.9ZrO$_2$·0.1HAp sample (Figure 3c), as proven by the modification of the FTIR spectrum, which assumes the typical shape of the HAp spectrum. The broad band in the region 700–400 cm^{-1} and the strong peak at 1060 cm^{-1} split, and the typical HAp doublets arise at 600–570 cm^{-1} and 1090–1040 cm^{-1}, respectively. Moreover, the signal of the –OH groups in HAp are also visible at 3570 and 630 cm^{-1}, whereas the carbonate signals disappear, confirming that such ions were not strictly incorporated in the materials structure (non apatitic carbonate). However, the peaks at 748 and 420 cm^{-1}, and the shoulder at 510 cm^{-1}, prove the presence of ZrO$_2$ in the sample. ZrO$_2$ peaks are still visible in the spectrum of the 1000° heated 0.7ZrO$_2$·0.3HAp sample (Figure 3d), but with lower intensity and as shoulders, whereas they are not detectable in the spectra of the 0.5ZrO$_2$·0.5HAp sample (Figure 3e) because they are overlapped by HAp signals. In the spectra of the 1000° heated 0.7ZrO$_2$·0.3HAp and 0.5ZrO$_2$·0.5HAp samples, the signal of CO_3^{2-} type B residues are also visible. The band at 1455 cm^{-1} and the weak peak of 875 cm^{-1}, indeed, are ascribable to residual apatitic carbonate [46,50]. Moreover, the formation of β-TCP, as a degradation product of the thermal degradation of the HAp phases, probably occurs also in those composites samples. In the literature, indeed, it is reported that zirconia can act as a catalyst of the decomposition reaction [61], which, thus, can also occur at a lower temperature [13]. However, β-TCP is undetectable in the spectra of the composites, because in those materials a lower HAp content is present which, thus, leads to the formation of a lower β-TCP amount.

2.2. Evaluation of Biological Properties

The biocompatibility of the synthesized samples after 600 °C and 1000 °C heating was assessed by evaluating both the materials' ability to absorb blood proteins on their surface and the materials' cytotoxicity. The biocompatibility of the samples after 120 °C heating was not tested, as in these samples toxic nitrate ions are present, as proved by FTIR analysis (Figure 1).

Figures 4 and 5 show FTIR spectra of all samples, heated to 600 °C and 1000 °C, respectively, after 24 h of exposure to bovine serum albumin (BSA) solution. The comparison between sample spectra (Figure 4, curves from b to e) and the BSA spectrum (Figure 4, curve a) showed that after heat treatment at 600 °C, the protein adsorption is low and occurs only on the surface of xZrO$_2$·(1 − x)HAp composites. Indeed, the main peak of albumin at 1654 cm^{-1}, due to the stretching of C-O in amide I [64], is visible with low intensity only in the FTIR spectra of the samples 0.9ZrO$_2$·0.1HAp (as a shoulder of the band at 1512 cm^{-1}), 0.7ZrO$_2$·0.3Hap, and 0.5ZrO$_2$·0.5HAp as weak peak.

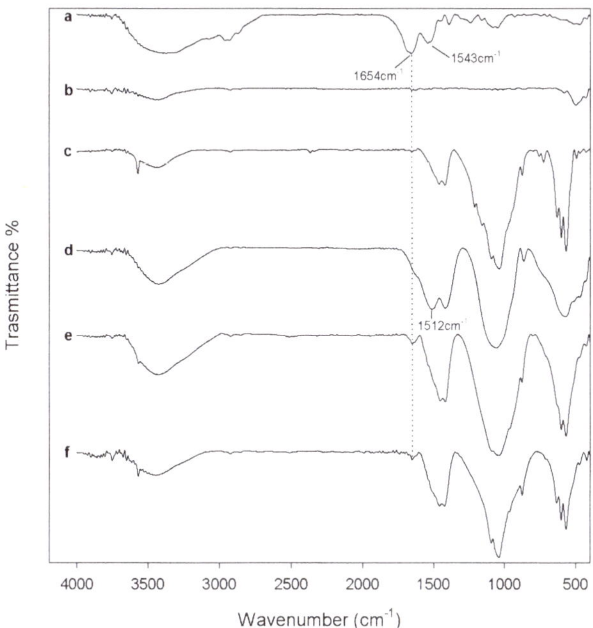

Figure 4. FTIR of (**a**) bovine serum albumin (BSA) and (**b**) ZrO$_2$; (**c**) Hap; (**d**) 9ZrO$_2$·1Hap; (**e**) 7ZrO$_2$·3Hap; and (**f**) 5ZrO$_2$·5HAp gels, heated to 600 °C, after 24 h of exposure to the BSA solution.

The presence of the adsorbed proteins is more evident in the samples heat-treated to 1000 °C (Figure 5). In the spectra of all samples, except in the FTIR spectrum of pure ZrO$_2$, an increase of the intensity of the albumin band at 1654 cm^{-1} is visible. Moreover, in the samples of 0.5ZrO$_2$·0.5HAp the BSA band at 1543 cm^{-1} also appears, due to N–H in-plane bending of amide II [64]. Therefore, the protein adsorption is affected by heat treatment carried out on the materials after synthesis, and is higher when the materials are heated to 1000 °C. This can be ascribed to a different degree of ions' release, from the materials heated to different temperatures. Mavropoulos et al. [65] proved that BSA adsorption on synthetic hydroxyapatite is affected by ions present in the solution containing the protein. In particular, the authors observed a decrease of the BSA adsorption with an increase of the phosphate concentration in the BSA solution. The presence of a high amount of PO$_4$$^{3-}$ in the diffusion layer at the HAp surface, indeed, results in an increase of the electrostatic repulsion force between HAp and BSA. Moreover, Catauro et al. [45] showed that calcium phosphates ternary systems, heated to 600 °C, release a higher amount of Ca^{2+} and PO$_4$$^{3-}$ ions compared to the same materials heated at 1000 °C. Therefore, the test results can be explained by a higher ion release from the materials heated to 600 °C, compared to those heated to 1000 °C (which modifies the pH of the solution, the charges of both the materials surface, and the BSA and, thus, affects their interaction). The higher ion release observed in the sol-gel materials heated to a lower temperature is due to their lower crystallinity degree, as reported in the literature [66].

Moreover, the presence of HAp improves the ZrO$_2$ ability of absorbing proteins. As the blood protein adsorption on the sample surface is the first step leading to cell adhesion and proliferation, the test results suggest that the materials heated to 1000 °C are more biocompatible than those heated to 600 °C, regardless of the HA amount in the samples.

In order to confirm this data, WST-8 assay was carried out to evaluate material cytotoxicity.

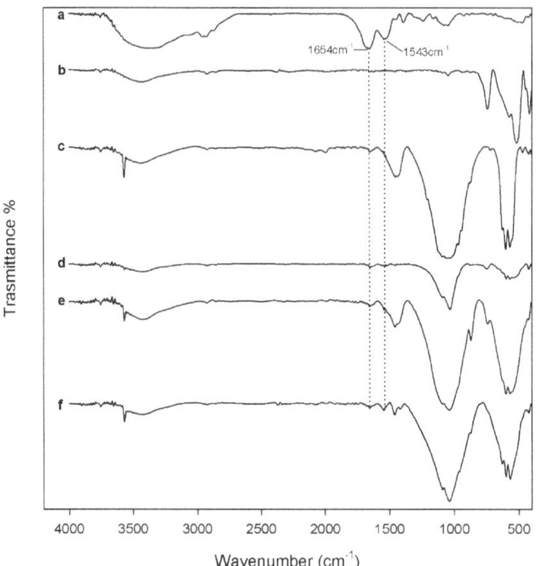

Figure 5. FTIR of (**a**) BSA and (**b**) ZrO_2; (**c**) Hap; (**d**) $9ZrO_2 \cdot 1Hap$; (**e**) $7ZrO_2 \cdot 3Hap$; and (**f**) $5ZrO_2 \cdot 5HAp$ gels, heated to 1000 °C, after 24 h of exposure to BSA solution.

The results of the cytotoxicity assay are reported in Figure 6. NIH-3T3 (National Institutes of Health—3 day transfer, inoculum 3×10^5 cells) murine fibroblast cell line, after contact with extracts of the samples heated to 600 °C, showed viability almost similar to control cells, regardless of the HAp amount. This result proves that both 600°-heated ZrO_2 and Hap, as well as $xZrO_2 \cdot (1 - x)HAp$ composites, are bioinert materials. An increase of cell viability with respect to the control cells was recorded after exposure to extracts of the samples heated to 1000 °C.

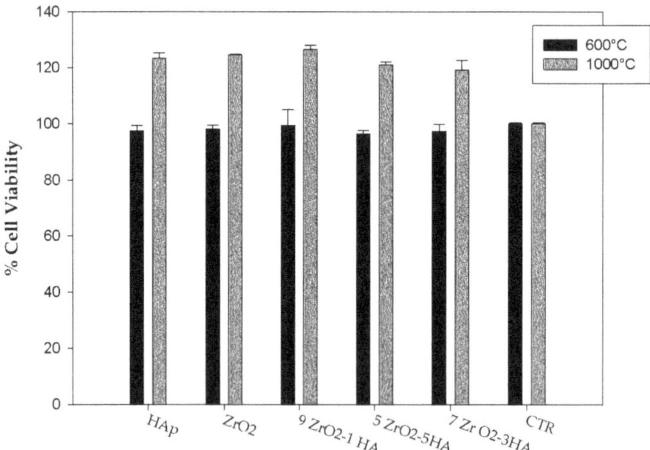

Figure 6. Cytotoxicity assay results. CTR: Control.

Therefore, the HAp content does not affect the biocompatibility of the sol-gel materials, whereas the heat treatment influences this biological property. All samples (pure ZrO_2 and Hap, as well as xZrO$_2$·(1 − x)HAp composites), heated to 1000 °C, are more biocompatible than those heated to 600 °C, in agreement with the results of the protein adsorption test. This improvement can be due to the higher protein adsorption ability of the materials heated to 1000 °C. Moreover, as the biocompatibility improvement occurs also in pure materials (ZrO$_2$ and HAp), it can be ascribed also to the zirconia and HAp microstructure modifications induced by heating. When the samples were heated to 1000 °C, an increase of crystallinity degree occurs and m-ZrO$_2$ and TCP are formed (as proved by FTIR analysis, Figure 3), which can contribute to biocompatibility improvement of the samples. It is known, indeed, that t-ZrO$_2$ is a bioinert material with good mechanical properties used in the dental field [1,3,5]. However, test results showed an increase of the viability of the cells seeded on m-ZrO$_2$, suggesting that its presence makes the composites more biocompatible (Figure 6). Moreover, it is proven that TCP has higher osteoconductivity than HAp [67]. Therefore, its presence in the materials heated to 1000 °C, as well as the presence of m-ZrO$_2$, can contribute to the improvement of the material's biological performance.

The osseointegration ability of the sol-gel materials was evaluated by studying their bioactivity in vitro. After soaking in SBF and drying, powders and disks of the synthesized materials after 600 °C and 1000 °C heating were analyzed by FTIR and scanning electron microscopy (SEM).

FTIR spectra of the pure HA and xZrO$_2$·(1 − x)HAp samples heated to 600 °C, recorded after 21 days of exposure to SBF (Figure 7 curves from b to e), did not show any new peak compared to the FTIR spectra of the same samples recorded before the test (Figure 2). Only an increase and a broadening of the HAp signals are visible, ascribable to the nucleation of new HAp on the samples' surfaces.

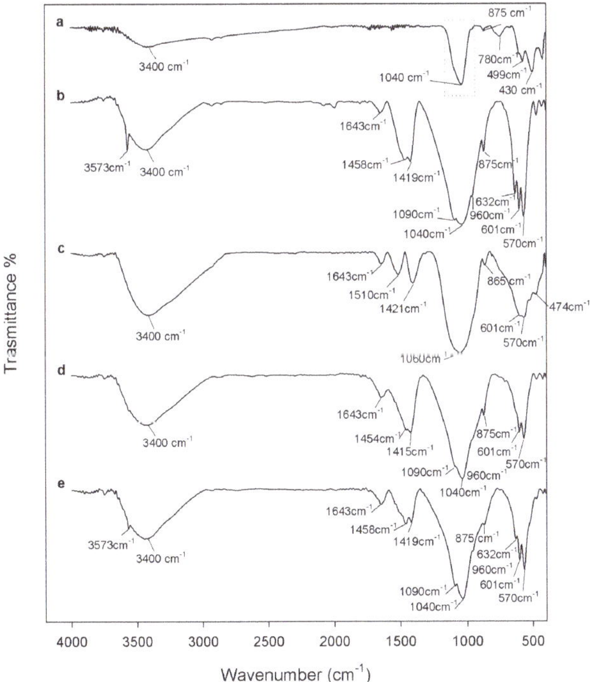

Figure 7. FTIR of (**a**) ZrO$_2$; (**b**) Hap; (**c**) 9ZrO$_2$·1Hap; (**d**) 7ZrO$_2$·3Hap; and (**e**) 5ZrO$_2$·5HAp gels, heated to 600 °C, after 21 days of exposure to Simulated Body Fluid (SBF).

In contrast, HAp nucleation on the pure ZrO$_2$ sample (Figure 7 curve a) caused the appearance of a new intense peak at 1040 cm^{-1}, due to P-O asymmetric stretching of the PO$_4$$^{3-}$ groups [54].

The change recorded in the spectra of the samples heated to 1000 °C after 21 days of exposure to SBF (Figure 8) are the same as those observed in the spectra of the samples heated to 600 °C. Therefore, a broadening and an intensity increase of the HAp signals is observable in all sample spectra and a peak at 1040 cm^{-1} [54], due to P-O asymmetric stretching in PO_4^{3-} groups, is visible in the spectrum of the pure ZrO_2 sample (Figure 8, curve a). Moreover, weak peaks at 1210 and 725 cm^{-1} are visible in the HAp spectrum (Figure 8, curve b), which can be assigned to P-O vibrations in pyrophosphate groups. The presence of pyrophosphate ion inclusions in the hydroxyapatite nucleated on the surface of the synthesized HAp can be due to the dissolution of the β-TCP (its formation was proved by FTIR analysis, Figure 3 curve b) in Simulated Body Fluid (SBF), which leads to the formation of HPO_4^{-2}, PO_4^{-3}, and OH^- ions [68]. As in the $xZrO_2 \cdot (1 - x)HAp$ composites, the amount of HAp is lower, while the content of β-TCP formed by HAp thermal decomposition is also lower than that present in HAp. Therefore, pyrophosphate signals are not visible in the composite spectra.

The presence of the carbonate signals proves that calcium-deficient hydroxyapatite was grown on the surface of all synthesized samples. The carbonatation is higher in the 600 °C heated samples, i.e., their spectra carbonate signals are more evident.

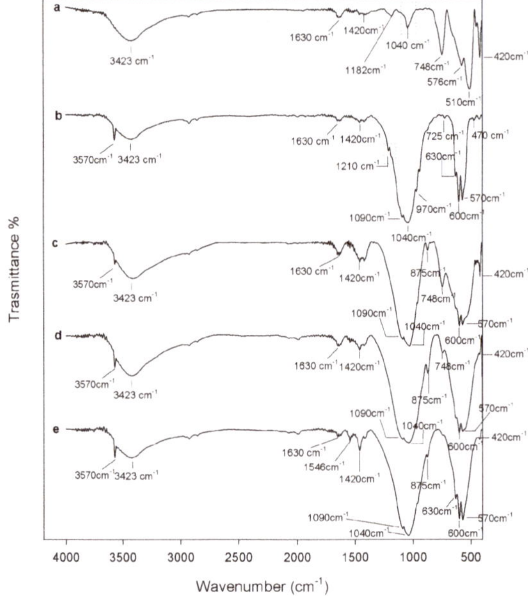

Figure 8. FTIR of (**a**) ZrO_2; (**b**) HAp; (**c**) $9ZrO_2 \cdot 1Hap$; (**d**) $7ZrO_2 \cdot 3Hap$; and (**e**) $5ZrO_2 \cdot 5HAp$ gels, heated to 1000 °C, after 21 days of exposure to SBF.

SEM/EDX (Energy Dispersive X-ray analysis) of the sample disks (Figure 9) confirmed FTIR results. On the surface of all samples the formation of a precipitate with a globular shape typical of HAp is visible (Figure 9). EDX analysis (Figure 10) of such globules show that they consist of an atomic ratio Ca/P < 1.67, identifying them as calcium-deficient HAp [69]. Therefore, all the synthesized materials are bioactive. However, on the surface of the 600°-heated samples only a few globules are visible, whereas the surface of the 1000°-heated samples, except to ZrO_2, appears entirely covered by the HAp globule, confirming that 1000 °C heating improves the material's bioactivity. The improvement of the bioactivity and the increase of the carbonatation process, recorded when the temperature of the heat treatment was increased from 600 °C to 1000 °C, can be ascribed to the different ion release degree caused by the different crystallinity degree of the materials. As already discussed above, in fact, the increase of crystallization, due to the 1000 °C heating, causes the decrease of the ion release in the SBF. The ion exchange, which occurs in the solution containing the 600 °C heated materials, is different from that which takes place in the solution containing the samples heated to 1000 °C. Therefore, the different ion exchange results in a different surface material charge and in a different pH of the SBF solution. The combination of those factors can modify the kinetic of both the nucleation reaction of HAp and of the CO_2 solubilization process in SBF and, thus, of the carbonatation reaction. In the literature [45,70], the presence of a relationship between the heat treatment of the materials and the nucleation of the carbonated HAp was already observed and ascribed mainly to the increase of solution pH, due to the release of Ca^{2+} and PO_4^{3-} ions [45].

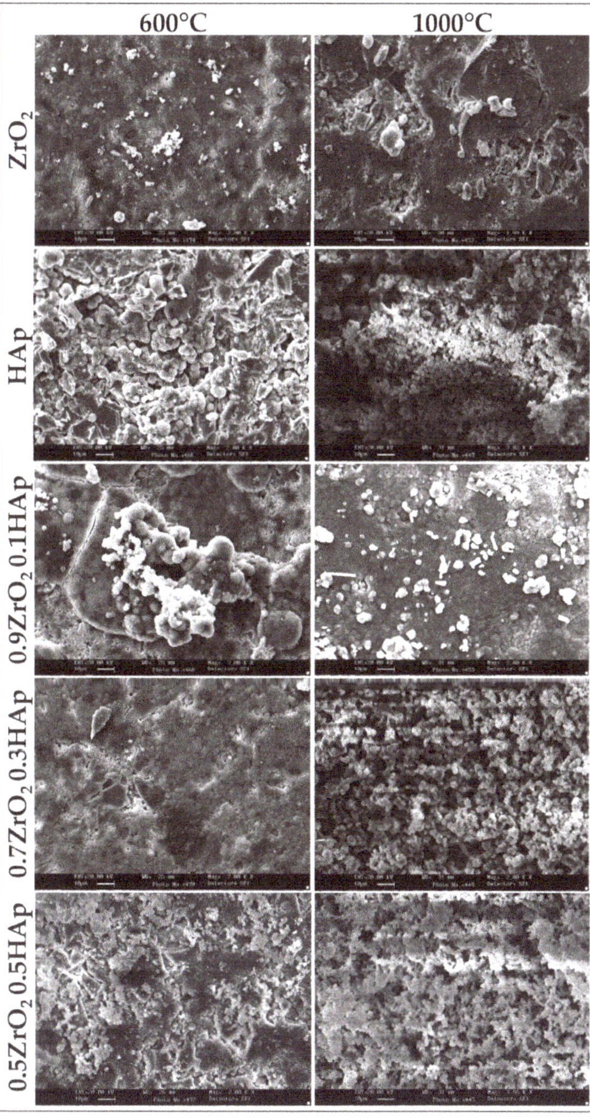

Figure 9. SEM images of all samples heated to 600 °C and 1000 °C after 21 days of exposure to SBF.

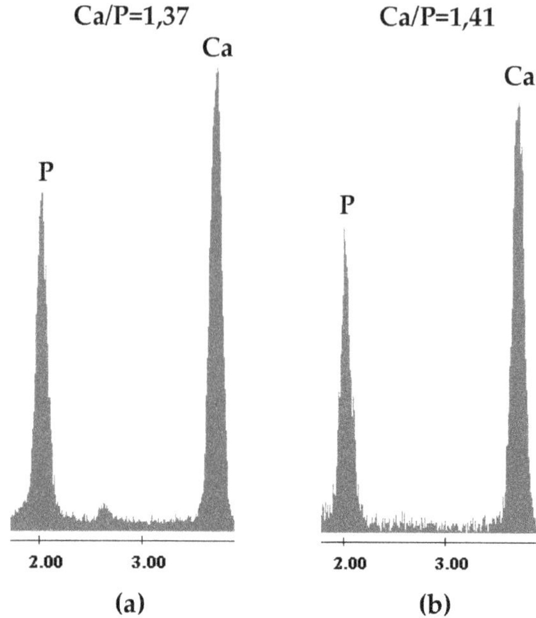

Figure 10. Representative Energy Dispersive X-ray analysis (EDX) images of the globular precipitate on the surface of the samples heated to (**a**) 600 °C and (**b**) 1000 °C after 21 days of exposure to SBF.

Moreover, the SEM images showed that the pure HAp sample is more bioactive than pure ZrO_2, and the higher the HAp content in the $xZrO_2 \cdot (1 - x)$HAp composites, the higher the material's bioactivity. This is ascribable mainly to the presence of Ca^{2+} ions. It is reported in the literature that a higher content of cation in the materials can induce HAp nucleation. The cations can be exchanged with the H^+ in the SBF solution. The increase of the pH value of the solution leads to the dissociation of the –OH groups on the surface of the sol-gel materials and a higher amount of negatively charged $-O^-$, able to induce HAp nucleation.

3. Materials and Methods

3.1. Sol-Gel Synthesis of the Composites

The sol-gel method was used to synthesize $xZrO_2$ $(1 - x)$HAp composites, where x is the mole fraction of ZrO_2 in the composites and is equal to 1, 0.9, 0.7, 0.5, and 0. All reagents were obtained from Sigma Aldrich (Milan, Italy). To synthesize the pure ZrO_2 sample (x = 1), a zirconium propoxide solution (70 wt. % in 1-propanol) was used as precursor. The metal alkoxide was added to a solution of acetyl acetone (AcAc) in ethanol 99.8%. The AcAc was added to act as inhibitor of the fast hydrolytic activity of zirconium propoxide. In the obtained sol the reagents had the following molar ratios: $Zr(OCH_2CH_2CH_3)_4$ / AcAc = 3 and $EtOH/Zr(OCH_2CH_2CH_3)_4$ = 6.

To synthesize the pure HAp sample (x = 0) two solutions were prepared, as follows:

1. Calcium nitrate tetrahydrate $(Ca(NO_3)_2 \cdot 4H_2O)$ was dissolved in ethanol 99.8% under stirring;
2. Phosphorus pentoxide was added to a solution of NH_4OH in Ethanol with pH = 11 under stirring.

When both the solutions were prepared, solution 1 was added to solution 2 under stirring. Within the obtained sol, the molar ratio Ca/P was equal to 1.67.

To synthesize the composites samples, the suitable amount of HAp sol (prepared as described above) was added drop by drop to ZrO_2 sol under stirring. All obtained sols, homogeneous and transparent, were left to gel at room temperature. The gels were dried at 120 °C for 2 h and then heat treated at 600 °C and 1000 °C for 2 h.

A schematic representation of the sol-gel synthesis is shown in Figure 11.

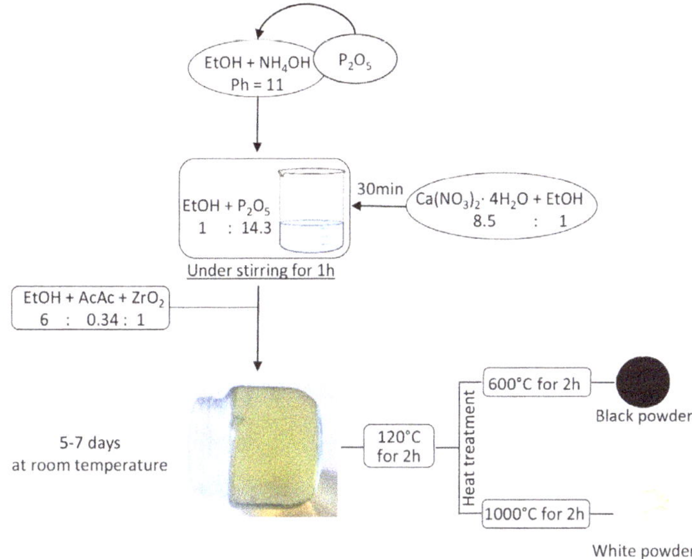

Figure 11. Flow chart of the sol-gel synthesis and heat treatments.

3.2. Composites Chemical Structure

The investigation of the chemical structure of the obtained composites was carried out on all the prepared samples after each heat treatment, in order to follow the evolution of the systems as a function of both the temperature and the relative amount of HAp in the composites.

Synthesized composites were analyzed using FTIR. Prestige 21 spectrophotometer (Shimadzu, Tokyo, Japan) was used to record transmittance spectra in the 400–4000 cm^{-1} region with a resolution of 4 cm^{-1} (45 scans). The instrument was equipped with a DTGS KBr (Deuterated Tryglycine Sulphate with potassium bromide windows) detector. Pelleted disks containing 2 mg of sample diluted with KBr (sample to KBr ratio = 1:100) were made. FTIR spectra were analyzed by Prestige software (IRsolution).

3.3. Biological Properties

3.3.1. Protein Adsorption Evaluation

In order to evaluate the ability of the materials to interact with the blood proteins (the first step of the cell adhesion process), 10 mg/mL of each of the sample powders were soaked for 24 h in a solution of BSA, in a phosphate buffer with 2 mg/mL concentration. After soaking, the powders were gently rinsed three times with distilled water, and were dried at room temperature in a glass desiccator for 24 h. To evaluate the effective adsorption of the albumin on the materials, the dried powders were then analyzed by means of FTIR spectroscopy to assess the presence of the main peak of the albumin, at 1654 cm^{-1} (stretching of C-O in amide I) and at 1543 cm^{-1} (N–H in-plane bending of amide II) [64].

3.3.2. WST-8 Assay

Cytotoxicity evaluation was performed using the WST-8 assay (Dojindo Molecular Technologies Inc., MD, USA), a colorimetric assay. The extracted materials were obtained by incubating 150.0 mg of the sample powders for 2 h, in 7.5 mL of a complete culture medium, at 37 °C under continuous stirring. The NIH-3T3 murine fibroblast cell line (ATCC, Manassas VA, USA) was used in this cytotoxicity test. The cell line was grown in DMEM medium (Gibco, Gaithersburg, MD, USA) with 10% (*v/v*) fetal bovine serum (1% pen-strep) in a humidified incubator, at 37 °C and 5% CO_2. Cells were seeded into 96 multiwell plates at a density of 5.0×10^3 cells/well. After 24 h of incubation, cells were treated with the extracts for 24 h. Afterwards, they were washed 3 times with PBS (phosphate buffered saline) and again incubated with 10% *v/v* of WST-8 [2-(2-methoxy-4-nitrophenyl)-3-(4-nitrophenyl)-5-(2,4-disulfophenyl)-2H-tetrazolium, monosodium salt] in a fresh medium for 2 h. (the water-soluble purple-coloured WST-8 tetrazolium salt is able to penetrate the cellular membrane, and is cleaved by mitochondrial dehydrogenases of the live cells, producing insoluble yellow-orange crystals of formazan). The quantification of the generated formazan can be carried out measuring the absorbance, at 450 nm, of the well-plates, which is proportional to the number of viable cells. The absorbance value was measured with a UV-visible spectrophotometer (Biomate 3, Thermo Scientific, Walkersville, MD, USA). A low absorbance value means that the materials in contact with the cells are cytotoxic agents, able to inhibit their mitochondrial activity. Cell viability was expressed as a percentage of mitochondrial redox activity of the cells directly exposed to material extracts, compared to that of an unexposed control. Therefore, the value of the cell viability has been expressed as the percentage of UV absorbance at 450 nm, recorded in the well where the cells treated with the sample extracts were seeded (compared to the absorbance recorded in the well where untreated control cells where seeded, considered as 100% of viability). The percentages of cell viability were calculated as an average of 3 determinations ± the standard deviation.

3.3.3. Apatite-Forming Ability Test

For evaluations of in vitro bioactivity, the apatite-forming ability test was carried out, as proposed by Kokubo et al. [19]. Both the sample powders and sample disks (13 mm of diameter and 2 mm of thickness), obtained by sample powder pressing, were soaked for 21 days in a SBF with an ion concentration nearly equal to that in the human blood plasma. Polystyrene bottles containing powder and SBF were placed in a water bath at 37.5 ± 0.5 °C. As the ratio between the exposed sample surface and the SBF volume affects the test, it was chosen for the powders and the disks in accordance with Catauro et al. [45] and Kokubo et al. [71], respectively, and kept constant. As the exposed surface area of the powders is higher than the one of the sample disks, a higher volume of SBS was used to test the bioactivity of the sample powders compared to the SBS volume used to test the sample disks.

The solution was replaced every 2 days to avoid depletion of the ionic species in the SBF, due to the nucleation of biominerals on the samples.

After each exposure time, the samples, as powders and as disks, were removed from SBF, gently washed with deionized water, and dried in a desiccator. The hydroxyapatite deposition on the powders was evaluated by FTIR spectroscopy, whereas on the disks it was evaluated by SEM(SEM Quanta 200, FEI, Eindhoven, The Netherlands), equipped with an EDX.

4. Conclusions

The sol-gel technique allowed the preparation of ZrO_2-based composites containing different amounts of HAp. Modification of the materials' structure was induced by heating (120 °C, 600 °C, and 1000 °C), and was followed by FTIR analysis of all samples after each heat treatment. Moreover, biological properties of the synthesized materials were tested in vitro as a function of heat treatment and HAp content. The results showed that HAp content does not influence the composites' ability to adsorb blood proteins or their cytotoxicity, but improves the materials' bioactivity. In contrast, the heat treatment acts on all the tested biological properties and, in particular, 1000 °C heating allowed a higher performance improvement than 600 °C heating.

Therefore, the results of the reported preliminary tests encourage performing, in the future, more extensive analysis on cells attachment, proliferation, and viability. A more extensive biological characterization will be needed to fully understand the behavior of the synthesized materials in vitro, and to identify the cell-materials' interaction mechanism and the effect of such interaction on the cell cycle.

Author Contributions: Flavia Bollino conceived and designed the experiments. Moreover, she coordinated the experimental activity, analyzed data and wrote the paper. Elisabetta Tranquillo and Emilia Armenia, coordinated by Flavia Bollino, performed the experiments. The former carried out the synthesis of the materials and the FTIR analyses while the latter was dealing with the cell culture and WST-8 assay.

Conflicts of Interest: The authors declare no conflict of interest.

References

1. Abd El-Ghany, O.S.; Sherief, A.H. Zirconia based ceramics, some clinical and biological aspects: Review. *Future Dent. J.* **2016**, *2*, 55–64. [CrossRef]
2. Catauro, M.; Bollino, F.; Papale, F.; Pacifico, S.; Galasso, S.; Ferrara, C.; Mustarelli, P. Synthesis of zirconia/polyethylene glycol hybrid materials by sol-gel processing and connections between structure and release kinetic of indomethacin. *Drug Deliv.* **2014**, *21*, 595–604. [CrossRef] [PubMed]
3. Harianawala, H.; Kheur, M.; Kheur, S.; Sethi, T.; Bal, A.; Burhanpurwala, M.; Sayed, F. Biocompatibility of Zirconia. *J. Adv. Med. Dent. Sci. Res.* **2016**, *4*, 35–39.
4. Catauro, M.; Papale, F.; Bollino, F.; Gallicchio, M.; Pacifico, S. Biological evaluation of zirconia/PEG hybrid materials synthesized via sol-gel technique. *Mater. Sci. Eng. C* **2014**, *40*, 253–259. [CrossRef] [PubMed]
5. Ramesh, T.R.; Gangaiah, M.; Harish, P.V.; Krishnakumar, U.; Nandakishore, B. Zirconia Ceramics as a Dental Biomaterial—An Over view. *Trends Biomater. Artif. Organs* **2012**, *26*, 154–160.
6. Hannink, R.H.J.; Kelly, P.M.; Muddle, B.C. Transformation toughening in zirconia-containing ceramics. *J. Am. Ceram. Soc.* **2000**, *83*, 461–487. [CrossRef]
7. Burger, W.; Richter, H.G.; Piconi, C.; Vatteroni, R.; Cittadini, A.; Boccalari, M. New Y-TZP powders for medical grade zirconia. *J. Mater. Sci. Mater. Med.* **1997**, *8*, 113–118. [CrossRef] [PubMed]
8. Ruiz, L.; Readey, M.J. Effect of Heat Treatment on Grain Size, Phase Assemblage, and Mechanical Properties of 3 mol% Y-TZP. *J. Am. Ceram. Soc.* **1996**, *79*, 2331–2340. [CrossRef]
9. Gupta, T.K.; Lange, F.F.; Bechtold, J.H. Effect of stress-induced phase transformation on the properties of polycrystalline zirconia containing metastable tetragonal phase. *J. Mater. Sci.* **1978**, *13*, 1464–1470. [CrossRef]
10. Piconi, C.; Maccauro, G. Zirconia as a ceramic biomaterial. *Biomaterials* **1999**, *20*, 1–25. [CrossRef]
11. Evans, A.G.; Heuer, A.H. Review—Transformation Toughening in Ceramics: Martensitic Transformations in Crack-Tip Stress Fields. *J. Am. Ceram. Soc.* **1980**, *63*, 241–248. [CrossRef]
12. Coli, P.; Karlsson, S. Fit of a New Pressure-Sintered Zirconium Dioxide Coping. *Int. J. Prosthodont.* **2004**, *17*, 59–64. [PubMed]
13. Salehi, S.; Fathi, M.H. Fabrication and characterization of sol–gel derived hydroxyapatite/zirconia composite nanopowders with various yttria contents. *Ceram. Int.* **2010**, *36*, 1659–1667. [CrossRef]
14. Qiu, D.; Wang, A.; Yin, Y. Characterization and corrosion behavior of hydroxyapatite/zirconia composite coating on NiTi fabricated by electrochemical deposition. *Appl. Surf. Sci.* **2010**, *257*, 1774–1778. [CrossRef]
15. Say, Y.; Aksakal, B. Effects of hydroxyapatite/Zr and bioglass/Zr coatings on morphology and corrosion behaviour of Rex-734 alloy. *J. Mater. Sci. Mater. Med.* **2016**, *27*, 105. [CrossRef] [PubMed]
16. Kong, D.-J.; Long, D.; Wu, Y.-Z.; Zhou, C.-Z. Mechanical properties of hydroxyapatite-zirconia coatings prepared by magnetron sputtering. *Trans. Nonferrous Met. Soc. China* **2012**, *22*, 104–110. [CrossRef]
17. Hench, L.L. Bioceramics. *J. Am. Ceram. Soc.* **1998**, *81*, 1705–1728. [CrossRef]
18. Catauro, M.; Bollino, F.; Papale, F.; Pacifico, S. Modulation of indomethacin release from ZrO_2/PCL hybrid multilayers synthesized via sol-gel dip coating. *J. Drug Deliv. Sci. Technol.* **2015**, *26*, 10–16. [CrossRef]
19. Catauro, M.; Bollino, F.; Papale, F. Biocompatibility improvement of titanium implants by coating with hybrid materials synthesized by sol-gel technique. *J. Biomed. Mater. Res. Part A* **2014**, *102*, 4473–4479. [CrossRef] [PubMed]
20. Catauro, M.; Bollino, F.; Papale, F. Preparation, characterization, and biological properties of organic-inorganic nanocomposite coatings on titanium substrates prepared by sol-gel. *J. Biomed. Mater. Res. Part A* **2014**, *102*, 392–399. [CrossRef] [PubMed]

21. Catauro, M.; Bollino, F.; Papale, F.; Giovanardi, R.; Veronesi, P. Corrosion behavior and mechanical properties of bioactive sol-gel coatings on titanium implants. *Mater. Sci. Eng. C* **2014**, *43*, 375–382. [CrossRef] [PubMed]
22. Catauro, M.; Bollino, F.; Papale, F.; Mozetic, P.; Rainer, A.; Trombetta, M. Biological response of human mesenchymal stromal cells to titanium grade 4 implants coated with PCL/ZrO_2 hybrid materials synthesized by sol-gel route: In vitro evaluation. *Mater. Sci. Eng. C* **2014**, *45*, 395–401. [CrossRef] [PubMed]
23. Faure, J.; Drevet, R.; Lemelle, A.; Ben Jaber, N.; Tara, A.; El Btaouri, H.; Benhayoune, H. A new sol–gel synthesis of 45S5 bioactive glass using an organic acid as catalyst. *Mater. Sci. Eng. C* **2015**, *47*, 407–412. [CrossRef] [PubMed]
24. Catauro, M.; Laudisio, G.; Costantini, A.; Fresa, R.; Branda, F. Low Temperature Synthesis, Structure and Bioactivity of $2CaO·3SiO_2$ Glass. *J. Sol-Gel Sci. Technol.* **1997**, *10*, 231–237. [CrossRef]
25. Pirayesh, H.; Nychka, J.A. Sol–Gel Synthesis of Bioactive Glass-Ceramic 45S5 and its in vitro Dissolution and Mineralization Behavior. *J. Am. Ceram. Soc.* **2013**, *96*, 1643–1650. [CrossRef]
26. Catauro, M.; Bollino, F. Anti-inflammatory entrapment in polycaprolactone/silica hybrid material prepared by sol-gel route, characterization, bioactivity and in vitro release behavior. *J. Appl. Biomater. Funct. Mater.* **2013**, *11*, 172–179. [CrossRef] [PubMed]
27. Balamurugan, A.; Sockalingum, G.; Michel, J.; Fauré, J.; Banchet, V.; Wortham, L.; Bouthors, S.; Laurent-Maquin, D.; Balossier, G. Synthesis and characterisation of sol gel derived bioactive glass for biomedical applications. *Mater. Lett.* **2006**, *60*, 3752–3757. [CrossRef]
28. Brinker, C.; Scherer, G. *Sol-Gel Science: The Physics and Chemistry of Sol-Gel Processing*; Academic Press: San Diego, CA, USA, 1989.
29. Gupta, R.; Kumar, A. Bioactive materials for biomedical applications using sol-gel technology. *Biomed. Mater.* **2008**, *3*. [CrossRef] [PubMed]
30. Martin, R.A.; Yue, S.; Hanna, J.V.; Lee, P.D.; Newport, R.J.; Smith, M.E.; Jones, J.R. Characterizing the hierarchical structures of bioactive sol-gel silicate glass and hybrid scaffolds for bone regeneration. *Philos. Trans. R. Soc. A Math. Phys. Eng. Sci.* **2012**, *370*, 1422–1443. [CrossRef] [PubMed]
31. Sepulveda, P.; Jones, J.R.; Hench, L.L. Characterization of melt-derived 45S5 and sol-gel–derived 58S bioactive glasses. *J. Biomed. Mater. Res.* **2001**, *58*, 734–740. [CrossRef] [PubMed]
32. Picquart, M.; López, T.; Gómez, R.; Torres, E.; Moreno, A.; Garcia, J. Dehydration and crystallization process in sol-gel zirconia. *J. Therm. Anal. Calorim.* **2004**, *76*, 755–761. [CrossRef]
33. Agrawal, K.; Singh, G.; Puri, D.; Prakash, S. Synthesis and Characterization of Hydroxyapatite Powder by Sol-Gel Method for Biomedical Application. *J. Miner. Mater. Charact. Eng.* **2011**, *10*, 727–734. [CrossRef]
34. Mutlu, H.I.; Hascicek, Y.S. Insulation for Wind and React High Temperature Superconducting Coils. In *Advances in Cryogenic Engineering Materials*; Balachandran, U.B., Gubser, D.G., Hartwig, K.T., Reed, R., Warnes, W.H., Bardos, V.A., Eds.; Springer Science + Business Media, LLC: New York, NY, USA, 1998.
35. Wachsman, E.D.; Henn, F.E.G.; Jiang, N.; Leezenberg, P.B.; Buchanan, R.M.; Frank, C.W.; Stevenson, D.A.; Wenckus, J.F. Luminescence of Anion Vacancies and Dopant-Vacancy Associated in Stabilized Zirconia. In *Science and Technology of Zirconia V*; Badwal, S.P.S., Bannister, M.J., Hannink, R.H.J., Eds.; Technomic Publishing Company: Lancaster, PA, USA, 1993; pp. 584–592.
36. Catauro, M.; Bollino, F.; Papale, F.; Mozzati, M.C.; Ferrara, C.; Mustarelli, P. ZrO_2/PEG hybrid nanocomposites synthesized via sol-gel: Characterization and evaluation of the magnetic properties. *J. Non-Cryst. Solids* **2015**, *413*, 1–7. [CrossRef]
37. Santos, V.; Zeni, M.; Bergmann, C.; Hohemberger, J. Correlation between thermal treatment and tetragonal/monoclinic nanostructured zirconia powder obtained by sol–gel process. *Rev. Adv. Mater. Sci.* **2008**, *17*, 62–70.
38. Petkova, N.; Dlugocz, S.; Gutzov, S. Preparation and optical properties of transparent zirconia sol–gel materials. *J. Non-Cryst. Solids* **2011**, *357*, 1547–1551. [CrossRef]
39. Georgieva, I.; Danchova, N.; Gutzov, S.; Trendafilova, N. DFT modeling, UV-Vis and IR spectroscopic study of acetylacetone-modified zirconia sol-gel materials. *J. Mol. Model* **2012**, *18*, 2409–2422. [CrossRef] [PubMed]
40. Elvira, M.R.; Mazo, M.A.; Tamayo, A.; Rubio, F.; Rubio, J.; Oteo, J.L. Study and characterization of organically modified silica-zirconia anti-Graffiti coatings obtained by sol-gel. *J. Chem. Chem. Eng.* **2013**, *7*, 120–131.
41. Hao, Y.; Li, J.; Yang, X.; Wang, X.; Lu, L. Preparation of ZrO_2–Al_2O_3 composite membranes by sol–gel process and their characterization. *Mater. Sci. Eng. A* **2004**, *367*, 243–247. [CrossRef]

42. Irish, D.E.; Walrafen, G.E. Raman and infrared spectral studies of aqueous calcium nitrate solutions. *J. Chem. Phys.* **1967**, *46*, 378–384. [CrossRef]

43. Brauer, D.S. Phosphate glass. In *Bio-Glasses: An Introduction*; Jones, J.R., Clare, A.G., Eds.; John Wiley & Sons, Ltd.: Oxford, UK, 2012; pp. 46–49.

44. De Oliveira, A.A.; de Souza, D.A.; Dias, L.L.; de Carvalho, S.M.; Mansur, H.S.; de Magalhães Pereira, M. Synthesis, characterization and cytocompatibility of spherical bioactive glass nanoparticles for potential hard tissue engineering applications. *Biomed. Mater.* **2013**, *8*. [CrossRef] [PubMed]

45. Catauro, M.; Bollino, F.; Renella, R.A.; Papale, F. Sol-gel synthesis of SiO_2-CaO-P_2O_5 glasses: Influence of the heat treatment on their bioactivity and biocompatibility. *Ceram. Int.* **2015**, *41*, 12578–12588. [CrossRef]

46. Brangule, A.; Gross, K.A. *Importance of FTIR Spectra Deconvolution for the Analysis of Amorphous Calcium Phosphates*; IOP Conference Series: Materials Science and Engineering; IOP Publishing Ltd.: Bristol, UK, 2015.

47. Kurajica, S.; Lozić, I.; Pantaler, M. Thermal decomposition of calcium(II) bis(acetylacetonate) n-hydrate. *Polimeri* **2015**, *35*, 4–9.

48. Li, P.; Ohtsuki, C.; Kokubo, T.; Nakanishi, K.; Soga, N.; Nakamura, T.; Yamamuro, T. Process of formation of bone-like apatite layer on silica gel. *J. Mater. Sci. Mater. Med.* **1993**, *4*, 127–131. [CrossRef]

49. Ohtsuki, C.; Kokubo, T.; Yamamuro, T. Mechanism of apatite formation on CaOSiO2P2O5 glasses in a simulated body fluid. *J. Non-Cryst. Solids* **1992**, *143*, 84–92. [CrossRef]

50. Beganskienė, A.; Dudko, O.; Sirutkaitis, R.; Giraitis, R. Water Based Sol-Gel Synthesis of Hydroxyapatite. *Mater. Sci.* **2003**, *9*, 383–386.

51. Vecchio Ciprioti, S.; Bollino, F.; Tranquillo, E.; Catauro, M. Synthesis, thermal behavior and physicochemical characterization of ZrO/PEG inorganic/organic hybrid materials via sol–gel technique. *J. Therm. Anal. Calorim.* **2017**, 1–6. [CrossRef]

52. Jayakumar, S.; Ananthapadmanabhan, P.V.; Perumal, K.; Thiyagarajan, T.K.; Mishra, S.C.; Su, L.T.; Tok, A.I.Y.; Guo, J. Characterization of nano-crystalline ZrO2 synthesized via reactive plasma processing. *Mater. Sci. Eng. B* **2011**, *176*, 894–899. [CrossRef]

53. Chen, S.; Yin, Y.; Wang, D.; Liu, Y.; Wang, X. Structures, growth modes and spectroscopic properties of small zirconia clusters. *J. Cryst. Growth* **2005**, *282*, 498–505. [CrossRef]

54. Berzina-Cimdina, L.; Borodajenko, N. Research of Calcium Phosphates Using Fourier Transform Infrared Spectroscopy. In *Infrared Spectroscopy—Materials Science, Engineering and Technology*; Theophile, T., Ed.; InTech: Rijeka, Croatia, 2012; pp. 123–148.

55. Gozalian, A.; Behnamghader, A.; Daliri, M.; Moshkforoush, A. Synthesis and thermal behavior of Mg-doped calcium phosphate nanopowders via the sol gel method. *Sci. Iran.* **2011**, *18*, 1614–1622. [CrossRef]

56. Meejoo, S.; Maneeprakorn, W.; Winotai, P. Phase and thermal stability of nanocrystalline hydroxyapatite prepared via microwave heating. *Thermochim. Acta* **2006**, *447*, 115–120. [CrossRef]

57. Fleet, M.E.; Liu, X.; King, P.L. Accommodation of the carbonate ion in apatite: An FTIR and X-ray structure study of crystals synthesized at 2–4 GPa. *Am. Mineral.* **2004**, *89*, 1422–1432. [CrossRef]

58. Eichert, D.; Drouet, C.; Sfihi, H.; Rey, C.; Combes, C. Nanocrystalline apatite-based biomaterials: Synthesis, processing and characterization. In *Biomaterials Research Advances*; Kendall, J.B., Ed.; Nova Science Publishers: New York, NY, USA, 2007; pp. 93–143.

59. Barralet, J.; Knowles, J.C.; Best, S.; Bonfield, W. Thermal decomposition of synthesised carbonate hydroxyapatite. *J. Mater. Sci. Mater. Med.* **2002**, *13*, 529–533. [CrossRef] [PubMed]

60. Latifi, S.M.; Fathi, M.; Varshosaz, J.; Ghochaghi, N. Mechanisms controlling ca ion release from sol-gel derived in situ apatite-silica nanocomposite powder. *Ceram. Silik.* **2015**, *59*, 64–69.

61. Mansour, S.F.; El-dek, S.I.; Ahmed, M.K. Physico-mechanical and morphological features of zirconia substituted hydroxyapatite nano crystals. *Sci. Rep.* **2017**, *7*. [CrossRef] [PubMed]

62. Mortier, A.; Lemaitre, J.; Rouxhet, P.G. Temperature-programmed characterization of synthetic calcium-deficient phosphate apatites. *Thermochim. Acta* **1989**, *143*, 265–282. [CrossRef]

63. Kalinkin, A.M.; Kalinkina, E.V.; Zalkind, O.A.; Makarova, T.I. Chemical interaction of calcium oxide and calcium hydroxide with CO_2 during mechanical activation. *Inorg. Mater.* **2005**, *41*, 1073–1079. [CrossRef]

64. Popescu, R.A.; Magyari, K.; Vulpoi, A.; Trandafir, D.L.; Licarete, E.; Todea, M.; Stefan, R.; Voica, C.; Vodnar, D.C.; Simon, S.; et al. Bioactive and biocompatible copper containing glass-ceramics with remarkable antibacterial properties and high cell viability designed for future in vivo trials. *Biomater. Sci.* **2016**, *4*, 1252–1265. [CrossRef] [PubMed]

65. Mavropoulos, E.; Costa, A.M.; Costa, L.T.; Achete, C.A.; Mello, A.; Granjeiro, J.M.; Rossi, A.M. Adsorption and bioactivity studies of albumin onto hydroxyapatite surface. *Coll. Surf. B Biointerfaces* **2011**, *83*, 1–9. [CrossRef] [PubMed]
66. Kumar, R.; Münstedt, H. Polyamide/silver antimicrobials: Effect of crystallinity on the silver ion release. *Polym. Int.* **2005**, *54*, 1180–1186. [CrossRef]
67. Ogose, A.; Hotta, T.; Kawashima, H.; Kondo, N.; Gu, W.; Kamura, T.; Endo, N. Comparison of hydroxyapatite and beta tricalcium phosphate as bone substitutes after excision of bone tumors. *J. Biomed. Mater. Res. Part B Appl. Biomater.* **2005**, *72B*, 94–101. [CrossRef] [PubMed]
68. Jalota, S.; Bhaduri, S.B.; Tas, A.C. In vitro testing of calcium phosphate (HA, TCP, and biphasic HA-TCP) whiskers. *J. Biomed. Mater. Res. Part A* **2006**, *78A*, 481–490. [CrossRef] [PubMed]
69. Catauro, M.; Papale, F.; Sapio, L.; Naviglio, S. Biological influence of Ca/P ratio on calcium phosphate coatings by sol-gel processing. *Mater. Sci. Eng. C* **2016**, *65*, 188–193. [CrossRef] [PubMed]
70. Radev, L. Influence of thermal treatment on the structure and in vitro bioactivity of sol-gel prepared CaO-SiO2-P2O5 glass-ceramics. *Process. Appl. Ceram.* **2014**, *8*, 155–166. [CrossRef]
71. Kokubo, T.; Takadama, H. How useful is SBF in predicting in vivo bone bioactivity? *Biomaterials* **2006**, *27*, 2907–2915. [CrossRef] [PubMed]

Article

Hydrophobic Coatings by Thiol-Ene Click Functionalization of Silsesquioxanes with Tunable Architecture

Sandra Dirè [1,*], Davide Bottone [1,*], Emanuela Callone [1], Devid Maniglio [1], Isabelle Génois [2] and François Ribot [2]

[1] Department of Industrial Engineering, University of Trento, via Sommarive 9, 30123 Trento, Italy; emanuela.callone@unitn.it (E.C.); devid.maniglio@unitn.it (D.M.)

[2] Sorbonne Universités, UPMC University Paris 06—CNRS—College de France, UMR 7574, Laboratoire de Chimie de la Matière Condensée de Paris, 4 place Jussieu, 75005 Paris, France; isabelle.genois@upmc.fr (I.G.); francois.ribot@upmc.fr (F.R.)

* Correspondence: sandra.dire@unitn.it (S.D.); davide.bottone@uzh.ch (D.B.); Tel.: +39-0461-282456 (S.D.); +41-44-63-54422 (D.B.)

Received: 28 July 2017; Accepted: 5 August 2017; Published: 8 August 2017

Abstract: The hydrolysis-condensation of trialkoxysilanes under strictly controlled conditions allows the production of silsesquioxanes (SSQs) with tunable size and architecture ranging from ladder to cage-like structures. These nano-objects can serve as building blocks for the preparation of hybrid organic/inorganic materials with selected properties. The SSQs growth can be tuned by simply controlling the reaction duration in the in situ water production route (ISWP), where the kinetics of the esterification reaction between carboxylic acids and alcohols rules out the extent of organosilane hydrolysis-condensation. Tunable SSQs with thiol functionalities (SH-NBBs) are suitable for further modification by exploiting the simple thiol-ene click reaction, thus allowing for modifying the wettability properties of derived coatings. In this paper, coatings were prepared from SH-NBBs with different architecture onto cotton fabrics and paper, and further functionalized with long alkyl chains by means of initiator-free UV-induced thiol-ene coupling with 1-decene (C10) and 1-tetradecene (C14). The coatings appeared to homogeneously cover the natural fibers and imparted a multi-scale roughness that was not affected by the click functionalization step. The two-step functionalization of cotton and paper warrants a stable highly hydrophobic character to the surface of natural materials that, in perspective, suggests a possible application in filtration devices for oil-water separation. Furthermore, the purification of SH-NBBs from ISWP by-products was possible during the coating process, and this step allowed for the fast, initiator-free, click-coupling of purified NBBs with C10 and C14 in solution with a nearly quantitative yield. Therefore, this approach is an alternative route to get sol-gel-derived, ladder-like, and cage-like SSQs functionalized with long alkyl chains.

Keywords: silsesquioxanes; thiol-ene click reaction; in situ water production; hydrophobic coatings; cotton fabric; paper; NMR; wettability

---

1. Introduction

Hybrid organic-inorganic (O/I) materials combine the characteristics of their organic and inorganic components to obtain mechanical, electrical, and chemical properties that would not be achievable with a simple mixture of the constituent parts, and are therefore relevant in a wide range of applications [1,2]. Silicon-based sol-gel chemistry is recognized as one of the main routes for the synthesis of hybrid materials owing to the wide availability of molecular precursors and the stability of the Si–C bond [3]. Silsesquioxanes (SSQs) are compounds with the general formula $(RSiO_{1.5})_n$, where R is an organic functional group or a H atom, and *n* is the number of repeating units that can be obtained from the hydrolysis-condensation of organotrialkoxysilanes (RSi(OR′)₃) [4,5]. While SSQs are routinely synthesized following conventional hydrolytic sol-gel processes, a higher degree of control on the structure of the final material can be achieved with the nanobuilding blocks (NBBs) approach [6–9]. Indeed, the assembly of pre-formed nano-objects based on different SSQs architectures, such as cage and ladder-like structures (Supplementary Materials, Figure S1), can be used for producing the final material, allowing for a bottom-up design of its properties [4,5,10].

Some of us previously reported the sol-gel synthesis of silsesquioxane NBBs by the in situ water production (ISWP) route, where the water necessary for the hydrolysis of the silane precursor is provided by the esterification reaction between a carboxylic acid and an alcohol [11–13]. The extent of hydrolysis–condensation in dependence on the kinetics of water production was assessed by time-dependent monitoring of the reactions, mainly through infrared spectroscopy (FTIR), nuclear magnetic resonance (NMR), and gel permeation chromatography (GPC) analyses, demonstrating the ability to tune SSQs' size and architecture by simply controlling the reaction duration. In particular, in the case of thiol-functionalized SSQs (SH-NBBs) [14], obtained by controlled hydrolysis-condensation of 3-mercaptopropyltrimethoxysilane (McPTMS) with chloroacetic acid (ClAAc) in 1-propanol solution at 100 °C (Figure 1a), the well-defined octakis (3-mercaptopropylsilsesquioxane) is the major species between 70 and 100 h reaction time; on the contrary, the ladder-like species appeared predominant for lower duration, according to ^{29}Si NMR spectroscopy [15]. However, the ISWP route is characterized by the occurrence of different reactions (Figure 1b,c), leading also to side products such as chloroacetic acid-derived compounds that cannot be completely removed from the SH-NBBs solution.

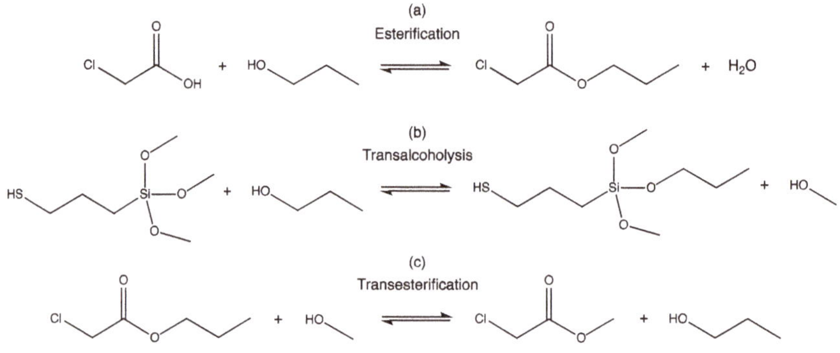

Figure 1. Reactions occurring in the ISWP reaction mixture (adapted with permission from [15] Borovin et al., 2016 © Wiley-VCH).

Despite the confirmed solution stability under prolonged storage and the substantial keeping of thiol functions in the final SH-NBBs, the presence of by-products can be detrimental towards both assembly and further SSQs functionalization in the final hybrid O/I materials. Attempts to purify sol-gel-prepared SSQs were reported in the literature generally with unsatisfactory results [9]; in the case of SH-NBBs, the elimination of by-products by sol evaporation under reduced pressure was unsuccessful [14].

The reactive SH- function is appealing for tuning the surface functionality of O/I hybrids, and can be exploited for different applications, such as immobilization and the sensing of biomolecules and trace metals [16–18], or the development of coatings with different wettability [19]. SH-NBBs sols demonstrated good filming ability onto glass and Si substrates [15], and this evidence prompted us to investigate their use for modifying the surface features of cotton fabrics and paper, for their possible application in oil-water separation. Microporous polymers, such as PU foams and PDMS sponges, and carbon-based materials present excellent adsorption characteristics due to oleophilic and hydrophobic properties [20]. However, adsorbent materials with low environmental impact in terms of waste disposal and eventual recyclability are envisaged, such as natural materials with improved oleophilic and hydrophobic behavior.

In this paper, a two-step functionalization process of both cotton fabrics and paper has been studied. In a first step, SH-NBBs coatings were prepared on cotton and paper substrates; in the second step, an initiator-free click thiol-ene coupling reaction was exploited for coating functionalization with long alkyl chains in order to further increase the hydrophobic character of the exposed surface. Scanning Electron Microscopy and Confocal Microscopy were used for coating characterization and, depending on the specific nature of the substrates, surface wettability was evaluated by shedding angle and dynamic contact angle measurements.

Moreover, according to the NMR study, the purification of SH-NBBs was also achieved during the coating procedure, thus enabling the preparation of long alkyl-chain functionalized NBBs solutions via click chemistry. This approach allows to overcome the difficulty in obtaining functionalized SSQs oligomers directly starting from trialkoxysilanes with long alkyl chains, such as octyltriethoxysilane, since these precursors own very low hydrolysis-condensation reactivity [21,22].

2. Results and Discussion

2.1. SH-NBBs Coatings

SH-NBBs solutions prepared via ISWP at a different reaction time were used for coating cotton and paper substrates. On the basis of the NMR study performed for elucidating the SH-NBBs growth in solution [15], three reaction times were selected as representative of different SSQs architectures. After 6 h reaction the condensation degree is limited and SH-NBBs are represented by small linear and cyclic species mainly based on T^2 units, i.e., according to the ^{29}Si NMR notation, $SiCO_3$ species with two bridging oxygen and one terminal OH function. These species after 16 h start to rearrange into fully condensed T^3 units (with three bridging oxygen) mainly in ladder–type architectures. At 80 h SH-NBBs are mainly composed of condensed units with closed cages, i.e., octakis (3-mercaptopropylsilsesquioxane), as the dominant structure.

The raw cotton and the cotton samples coated with SH-NBBs sols obtained after 6, 16, and 80 h reaction were characterized by Scanning Electron Microscopy, in order to assess the morphological changes undergone by the material surface during the first step of the process (Figure 2). In raw cotton, both the tubular and the secondary micro-fibrillar structures of the cotton fibers [23] can be observed. In *Cotton-NBB* samples the presence of a coating appears superimposed to the wrinkled surface textures and the appearance of sub-micrometric bumps, clumps, and bridge-like structures among the fibers can be clearly observed. From Figure 1, the filming ability of SH-NBBs solutions appear good independently of the duration of the ISWP reaction, thus assuring the effective coating of the cotton fibers in all samples.

To ascertain the presence and homogeneity of SH-NBBs coating on the fibers, EDX spectra were acquired on raw and coated cotton samples. In the case of SH-NBBs-coated samples, the signals due to Si and S, which are indicative of the presence of thiol-functionalized silsesquioxanes, can be clearly detected in the survey EDX spectra. On the contrary, the amount of Cl, which is indicative of chloroacetic acid and its esters i.e., the side products of the ISWP synthesis, appears very low and a clear correlation with the coating cannot be made on the basis of the acquired maps.

In *Cotton-NBB16h*, from the element maps obtained analyzing different selected zones, the Si and S signals do not appear uniform but are stronger in conjunction with the bridge-like structures. The SEM-EDX analysis of *Cotton–NBB80h* (Figure 3a,b) points out that, in comparison with *Cotton-NBB16h*, S and Si are more visible in the areas of the fibers where bridge-like structures or lumps are not present. These results can be probably explained by taking into account the viscosity of the SH-NBBs solutions, which increases with an increasing reaction time as a result of oligomers growth.

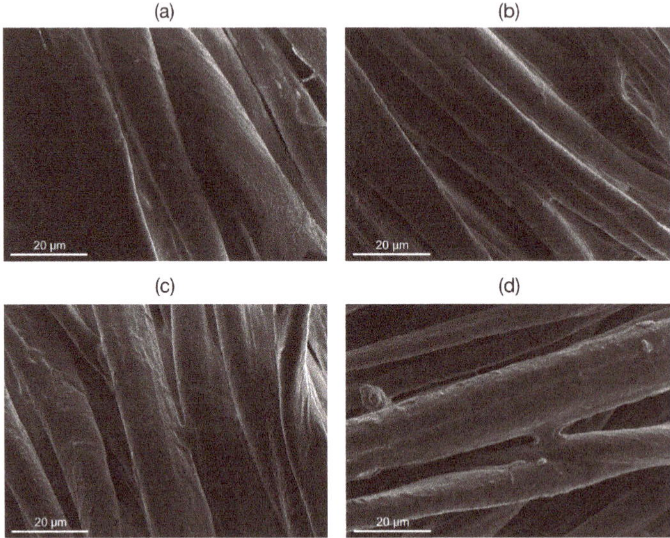

Figure 2. Scanning Electron Microscopy (SEM) images of raw cotton (**a**); *Cotton-NBB6h* (**b**); *Cotton-NBB16h* (**c**) and *Cotton-NBB80h* (**d**). Images were acquired at the same level of magnification.

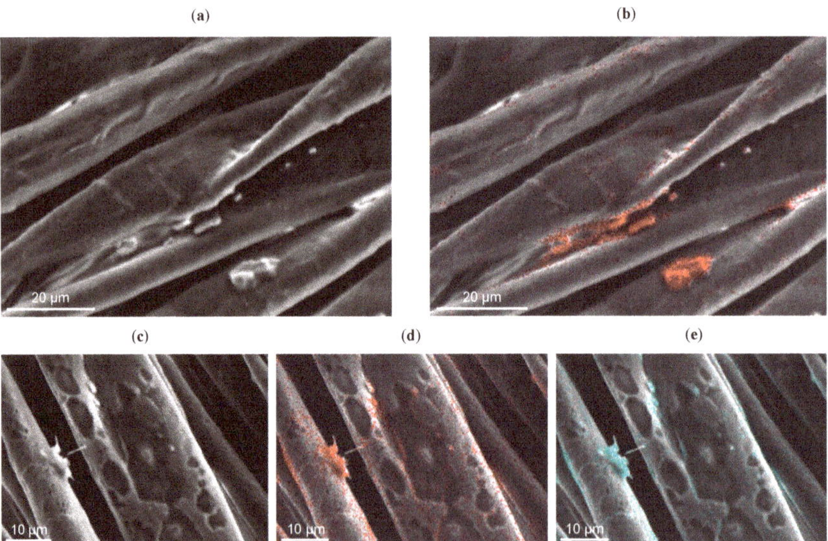

Figure 3. (**a**) SEM image of *Cotton-NBB80h*; (**b**) Si map of the corresponding area; (**c**) damaged *Cotton-NBB80h* specimen after prolonged SEM observation, and elemental maps for Si (**d**) and S (**e**) of the sample zone shown in (**c**).

This evidence is confirmed by the EDX results (that were recorded on a sample, exposed for a long time to the focused electron beam, and showed clear signs of damage compatible with coating degradation). Figure 3c,d and e suggest that the coating prior to degradation was more homogeneous than what was concluded on the basis of Figure 3a,b.

The SEM images of the raw paper substrate and *Paper-NBB* samples are shown in Figure 4. Owing to the high heterogeneity of the substrate, it is difficult to clearly distinguish signs of the NBB coating from the low magnification images (Figure 4 left). However, the images recorded at higher magnification (Figure 4 right) show the SH-NBBs layer covering the disordered fibrillary surface texture of the raw paper substrate. This layer exhibits a sub-micrometric roughness that persists even after rinsing of the samples in the ultrasound bath (see Section 3.2).

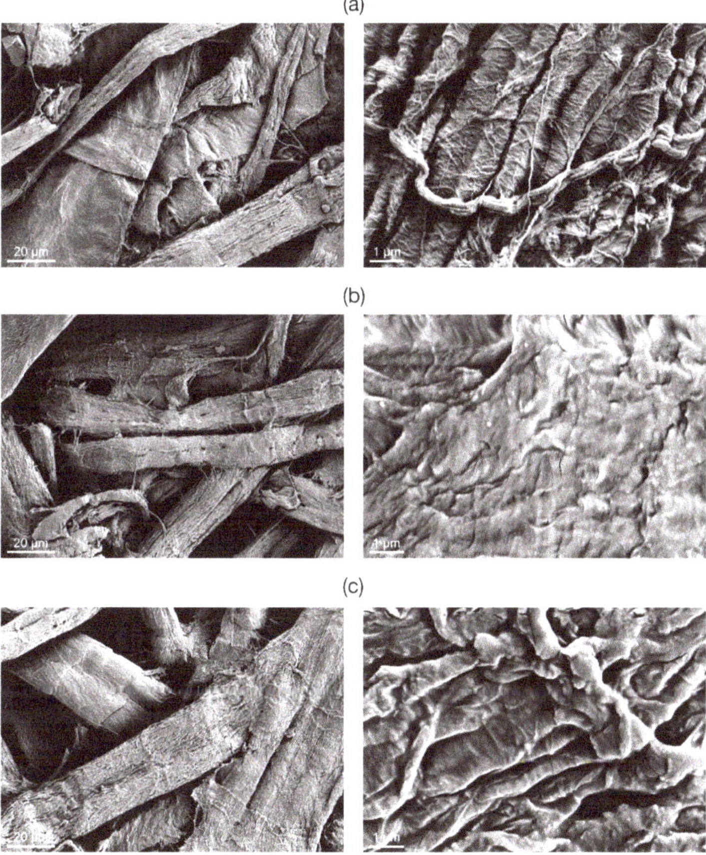

Figure 4. Low (left) and high (right) magnification SEM images of (**a**) the raw paper substrate and *Paper-NBB80h* samples, before (**b**) and after (**c**) rinsing in ultrasound bath. Images on the left were taken at the same level of magnification of those in Figure 1; images on the right were taken at a constant higher level of magnification.

From the SEM study, both *Cotton-NBB* and *Paper-NBB* samples show that coating with SH-NBBs leads to changes in surface morphology with the appearance, in the case of cotton fabrics, of bridge-like structures connecting the fibers. However, from EDX spectra the presence of SH-NBBs is detected also in the other regions, which suggests that a fairly homogeneous layer is present on the majority of the natural substrates. Therefore, a quite good distribution of the thiol anchoring sites for the subsequent alkene functionalization is also expected.

In the case of *Paper-NBB*, it is worth noting that the coating roughness is enhanced as a consequence of its rinsing using an ultrasound bath to remove the SH-NBBs excess. While the substrate fibers coating is still preserved, the change in morphology could be beneficial for the wetting behavior. Indeed, the sub-micrometric texture coupled with the microstructure given by the cellulosic fibers of cotton and paper might give rise to hierarchical multi-scale roughness, suitable for achieving extreme wetting properties. Finally, it should be mentioned that, despite the presence of the SH-NBBs coating, both cotton and paper samples keep their macroscopic features, such as color and mechanical behavior, and exhibit quite negligible increases in weight.

The very low amount of deposited SSQs makes the structural characterization of the SH-NBBs coating and the study of the interface between substrate and coating very difficult. However, some hints on the structural features displayed by the coating were obtained by solid state NMR experiments. In particular, a ^{29}Si NMR spectrum was recorded on a cotton sample, prepared ad hoc by repeated immersion steps in the SH-NBBs solution reacted for 80 h (Supplementary Materials, Figure S2); 36 k scans were acquired in a cross-polarization experiment (CPMAS). The main signal, due to fully condensed T^3 units with the second resonance related to T^2 units, can be observed in the spectrum. The CPMAS sequence does not lead to quantitative results since it is well known that the magnetization transfer enhances the intensity of proton-rich environments. Accordingly, the low intensity of the T^2 signal is an indication of the high crosslinking degree of the SSQs coating.

2.2. SH-NBBs Purification and Thiol-Ene Click Functionalization

As reported in the experimental section (Section 3.2), after the immersion of cotton and paper substrates in the SH-NBBs solution the coated samples were rinsed in chloroform in order to remove the ungrafted SH-NBBs, the solutions were kept to dryness, and the obtained solids were dissolved in deuterated chloroform (*SH-NBB_c* and *SH-NBB_p* solutions) for the NMR analyses.

The structural features of SH-NBBs and ISWP by-products were already described in Borovin et al. [15]. Accordingly, it has been proved that the synthesis of SH-NBBs leads to a final solution containing propyl- (PrClA) and methyl-chloroacetate (MeClA) residues, as a result of esterification and transalcoholysis reactions (Figure 1), and chlorothioacetate derivatives. The latter compounds are formed both through direct reaction between McPTMS derivatives and ClAAc, or by transesterification between McPTMS derivatives and propyl or methyl chloroacetate [15]. In order to remove the side-products from the pristine SH-NBBs solution, purification methods (as extraction or volatilization under reduced pressure) were proved to be not effective, representing an obstacle for further reactions. As a matter of fact, the attempt at running the thiol-ene coupling using the as prepared SH-NBBs solution was unsuccessful, contrary to the observed reactivity of the McPTMS precursor in 1-propanol solution.

Therefore, the NMR spectroscopy was used not only to verify the structural integrity of SH-NBBs recovered by rinsing in chloroform the coated samples, but also to assess both the purification process and the effectiveness of thiol-ene functionalization of the purified SH-NBBs. Figure 5a shows the ^{1}H NMR spectrum of the *SH-NBB_c* solution prepared by rinsing *Cotton-NBB80h*, as an example of the results obtained with the different *SH-NBB_c* and *SH-NBB_p* solutions. The peak assignment is based on the numbering scheme shown in Figure 5.

The most intense signals are attributed to α, β, and γ resonances of the mercaptopropyl chain, and together with α′, due to the chlorothioacetate derivative of the pristine propylsilane group [15], confirm the presence of SH-NBBs in the rinsing residue and, according to the chemical shifts, prove the maintenance of the structural features displayed by the pristine SSQs. Moreover, the integrated areas of (α + α′), β, and γ resonances are an indication that the mercaptopropyl functionalities of the recovered SH-NBBs do not undergo any degradation.

The resonances due to the ISWP side products are still observable in the spectrum. In detail, signals c, e, e′, and e″ are due to chloroacetic acid and its esters; in addition, PrClA gives rise to a′ and b′ resonances, the latter appearing as a shoulder of the signal, while MeClA gives rise to g′ resonance.

The low intensity of c and g signals suggest the large removal of 1-propanol and methanol during cotton fabrics processing. Finally, other signals not ascribable to SH-NBBs or to ISWP side-products can be observed and have been tentatively assigned to products of cotton degradation.

In order to evaluate the amount of residual ClAAc derived compounds with respect to the SH-NBBs, the following parameter can be calculated (Equation (1)):

$$R_{Cl} = \frac{c' + e + e' + e''}{\alpha + \alpha'}$$

(1)

where the different quantities are the integral areas of the specified signals. The parameter calculated according to Equation (1) slightly overestimates the molar ratio of ClAAc and its esters with respect to SH-NBBs, since the resonances due to SH-NBBs (α and α'), ClAAc (e), and MeClA (e'') each contribute two protons to the signals, while PrClA contributes four protons (c' and e'). The calculated R_{Cl} values (Table 1) for *SH-NBB_c* and *SH-NBB_p* solutions indicate the effectiveness of the SH-NBBs purification following the adopted procedure. Indeed, taking into account the 6:1 molar ratio between ClAAc and McPTMS used for the SH-NBBs synthesis, the nominal R_{Cl} for the as-prepared solutions would assume values of 6 and 12, if only ClAAc and MeClA or PrClA were present, respectively. Conversely, in *SH-NBB_c* and *SH-NBB_p*, R_{Cl} assume values ranging from 0.061 to 0.165, resulting in a more than fifty-fold reduction in the relative amount of ClAAc and its esters.

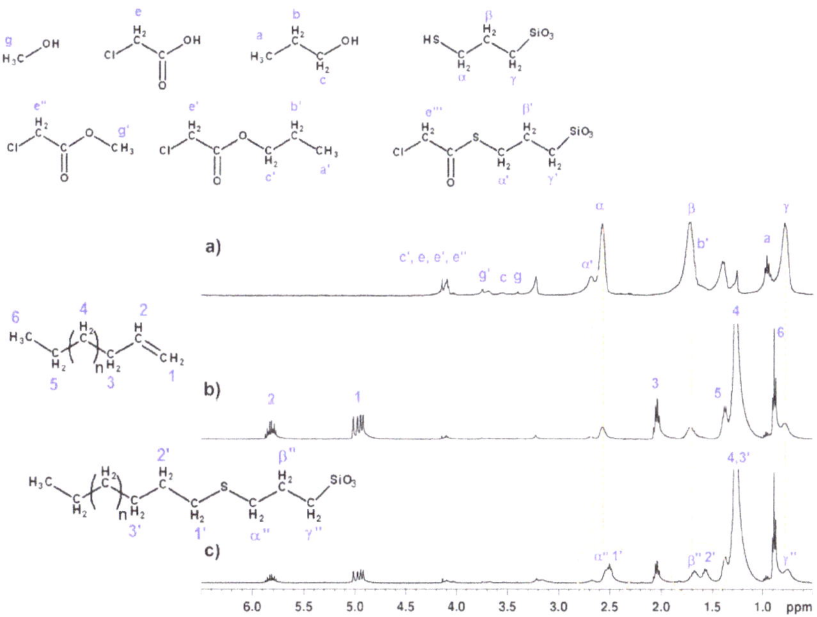

Figure 5. ^{1}H NMR spectra and peak assignment of (**a**) *SH-NBBs_c* solution obtained by rinsing *Cotton-NBB80h*; a mixture of *SH-NBBs_c* and C14 before (**b**); and after (**c**), 30 min irradiation at λ_{max} = 254 nm.

In order to quantitatively evaluate the reliability of the purification step, the percentage of unreacted available thiol groups can be calculated according to Equation (2):

$$SH_{av}(\%) = \frac{\alpha}{\alpha + \alpha'}$$

(2)

Table 1. Summary of the values of R_{Cl} and SH_{av} (%), as calculated from the ^{1}H NMR spectra of *SH-NBB_c* and *SH-NBB_p* solutions.

SH-NBBs	6 h		16 h		80 h	
Substrate	Cotton	Paper	Cotton	Paper	Cotton	Paper
R_{Cl}	0.108	0.042	0.165	0.061	0.145	0.134
SH_{av} (%)	96.4	100	95.6	92.3	78.6	78.1

The SH_{av} values (Table 1) are generally close to 100% for both paper and cotton supports, with the exception of the purified SH-NBBs 80 h that present a loss of thiol functionalities higher than what was previously reported for the as-prepared SH-NBBs solutions [15]. However, this result can be related to the fact that the solutions were kept at room temperature during the coating procedure, thus allowing the thioesterification reaction to proceed.

From the above results, the SH-NBBs purification appears effective, and the process is independent of the coated substrate and highly reproducible. In conclusion, the employed NBB coating procedure of cotton and paper produces recyclable wastes, from which the purified SH-NBBs can be easily collected and further reacted.

The *SH-NBB_c* and *SH-NBB_p* solutions were mixed with 1-decene or 1-tetradecene to study the NBBs reactivity in the thiol-ene coupling run without any photoinitiator addition. Figure 5b,c shows the ^{1}H NMR spectra of *SH-NBBs_c* (obtained from *Cotton-NBB80h*) and C14 mixture before and after irradiation for 30 min with λ_{max} = 254 nm. According to the signal labeling presented in Figure 5, in the mixture before irradiation C14 can be easily identified by the strong peaks at 1.27, 2.04, and 1.37 ppm due to the methylene protons 4, 3, and 5, respectively, in the alkyl chain, the methyl signal (6) at 0.88 ppm, and the two multiplets at 5.80 and 5.00 ppm due to the double bond end group (1 and 2). After UV irradiation, it is possible to observe a decrease in signal intensity of 1, 2, and 3 resonances and the corresponding appearance of signals 1′ at 2.50, 2′ at 1.57, and of resonance 3′ at 1.26 (overlapped with 4), which is proof of the conversion of C14 into the thioether product of thiol-ene coupling. Moreover, the formation of the thioether group leads to shielding of the α, β, and γ nuclei, with a consequent up-field shift of 0.03 ppm of the mercaptoproyl signals (α″, β″, and γ″). The reaction gives rise to methylene groups with similar neighboring, thus leading resonances 1′ and 2′ to appear overlapped to α″ and β″, respectively.

Since deconvolution of α″ and 1′ signals with Gaussian-Lorentzian line-shapes failed to obtain an acceptable fit, and integral areas of β″ and 2′ are not reliable due to the presence of other signals in the same range (the yield of the reaction was calculated from the comparison of the spectra before and after irradiation). The intensities of α, α″, and 1′ resonances were normalized in the spectra with respect to the signal 6 at 0.88 ppm, which is unaffected by the reaction. The integrated areas are used in Equation (3) that gives the thiol-ene click reaction yield in %:

$$\text{Yield}_\% = 100 \times \left[\left(\alpha + \alpha' + 1'\right)_{after} - \alpha_{before}\right] \tag{3}$$

Upon irradiation at λ_{max} = 254 nm in the presence of C14, an almost complete conversion is detected for *SH-NBB_c* and *SH-NBB_p* at 6, 16, and 80 h, which gives a yield above 99% with the only exception of *SH-NBB_p* at 16 h that reaches 93%. The same results are obtained for the reaction with 1-decene, indicating that the conversion reaction is achieved independently from the alkyl chain length. On the contrary, from the ^{1}H NMR spectra no reaction is observed with the same samples irradiated at 365 nm or kept in the darkness (control samples). All the obtained results are summarized in Table S1.

The ^{1}H NMR study of purified SH-NBBs/alkene mixtures demonstrates that initiator-free thiol-ene coupling between SH-NBBs and selected alkenes in over-stoichiometric amounts occurs with near to quantitative conversion after irradiation with light in the middle UV range with λ_{max} = 254 nm. Moreover, in accordance with its click character, the observed thiol-ene reaction is very fast, achieving almost complete completion within 15 min, as demonstrated for *SH-NBBs_p*/C14 mixtures. To the best of our knowledge, this constitutes the fastest and most efficient example of initiator-free thiol-ene coupling of silsesquioxane-based materials ever achieved [24,25].

While reactions conducted under irradiation at λ_{max} = 254 nm achieved excellent yields typical of click reactions, samples irradiated with λ_{max} = 365 nm did not react at all for the given exposure times (although, from this result it cannot be concluded that thiol-ene coupling cannot occur in those conditions; the reaction would certainly proceed much more slowly than for a λ_{max} = 254 nm irradiation, in accordance with the literature [26,27]).

2.3. Thiol-Ene Functionalization of SH-NBBs Coated Cotton and Paper

In the second functionalization step, *Cotton-NBB* and *Paper-NBB* samples were soaked in the pure alkene without the addition of any photoinitiator, and subsequently irradiated with λ_{max} = 254 nm. As reported above, the structural characterization of the coating by solid state NMR was possible only using the sample ad hoc prepared. With the purpose to check the occurrence of the click coupling between the alkene and the available thiol functions of the coatings, ^{1}H MAS NMR spectra were recorded on the previous sample before and after the reaction. Comparing the spectra of raw and coated cotton (Figure S3a,b), the overlapping of the relatively sharp peaks belonging to the NBBs and the broad cellulose signal can be appreciated. Moreover, the signal due to the substrate further broadens as a consequence of the interaction with the NBBs. The spectrum recorded on the coated sample soaked in C14 after UV irradiation (Figure S3c) presents two sharp and intense peaks at 1.3 and 0.9 ppm, which prove the presence of the long alkyl chain; the absence of sharp resonances in the region 5–6 ppm, attributable to the double bond protons (Figure S3c), confirms the occurrence of the click reaction.

In order to observe the effect of the click functionalization step on the coating, the SEM analysis was run on *Cotton-NBB click* (Figure S4) and *Paper-NBB click* (Figure S5) samples. *Cotton-NBB click* samples appear to have lost the bridge-like connections between fibers observed in their parent coated samples, but at higher magnification the sub-micrometric features previously observed on the coatings (Figure 2) appear clearly visible. Similar results are obtained for *Paper-NBB click* samples. Therefore, even if some large-scale features are lost with the click functionalization, the coating integrity is generally assessed on the basis of the SEM analysis.

Aiming to estimate the spatial distribution of available thiol groups and, consequently, the extent of alkene functionalization, the confocal microscopy analysis of *Cotton-NBB80h*, *Cotton-NBB80h C14 click* and raw cotton reference samples was conducted after binding with 7-diethylamino-3-(4-maleimidophenyl)-4-methylcoumarin fluorophore (CPM). An important characteristic of CPM is that it becomes fully fluorescent in the blue spectral range only when bound to thiols, while it is only weakly fluorescent in the absence of binding.

The raw cotton sample (shown in Figure S6a) appears brighter and with a more homogeneous emission than the *Cotton-NBB80h* sample (Figure 6a), while practically no fluorescence is visible in *Cotton-NBB80h C14 click* (Figure 6b). The brightest emission of raw cotton is produced by the cotton autofluorescence, which lies in the same spectral region of CPM emission. As it is common for natural fibers, cotton fluorescence has its origin in residual nucleotides, such as nicotinamide adenine dinucleotide phosphate (NADPH), and in the structure of cellulose, though the molecular origin of the latter phenomenon is still poorly understood [28,29]. Therefore, while it is not possible to determine the extent of CPM binding to SH-NBBs, the lower emission of *Cotton-NBB80h* can be reasonably ascribed to the quenching of cotton self-fluorescence resulting from the surface functionalization with SH-NBBs. Moreover, the evident reduced emission after click coupling (Figure 6b) can be taken as proof of the presence of a further functionalization of the surface, which derives from the reaction of the thiol groups available on the coating with the alkene.

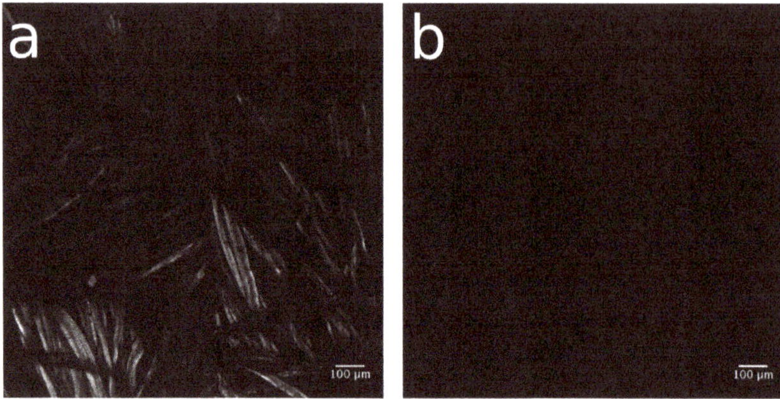

Figure 6. Confocal microscopy image of (**a**) *Cotton-NBB80h* and (**b**) *Cotton-NBB80h C14 click* samples after CPM treatment.

Even if raw paper exhibits a strong autofluorescence signal (Figure S6b), the spatial distribution of available thiol groups in the NBB coating before and after the click coupling was evaluated by means of confocal microscopy, after reaction with fluorescein diacetate 5-maleimide (FDM). As visible in Figure 7a, *Paper-NBB* exhibits diffused fiber fluorescence, with high intensity bright spots that could be addressed to a high local confinement of the FDM dye. After the click reaction (15′ and 30′, Figure 7b,c, respectively) the bright spots are switched off. Moreover, the overall fluorescence emission evidences a tendency to reduce with the UV exposure time and, thus, with the progression of the click reaction that reduces the available SH groups at the surface.

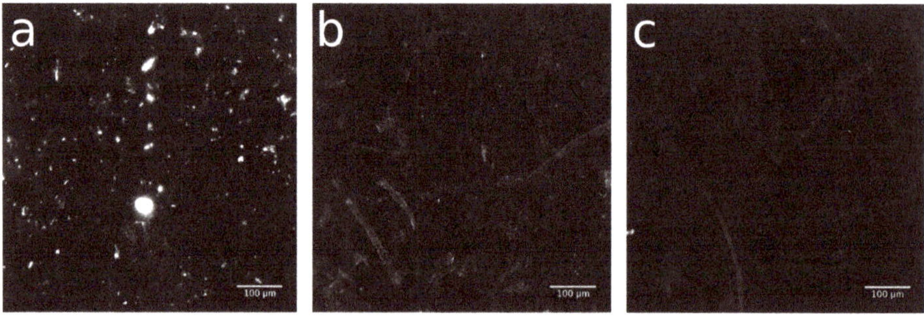

Figure 7. Confocal microscopy image of (**a**) *Paper-NBB80h*, (**b**) *Paper-NBB80h click 15′*, and (**c**) *Paper-NBB80h click 30′* samples after CPM treatment.

Owing to the different characteristics of the substrates, the performance of the two-step surface functionalization as a hydrophobization treatment was evaluated with different techniques for cotton and paper substrates.

Wettability of cotton samples was evaluated by the means of water shedding angle (ω) measurements, a technique specifically developed for rough, non-reflective textile surfaces such as cotton [30]. The results of the analysis are summarized in Figure 8; raw cotton samples showed ω values higher than 70° and are not reported in the figure. ω decreases both after NBB coating and click functionalization of the samples. Only a small difference was observed among coated samples, with *Cotton-NBB6h* and *Cotton-NBB80h* showing better performance than *Cotton-NBB16h*, but the difference disappeared after click functionalization. It is worth of noting that C14 functionalized samples appear to be more hydrophobic than those functionalized with C10.

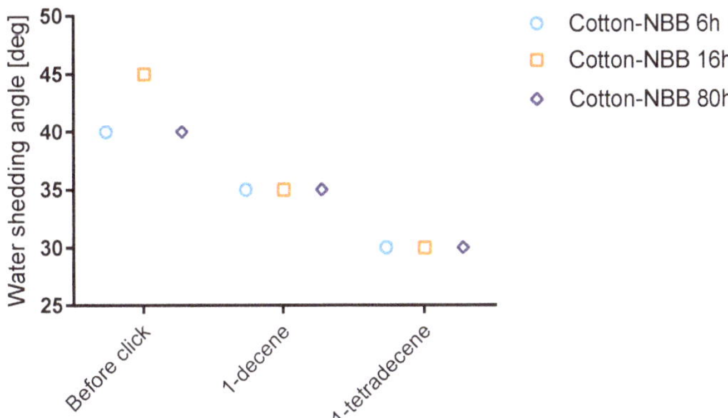

Figure 8. Results of the water shedding angle measurements of *Cotton-NBB* samples before and after click functionalization with either C10 or C14.

However, when water droplets are pinned on *Cotton-NBB* samples, after a very short time they are absorbed into the fabric. On the other hand, droplets pinned on the click functionalized samples remain indefinitely on the fabric surface. The behavior of *Cotton-NBB* samples that show an instaneous hydrophobic character suggests that the coating, although characterized by sub-micrometric surface roughness, is not sufficient to permit a stable surface hydrophobicity. On the other hand, after click functionalization, the introduced long alkyl chains, together with the hierarchical roughness conferred by the coating, can satisfy both the chemical and morphological conditions for stable hydrophobicity of the surface. Indeed, at least for the samples reacted with C14, the obtained ω values are comparable with those reported in the literature for superhydrophobic cotton fabrics obtained by silicon nanofilament coating [31].

The higher hydrophobicity of C14 functionalized samples could be related to the interactions occurring among the alkyl chains, which can increase with the functionalization density [32]. The grafted alkyl chains are free to rotate and can fold to find the best accommodation over the material surface, thus offering a degree of hydrophobic coverage that is probably higher for C14 than for C10.

Water contact angle on paper samples was evaluated by the means of the Wilhemly technique. Force curves were analyzed to evaluate the advancing contact angle (Figure 9). The receding angle (ϑ_{rec}) was impossible to evaluate because samples resulted completely wet after the advancing phases. This fact is explained by water adsorption due to defects on the sample edges and the capillary rise into the paper fibers network.

Raw paper samples, as expected, appeared to be completely hydrophilic with ϑ_{adv} compatible with 0° (not reported in the Figure 9). ϑ_{adv}, which slightly increased after NBB coating, was found to significantly increase after click functionalization, reaching values greater than 140°.

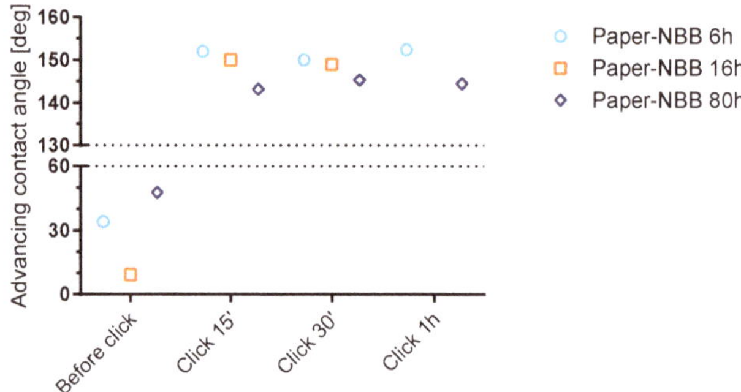

Figure 9. Results of the advancing contact angle of *Paper-NBB* samples before and after click functionalization with C14 with increasing irradiation time. Error bars are smaller than the used data symbols.

The high values of ϑ_{adv} observed in *Paper-NBB click* samples, typical of highly hydrophobic material, confirm the effectiveness of the proposed two-step surface functionalization process as a hydrophobization procedure. As observed for cotton samples, it appears that no substantial difference exists between samples coated with SH-NBBs prepared at different reaction times, although *Paper-NBB6h click* shows the highest ϑ_{adv} values. Moreover, it can be clearly seen that no remarkable change is observable with increasing irradiation times, and 15 min is a sufficient time to achieve the maximum observed hydrophobicity. This result would be in good accordance with the fast kinetics of the thiol-ene coupling between SH-NBBs and the C14 observed in solution.

According to the results obtained on cotton and paper, the click functionalization step is effective in introducing hydrophobic alkyl chains on the material surface while keeping the surface roughness obtained as a result of the NBB coating step. Consequently, at the end of the two-step surface functionalization, samples possess both the chemical and morphological requirements for extreme wettability. Therefore, the two-step surface functionalization process here proposed is a viable process for obtaining the efficient hydrophobization of natural materials by combined sol-gel and initiator free thiol-ene chemistry. In perspective, the preparation of superhydrophobic materials without employing fluorinated compounds and based on cellulose appears as a good response to the request of low environmental impact materials for water remediation.

3. Materials and Methods

3.1. Materials

Chloroacetic acid (ClAAc, pure), alkenes: 1-decence (C10, >97%), 1-tetradecene (C14, >97%); chloroform (purum, >99.5%), chloroform-d (>99.8 atom %D, with 0.05% (v/v) tetramethylsilane), dimethyl sulfoxide (DMSO, >99.5%), 7-diethylamino-3-(4-maleimidophenyl)-4-methylcoumarin (CPM, >90%), and fluorescein diacetate 5-maleimide (FDM, for fluorescence) were acquired from Sigma-Aldrich (St. Louis, MO, USA); 3-mercaptopropyltrimethoxysilane (McPTMS, >95%) and dibutyltin dilaurate (DBTL) were acquired from ABCR (Karlsruhe, Germany); 1-propanol (PrOH, Baker analysed >99%) was acquired from Fischer Scientific (Hampton, NH, USA). White, commercially available plain-weave cotton fabrics were used without any further treatment; 70 mm diameter Whatman Schleicher & Schuell ashless 589/2 white ribbon filter paper (Maidstone, UK) was used as the paper substrate without any further treatment.

3.2. SH-NBB Coatings and Purification

Thiol-functionalized silsesquioxane NBBs (SH-NBBs) were synthesized according to previous works [14,15]. ClAAc (1275.8 mg) was first brought to 100 °C and melted in an open borosilicate glass tube under N2 fluxing. PrOH (3.027 mL) was then added as a solvent, followed by the trialkoxysilane precursor McPTMS, (418 µL) and finally by DBTL (20 µL) as a condensation promoter. The tube was then sealed, obtaining a self-condensing reaction vial, and the mixture was left to react for a given time (6 h, 16 h or 80 h). The obtained clear, faintly yellow solutions (SH-NBB solutions) were stored at 4 °C in order to prevent any further reaction.

Raw cotton fabrics were immersed for 24 h into the SH-NBB solutions and subsequently left to dry overnight in air at room temperature on a horizontal stainless steel drying rack. The fabrics were then rinsed in chloroform for 1 h in order to remove the ungrafted SH-NBBs. The obtained samples were labeled *Cotton-NBB*. NBB coating of paper samples was conducted following the same procedure as *Cotton-NBB* samples, with an additional rinsing with chloroform in an ultrasound bath for 7 min. The obtained samples were labeled *Paper-NBB*.

The chloroform solutions obtained by rinsing both *Cotton-NBB and Paper-NBB* samples were left to evaporate at room temperature under a fume hood for several days. The whitish precipitate was dissolved in chloroform-d, containing 0.05% (*v/v*) tetramethylsilane (TMS) as an internal standard for the ^{1}H NMR characterization. The solutions obtained from *Cotton-NBB and Paper-NBB* samples were labeled *SH-NBBs_c* and *SH-NBBs_p*, respectively.

3.3. Thiol-Ene Click Functionalization

400 µL of *SH-NBB_c* or *SH-NBB_p* solutions were transferred in a 5 mm NMR tube and an appropriate alkene (C14 or C10) volume was added so that [SH-NBBs]:[alkene] = 1:1.75. Tubes were sealed and both irradiated (at λ_{max} = 254 nm or λ_{max} = 365 nm, 4 W, 2 cm distance from lamp) for 15 or 30 min, or kept in the darkness for 2 days (control). ^{1}H NMR spectra were acquired before and after each procedure.

Coated cotton fabrics and paper were soaked in the pure alkene (C14 or C10) for 5 min, without any photoinitiator, and immediately transferred in a black box where they were irradiated continuously with λ_{max} = 254 nm for 1 h tilting the exposed side every 15 min. The samples were then rinsed in chloroform and left to dry in air at room temperature. Cotton samples were labelled *Cotton-NBB* C10 *click* or *Cotton-NBB* C14 *click*, respectively. Coated paper samples, soaked in C14 and irradiated for 15 and 30 min, and 1 h, were labelled *Paper-NBB click 15′, Paper-NBB click 30′, and Paper-NBB 1h*.

3.4. Characterization Techniques

Raw cotton and *Cotton-NBB* carbon coated samples were analysed with an Hitachi S-3400N microscope (Hitachi Ltd., Tokyo, Japan) with an accelerating voltage of 3 kV; EDX analysis were conducted with the built-in Oxford Instruments EDX instrument with an accelerating voltage of 10 kV. *Cotton-NBB click* samples were analysed after Pt/Pd sputtering with a Jeol JSM-5510 microscope (Jeol Ltd., Tokyo, Japan). Images at low magnification were taken with an accelerating voltage of 3 kV; images at a high magnification were taken with an accelerating voltage of 10 kV. *Paper-NBB* samples were analysed after Pt/Pd sputtering with a Supra 40 microscope (Carl Zeiss AG, Oberkochen, Germany). Images were taken with an accelerating voltage of 2 or 3 kV.

Solution-state NMR spectra were collected on a Bruker 400 WB spectrometer, equipped with a BBO liquid probe, or a Bruker 300 AvanceIII spectrometer, equipped with a 5 mm BBFO probe (Bruker Corporation, Billerica, MA, USA). The carrier frequencies were 400.13 and 300.13 MHz, respectively. ^{1}H single pulse sequence with pi/6 flip angle was used (400: pulse = 10 µs/recycle delay = 5 s/32 scans; 300: pulse = 4 µs/recycle delay = 8 s/32 scans). The Bruker Topspin software was used for spectra analysis.

Confocal microscopy analyses were run on cotton samples treated with CPM florophore, whereas paper samples were reacted with FDM. 10 mg of CPM was dissolved in 1.6 mL of chloroform, obtaining a bright orange solution. Cotton fabrics were immersed in 500 μL of the obtained solution and samples were left to react for 2 h at 4 °C in the darkness. The bright orange cotton samples were removed from the solution and dried at ambient conditions in the darkness. Paper samples were stained with FDM [33]. 3 mg of FDM were first dissolved in 25 mL of DMSO and 275 μL of sodium phosphate buffer solution (pH = 7) were added. Paper samples were immersed in the obtained solution (5 mL) and left to react for 21 h at room temperature away from any light source. The samples were rinsed in chloroform for 30 min to remove excess solvent. Analysis on CPM treated samples was carried out with a Nikon A1 laser confocal microscope (Nikon Corporation, Tokyo, Japan), exciting the samples at 405 nm wavelength and collecting fluorescence emission at 450/50 nm. Conversely, confocal microscopy on FDM-treated samples was obtained exciting samples with λ = 488 nm laser source and collecting emission at 525/50 nm.

Multi-stack pictures were collected for each sample and then were post elaborated using Fiji image processing software [34]. Multi layers where Z-projected to get single maximum intensity images for each sample. Picture brightness was finally normalized to compensate for different pinhole apertures and photomultiplier gain settings, and to allow image comparison.

Water shedding angle measurements on cotton samples were conducted following the procedure described in the literature [30] on an apparatus built using a Standa 8MR151 motorized rotation stage (Standa Ltd., Vilnius, Lithuania) with automatic droplet dispensation constituted by a Harvard Instruments Nanomite MA1 70-2217 syringe pump (Harvard Apparatus Inc., Holliston, MA, USA) and a 100 μL volume Hamilton syringe (Hamilton Company, Switzerland). The following parameters were used: droplet volume equal to 10.0 ± 0.1 μL, needle-to-substrate distance set to 10.0 ± 0.5 mm, and minimum droplet path on sample equal to 20.0 ± 0.5 mm. Fifteen measurements were conducted for each inclination angle, with care taken that the impact of the drops on the samples occurred in a different place. If all the droplets rolled off the sample surface, inclination was decreased by 5°; the smallest angle for which all droplets rolled off the sample surface was addressed as the shedding angle.

Contact angle measurements on paper samples were conducted using a Dynamic contact angle analyzer (*Cahn DCA-322*) on rectangular specimens (10×25 mm) obtained by cutting the samples, setting an advancing/receding speed of 40 μm/s.

4. Conclusions

The in situ water production route provides the opportunity to control the size and architecture of silsesquioxanes, obtained through sol-gel reactions of trialkoxysilanes bearing reactive organic groups like the thiol function in the case of SH-NBBs [15]. Owing to the presence of side-products that cannot be efficiently removed from the solutions [14], further SSQs functionalization can present difficulties that limit the versatility of the synthesis route. The filming ability of SH-NBBs solutions obtained by ISWP at different reaction times has been exploited for coating both cotton fabrics and paper. A good degree of coating was generally achieved and the fairly homogenous deposited layers imparted to the natural materials a hierarchical roughness, necessary but not sufficient to extreme hydrophobicity. The ^{1}H NMR analysis of the rinsing residues obtained during the SH-NBBs coating of both cotton and paper pointed out that the treatment yields recyclable wastes, which are constituted of purified SSQs. Therefore, the rinsing of coatings was proved to be a viable route to remove side-products as ClAAc and its derivatives, and gave the opportunity for the preparation of long alkyl chains substituted SSQs by means of fast, initiator-free thiol-ene click reactions with a nearly quantitative yield. To the best of our knowledge, this constitutes the fastest and most efficient example of initiator-free thiol-ene coupling of silsesquioxane-based materials ever achieved [24,25]. The functionalization of coated cotton and paper by thiol-ene click coupling was obtained via a simple immersion and UV-irradiation procedure, preserving the coating structure. The combination of multi-scale roughness, ensured by the sol-gel coating, and the hydrophobic surface character, imparted by the C10 and C14 chains, led the natural substrates to display stable highly hydrophobic behavior.

Supplementary Materials: The following are available online at www.mdpi.com/1996-1944/10/8/913/s1, Table S1. Reaction yield for each purified SH-NBB/alkene mixture and irradiation procedure; Figure S1. Cross-linking morphologies of general silsesquioxane networks (reprinted with permission from V. Tagliazucca, E. Callone, S. Dirè, "Influence of synthesis conditions on the cross-link architecture of silsesquioxanes prepared by in situ water production route" J Sol-Gel Sci Technol (2011) 60:236–245, © Springer); Figure S2. ^{29}Si CPMAS NMR spectrum of *Cotton-NBB80h* sample; Figure S3. ^{1}H MAS NMR spectra of (a) raw cotton, (b) *Cotton-NBB80h* and (c) the product of click reaction with 1-tetradecene (*Cotton-NBB80h click*); Figure S4. SEM images of *Cotton-NBB6h C14 click* (a), *Cotton-NBB16h C14 click* (b), and (c) *Cotton-NBB80h C14 click*; Figure S5. SEM images of *Paper-NBB80h C14 click* samples after 15′ (a), 30′ (b) and 1h (c) exposure to UV radiation; Figure S6. Confocal microscopy pictures of autofluorescence emission from raw cotton (a) and raw paper (b).

Acknowledgments: The research was performed within the frame of COST Action MP1202 (HINT) "Rational design of hybrid organic/inorganic interfaces: the next step towards advanced functional materials".

Author Contributions: D.B. prepared the coated samples and performed the click functionalization; D.B. and F.R. studied the NBBs purification; D.B., F.R., and E.C. performed the NMR experiments; E.C. and S.D. analyzed and discussed the NMR data; D.B., I.G., and D.M. acquired the SEM images; D.B. and D.M. performed the confocal microscopy and wettability analyses. All data were analyzed by D.B. with the assistance of F.R., E.C., D.M., and S.D.; D.B. and S.D. wrote the draft paper, but all authors have contributed and given approval to the final version of the manuscript.

Conflicts of Interest: The authors declare no conflict of interest.

References

1. Gómez-Romero, P.; Sanchez, C. Hybrid Materials, Functional Applications. An Introduction. In *Functional Hybrid Materials*; Gómez-Romero, P., Sanchez, C., Eds.; Wiley-VCH Verlag GmbH & Co. KGaA: Weinheim, Germany, 2004; pp. 1–14, ISBN 978-3-527-30484-4.

2. Sanchez, C.; Belleville, P.; Popall, M.; Nicole, L. Applications of advanced hybrid organic-inorganic nanomaterials: From laboratory to market. *Chem. Soc. Rev.* **2011**, *40*, 696–753. [CrossRef] [PubMed]

3. Kickelbick, G. Introduction to hybrid materials. In *Hybrid Materials: Synthesis, Characterization, and Applications*; Kickelbick, G., Ed.; John Wiley & Sons: Weinheim, Germany, 2007; pp. 1–46, ISBN 978-3-527-31299-3.

4. Kickelbick, G. Silsesquioxanes. In *Functional Molecular Silicon Compounds I: Regular Oxidation States*; Scheschkewitz, D., Ed.; Springer: Cham, Switzerland, 2014; pp. 1–28, ISBN 978-3-319-03620-5.

5. Ribot, F.; Sanchez, C. Organically Functionalized Metallic Oxo-Clusters: Structurally Well-Defined Nanobuilding Blocks for the Design of Hybrid Organic-Inorganic Materials. *Comments Inorg. Chem.* **1999**, *20*, 327–371. [CrossRef]

6. Sanchez, C.; de Soler-Illia, G.J.A.A.; Ribot, F.; Lalot, T.; Mayer, C.R.; Cabuil, V. Designed Hybrid Organic—Inorganic Nanocomposites from Functional Nanobuilding Blocks. *Chem. Mater.* **2001**, *13*, 3061–3083. [CrossRef]

7. Kickelbick, G. Concepts for the incorporation of inorganic building blocks into organic polymers on a nanoscale. *Prog. Polym. Sci.* **2003**, *28*, 83–114. [CrossRef]

8. Mammeri, F.; Bonhomme, C.; Ribot, F.; Babonneau, F.; Dirè, S. New Monofunctional POSS and Its Utilization as Dewetting Additive in Methacrylate Based Free-Standing Films. *Chem. Mater.* **2009**, *21*, 4163–4171. [CrossRef]

9. Dirè, S.; Borovin, E.; Ribot, F. Architecture of Silsesquioxanes. In *Handbook of Sol-Gel Science and Technology*; Klein, L., Aparicio, M., Jitianu, A., Eds.; Springer: Cham, Switzerland, 2016; pp. 1–34, ISBN 978-3-319-19454-7.

10. Pescarmona, P.P.; Maschmeyer, T. Review: Oligomeric Silsesquioxanes: Synthesis, Characterization and Selected Applications. *Aust. J. Chem.* **2002**, *54*, 583–596. [CrossRef]

11. Dirè, S.; Tagliazucca, V.; Brusatin, G.; Bottazzo, J.; Fortunati, I.; Signorini, R.; Dainese, T.; Andraud, C.; Trombetta, M.; Di Vona, M.; et al. Hybrid organic/inorganic materials for photonic applications via assembling of nanostructured molecular units. *J. Sol-Gel Sci. Technol.* **2008**, *48*, 217–223. [CrossRef]

12. Dirè, S.; Licoccia, S. Sol-gel hybrid organic/inorganic nanocomposites by condensation reactions of diphenylsilanediol and organo-alkoxysilanes for photonic applications. In *Organosilanes: Properties, Performance and Applications*; Wyman, E., Skief, M., Eds.; Nova: New York, NY, USA, 2010; pp. 131–155, ISBN 978-1-60876-452-5.

13. Tagliazucca, V.; Callone, E.; Dirè, S. Influence of synthesis conditions on the cross-link architecture of silsesquioxanes prepared by in situ water production route. *J. Sol-Gel Sci. Technol.* **2011**, *60*, 236–245. [CrossRef]

14. Borovin, E.; Callone, E.; Papendorf, B.; Guella, G.; Dirè, S. Influence of Sol-Gel Conditions on the Growth of Thiol-Functionalized Silsesquioxanes Prepared by In Situ Water Production. *J. Nanosci. Nanotechnol.* **2016**, *16*, 3030–3038. [CrossRef] [PubMed]

15. Borovin, E.; Callone, E.; Ribot, F.; Dirè, S. Mechanism and Kinetics of Oligosilsesquioxane Growth in the In Situ Water Production Sol-Gel Route: Dependence on Water Availability. *Eur. J. Inorg. Chem.* **2016**, 2166–2174. [CrossRef]

16. Pierre, A.C. The sol-gel encapsulation of enzymes. *Biocatal. Biotransform.* **2004**, *22*, 145–170. [CrossRef]

17. Hendan, B.J.; Marsmann, H.C. Silsesquioxanes as models of silica supported catalyst I. [3-(Diphenylphosphino)propyl]-hepta[propyl]-[octasilsesquioxane] and [3-mercapto-propyl]-hepta[propyl]-[octasilsesquioxane] as ligands for transition-metal ions. *Appl. Organomet. Chem.* **1999**, *13*, 287–294. [CrossRef]

18. Anker, J.N.; Hall, W.P.; Lyandres, O.; Shah, N.C.; Zhao, J.; Van Duyne, R.P. Biosensing with plasmonic nanosensors. *Nat. Mater.* **2008**, *7*, 442–453. [CrossRef] [PubMed]

19. Xue, C.-H.; Guo, X.-J.; Zhang, M.-M.; Ma, J.-Z.; Jia, S.-T. Fabrication of robust superhydrophobic surfaces by modification of chemically roughened fibers via thiol-ene click chemistry. *J. Mater. Chem. A* **2015**, *3*, 21797–21804. [CrossRef]

20. Tran, D.N.H.; Kabiri, S.; Sim, T.R.; Losic, D. Selective adsorption of oil-water mixtures using polydimethylsiloxane (PDMS)-graphene sponges. *Environ. Sci. Water Res. Technol.* **2015**, *1*, 298–305. [CrossRef]

21. Delattre, L.; Babonneau, F. Influence of the nature of the R Group on the Hydrolysis and Condensation Process of Trifunctional Silicon Alkoxides, R-Si(OR′)$_3$. *MRS Proc.* **1994**, *346*. [CrossRef]

22. Brochier Salon, M.-C.; Bayle, P.-A.; Abdelmouleh, M.; Boufi, S.; Belgacem, M.N. Kinetics of hydrolysis and self condensation reactions of silanes by NMR spectroscopy. *Colloids Surf. A Physicochem. Eng. Asp.* **2008**, *312*, 83–91. [CrossRef]

23. Wakelyn, P.J.; Bertoniere, N.R.; French, A.D.; Thibodeaux, D.P.; Triplett, B.A.; Rousselle, M.-A.; Goynes, W.R.J.; Edwards, J.V.; Hunter, L.; McAlister, D.D.; et al. Cotton Fibers. In *Handbook of Fiber Chemistry*; Lewin, M., Ed.; CRC Press: Boca Raton, FL, USA, 2006; ISBN 978-0-8247-2565-5.

24. Göbel, R.; Hesemann, P.; Friedrich, A.; Rothe, R.; Schlaad, H.; Taubert, A. Modular thiol–ene chemistry approach towards mesoporous silica monoliths with organically modified pore walls. *Chem. Eur. J.* **2014**, *20*, 17579–17589. [CrossRef] [PubMed]

25. Strauch, H.; Engelmann, J.; Scheffler, K.; Mayer, H.A. A simple approach to a new T 8-POSS based MRI contrast agent. *Dalton Trans.* **2016**, *45*, 15104–15113. [CrossRef] [PubMed]

26. Cramer, N.B.; Scott, J.P.; Bowman, C.N. Photopolymerizations of thiol- ene polymers without photoinitiators. *Macromolecules* **2002**, *35*, 5361–5365. [CrossRef]

27. Cramer, N.B.; Reddy, S.K.; Cole, M.; Hoyle, C.; Bowman, C.N. Initiation and kinetics of thiol-ene photopolymerizations without photoinitiators. *J. Polym. Sci. Part Polym. Chem.* **2004**, *42*, 5817–5826. [CrossRef]

28. Olmstead, J.A.; Gray, D.G. Fluorescence emission from mechanical pulp sheets. *J. Photochem. Photobiol. Chem.* **1993**, *73*, 59–65. [CrossRef]

29. Malinowska, K.H.; Rind, T.; Verdorfer, T.; Gaub, H.E.; Nash, M.A. Quantifying Synergy, Thermostability, and Targeting of Cellulolytic Enzymes and Cellulosomes with Polymerization-Based Amplification. *Anal. Chem.* **2015**, *87*, 7133–7140. [CrossRef] [PubMed]

30. Zimmermann, J.; Seeger, S.; Reifler, F.A. Water Shedding Angle: A new technique to evaluate the water repellent properties of superhydrophobic surfaces. *Text. Res. J.* **2009**. [CrossRef]

31. Zimmermann, J.; Reifler, F.A.; Fortunato, G.; Gerhardt, L.-C.; Seeger, S. A Simple, One-Step Approach to Durable and Robust Superhydrophobic Textiles. *Adv. Funct. Mater.* **2008**, *18*, 3662–3669. [CrossRef]

32. Heinz, H.; Vaia, R.A.; Farmer, B.L. Relation between packing density and thermal transitions of alkyl chains on layered silicate and metal surfaces. *Langmuir* **2008**, *24*, 3727–3733. [CrossRef] [PubMed]

33. Navarro, J.R.G.; Conzatti, G.; Yu, Y.; Fall, A.B.; Mathew, R.; Edén, M.; Bergström, L. Multicolor Fluorescent Labeling of Cellulose Nanofibrils by Click Chemistry. *Biomacromolecules* **2015**, *16*, 1293–1300. [CrossRef] [PubMed]

34. Schindelin, J.; Arganda-Carreras, I.; Frise, E.; Kaynig, V.; Longair, M.; Pietzsch, T.; Preibisch, S.; Rueden, C.; Saalfeld, S.; Schmid, B.; et al. Fiji: An open-source platform for biological-image analysis. *Nat. Methods* **2012**, *9*, 676–682. [CrossRef] [PubMed]

 *materials*

Article

Synthesis, Characterization, and Electrochemical Behavior of LiMn$_x$Fe$_{(1-x)}$PO$_4$ Composites Obtained from Phenylphosphonate-Based Organic-Inorganic Hybrids

Alessandro Dell'Era [1,2,*], Mauro Pasquali [1], Elvira Maria Bauer [2], Stefano Vecchio Ciprioti [1,*], Francesca A. Scaramuzzo [1] and Carla Lupi [3]

[1]   Department of Basic and Applied Sciences for Engineering (SBAI), Sapienza University of Rome, Via del Castro Laurenziano 7, 00161 Rome, Italy; mauro.pasquali@uniroma1.it (M.P.); francesca.scaramuzzo@uniroma1.it (F.A.S.)

[2]   Istituto di Struttura della Materia ISM—CNR, Via Salaria, km. 29.300, C.P. 10, 00015 Rome, Italy; elvira.bauer@ism.cnr.it

[3]   Department Chemical Engineering Materials Environment DICMA, University Sapienza Rome, Via Eudossiana 18, 00184 Rome, Italy; carla.lupi@uniroma1.it

*   Correspondence: alessandro.dellera@uniroma1.it (A.D.); stefano.vecchio@uniroma1.it (S.V.C.); Tel.: +39-064-976-6906 (S.V.C.)

Received: 14 November 2017; Accepted: 27 December 2017; Published: 30 December 2017

Abstract: The synthesis of organic-inorganic hybrid compounds based on phenylphosphonate and their use as precursors to form LiMn$_x$Fe$_{(1-x)}$PO$_4$ composites containing carbonaceous substances with sub-micrometric morphology are presented. The experimental procedure includes the preliminary synthesis of Fe^{2+} and/or Mn^{2+} phenylphosphonates with the general formula Fe$_{(1-x)}$Mn$_x$[(C$_6$H$_5$PO$_3$)(H$_2$O)] (with $0 < x < 1$), which are then mixed at different molar ratios with lithium carbonate. In this way the carbon, obtained from in situ partial oxidation of the precursor organic part, coats the LiMn$_x$Fe$_{(1-x)}$PO$_4$ particles. After a structural and morphological characterization, the electrochemical behavior of lithium iron manganese phosphates has been compared to the one of pristine LiFePO$_4$ and LiMnPO$_4$, in order to evaluate the doping influence on the material.

Keywords: lithium-ion battery; LiMn$_x$Fe$_{(1-x)}$PO$_4$; carbon coating; pseudo-diffusion coefficient; potential step voltammetry; electrochemical impedance spectroscopy

1. Introduction

Nowadays, lithium-ion batteries are the most developed energy sources for modern portable electronics and their use in automotive application is also increasing [1–8]. So far, several materials, such as LiCoO$_2$ and LiMn$_2$O$_4$, have been used as cathode, but recently LiFePO$_4$ has attracted researchers' interest due to its high specific energy, which may reach 580 Wh/kg, and relatively low production cost [9–14]. As a drawback, LiFePO$_4$ has low ionic diffusivity and conductibility [15–17], which limits its use as cathode. The electronic conductibility of LiFePO$_4$ can be enhanced by using several materials processing methods such as in situ carbon synthesis, or by particle coating with conductive carbons [18], or by an ion doping approach [19–24]. In the latter case, the oxidized form of LiFePO$_4$ should be modified with cations having ionic radius slightly higher than Fe^{2+} and Fe^{3+}, such as manganese, facilitating a wider channel for lithium-ion diffusion, increasing the mobility of lithium ion but, at the same time, avoiding the structure to be stressed [25]. Moreover, the LiFePO$_4$ particles size should be reduced to decrease the average free lithium pathway in insertion/de-insertion process and raise the performances. Indeed, in this way, all the material can be effectively used, consequently enhancing the specific capacity. The aim of this work has been to analyze the electrochemical performance of lithium iron phosphate with the addition of manganese $LiMn_x Fe_{(1-x)}PO_4$ starting from Fe^{2+} and Mn^{2+} phenylphosphonates (general formula $Fe[(C_6H_5PO_3)(H_2O)]$ or $Mn[(C_6H_5PO_3)(H_2O)])$) in appropriate ratio as metal-organic precursor and Li_2CO_3 as inorganic precursor. Moreover, in order to verify possible differences in the electrochemical performances, one of the lithium iron manganese phosphates, i.e., $LiFe_{0.9}Mn_{0.1}PO_4$, has been also synthesized by using the precursor $Fe_{0.9}Mn_{0.1}[(C_6H_5PO_3)(H_2O)]$. Thermal, structural, and morphological analyses have been performed on both precursors and final materials; finally, an electrochemical characterization has been carried out on all prepared samples to evaluate if the synthesis process and the hetero-metal adding degree can influence their specific capacity.

2. Materials and Methods

2.1. Synthesis of Precursors

Analytical grade (Sigma Aldrich Chemical Co., Darmstadt, Germany) phenylphosphonic acid ($H_2C_6H_5PO_3$), ammonium hydroxide (NH_4OH), iron(II) sulphateheptahydrate ($FeSO_4 \cdot 7H_2O$), and manganese(II) sulphatemonohydrate ($MnSO_4 \cdot H_2O$) were used for the synthesis without further purification. HPLC (High Pressure Liquid Chromatography) water (Carlo Erba, Milan, Italy) was used as a solvent. Usual Schlenck techniques were used to prepare the phenylphosphonate precursor materials.

All metal(II) phosphonate precursors, i.e., $Fe_{0.9}Mn_{0.1}(C_6H_5PO_3)(H_2O)$ (**P1**), $Fe[(C_6H_5PO_3)H_2O]$ (**P2**) and $Mn(C_6H_5PO_3)(H_2O)$ (**P3**),were obtained following the synthetic procedure described previously for iron(II) phenylphosphonate monohydrate $Fe[(C_6H_5PO_3)H_2O]$ [21,22]: 10 g (63.25 mmol) of $H_2C_6H_5PO_3$ were suspended under continuous stirring in 50 mL of water in a 100 mL two-necked flask (flask 1). NH_4OH (about 9.5 mL, 30% in H_2O) was added, up to pH = 7, to the white colloidal suspension obtained, thus giving the water soluble ammonium salt of the phenylphosphonic acid $(NH_4)_2(C_6H_5PO_3)$. In another 100 mL two-necked flask (flask 2), 7 g (25.17 mmol) of $FeSO_4 \cdot 7H_2O$ were dissolved in 35 mL of degassed water. After the complete dissolution of ferrous sulphate, the degassed aqueous solution of $(NH_4)_2(C_6H_5PO_3)$ was transferred from flask 1 to flask 2 under a stream of inert gas and with a filtration system. The filtration system guarantees the transfer of a filtered and clear solution of the ammonium salt of phenylphosphonic acid to iron(II) sulphate. During the transfer, iron(II) phenylphosphonate, $Fe[(C_6H_5PO_3)H_2O]$ formed instantaneously as a white flaked precipitate. The white colloidal suspension thus obtained was maintained under continuous stirring under flowing nitrogen for approx. 2 h (pH_{fin} = 6.14). The precipitate was then filtered in air, washed with water to neutrality with acetone, and finally air-dried. Three different metal(II) phenylphosphonate precursors, reported in Table 1, were isolated by the former preparation method.

Table 1. Molecular formulas of precursors.

Material	Formula
P1	$Fe_{0.9}Mn_{0.1}(C_6H_5PO_3)(H_2O)$
P2	$Fe(C_6H_5PO_3)(H_2O)$
P3	$Mn(C_6H_5PO_3)(H_2O)$

2.2. Synthesis of $LiMn_{0.1}Fe_{0.9}PO_4$

In particular, $LiMn_{0.1}Fe_{0.9}PO_4$ was synthesized starting from different precursors following different procedures, and the final products obtained were compared in terms of morphology and electrochemical performances. In detail, $LiMn_{0.1}Fe_{0.9}PO_4$ was synthesized by mixing in a mechanical mill Li_2CO_3 (analytical grade by Sigma Aldrich Chemical Co.) and either the precursor **P1** $Fe_{0.9}Mn_{0.1}(C_6H_5PO_3) \cdot H_2O$ (sample **S1**) or the precursors **P2** $Fe(C_6H_5PO_3) \cdot H_2O$ and

P3 $Mn(C_6H_5PO_3) \cdot H_2O$ in molar ratio 0.9/0.1 (sample **S2**). After the grinding process the light-grey powder homogeneous mixture of the reagents was placed in an alumina crucible and transferred into the central zone of a tubular furnace for calcination. In order to maintain an inert environment, the mixture of reagents was degassed for 1 h at room temperature under nitrogen flowing. Successively the powder underwent a calcination at 600 °C for 16 h under nitrogen flowing. The calcined product was then cooled under inert gas to room temperature, thus obtaining a fine black powder.

2.3. Synthesis of $LiMn_xFe_{(1-x)}PO_4$ (with x = 0.05, 0.1, 0.5, 0.9, and 0.95)

The synthesis was performed starting from Li_2CO_3, **P2** and **P3** precursors, as described in the previous paragraph. In order to obtain all the desired compounds, **P2** and **P3** were mechanically mixed with a molar ratio (1−x):x respectively, where x = 0.05, 0.1, 0.5, 0.9, and 0.95, followed by thermal treatment under inert atmosphere as reported above.

3. Results and Discussion

3.1. Precursors Characterization

Infrared spectra of the different monohydrate precursors appear quite similar and present several bands, as shown in Figure 1. The bands between 3420 and 3470 cm^{-1} and the band at 1604 cm^{-1} correspond respectively to the stretching and bending vibrations of the water molecule of M(II) phenylphosphonate monohydrate (M = Fe, Mn, $Fe_{0.9}Mn_{0.1}$). Other characteristic bands of this compounds are located between 3074 and 3054 cm^{-1} and are associated with the stretching vibrations of the C-H bond of the phenyl group, while the band at 1438 cm^{-1} corresponds to C-C bond stretching of the same group. Finally, in the region between 1200–970 cm^{-1}, the characteristic stretching vibrations of the P-O bond of the anion $(PO_3)^{2-}$ are observed. The complete conversion of phenylphosphonic acid to metal(II) phosphonate is confirmed by the absence of the typical OH-binding strain vibrations of the P-OH group, generally observed as wide bands between 2900 and 2300 cm^{-1}.

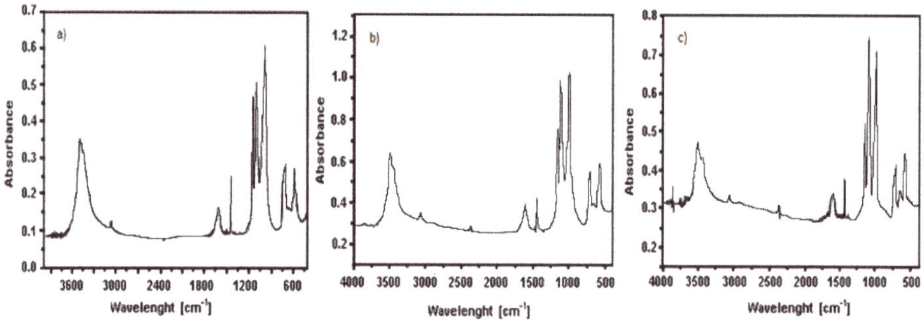

Figure 1. FT-IR spectra of precursors: (**a**) $Fe_{0.9}Mn_{0.1}(C_6H_5PO_3)(H_2O)$ (**P1**); and (**b**) $Fe[(C_6H_5PO_3)H_2O]$ (**P2**) and (**c**) $Mn(C_6H_5PO_3)(H_2O)$ (**P3**).

In Figure 2, the X-ray diffractograms and the refining results obtained by the Rietveld method are reported. The only crystalline phase present in the analyzed precursor powders corresponds to the expected M(II)phenylphosphonate (M = Fe, Mn, $Fe_{0.9}Mn_{0.1}$) [26,27]. All the three phenylphosphonate precursors crystallize in the orthorhombic spatial group $Pmn2_1$. In Table 2 the cell parameters for the different samples are reported. Excluding lattice parameter "b", it is possible to state that passing from iron to manganese produces an increasing of the cell size.

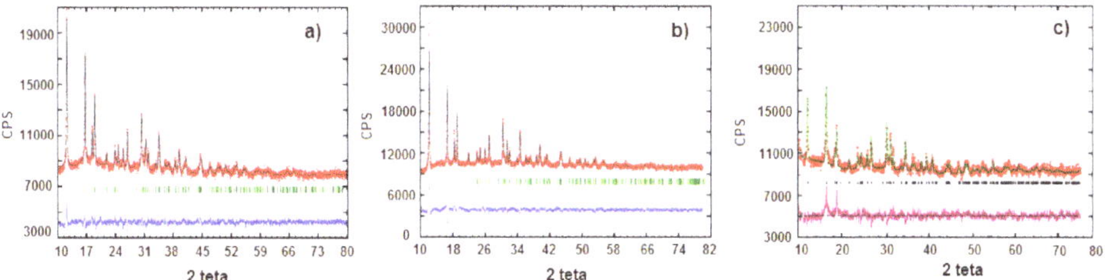

Figure 2. X-ray Diffraction pattern of precursors: (**a**) $Fe_{0.9}Mn_{0.1}(C_6H_5PO_3)(H_2O)$ (**P1**); and (**b**) $Fe[(C_6H_5PO_3)H_2O]$ (**P2**) and (**c**) $Mn(C_6H_5PO_3)(H_2O)$ (**P3**).

Table 2. Lattice parameters of precursors (in Å), where $\alpha = \beta = \gamma = 90°$.

Material (Symbol)	a	b	c
$Fe_{0.9}Mn_{0.1}(C_6H_5PO_3)\cdot H_2O$ (**P1**)	5.680	14.410	4.900
$Fe(C_6H_5PO_3)\cdot H_2O$ (**P2**)	5.652	14.404	4.882
$Mn(C_6H_5PO_3)\cdot H_2O$ (**P3**)	5.751	14.401	4.953

3.2. Characterization of $LiFe_{0.9}Mn_{0.1}PO_4$

DSC-TG (Differential Scanning Calorimetry–Thermo-Gravimetry) curves obtained under nitrogen flow for $Fe_{0.9}Mn_{0.1}(C_6H_5PO_3)\cdot H_2O$ (**P1**)/Li_2CO_3 mixture (sample **S1**) and for $Fe(C_6H_5PO_3)\cdot H_2O$ (**P2**) and $Mn(C_6H_5PO_3)\cdot H_2O$ (**P3**) (**P2**:**P3** = 0.9:0.1)/Li_2CO_3 mixture (sample **S2**) are reported, respectively, in Figure 3a,b. The thermal behaviour of both mixtures resulted to be quite similar.

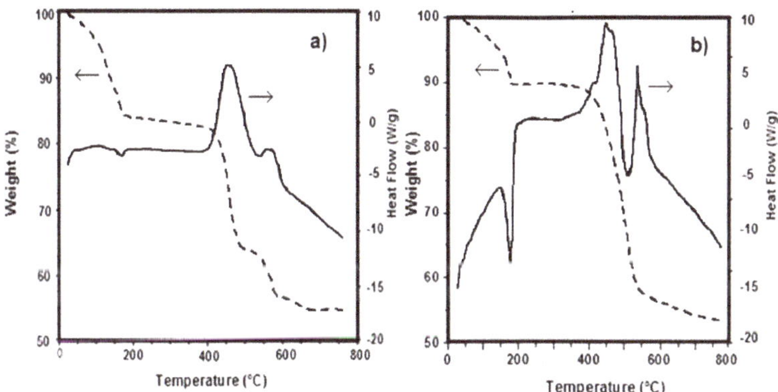

Figure 3. DSC-TG curves of: (**a**) sample **S1**: **P1**/Li_2CO_3; and (**b**) sample **S2**: (**P2**:**P3** = 0.9:0.1)/Li_2CO_3.

Part of the water, probably physically adsorbed on the sample, is lost at temperatures lower than 100 °C. The remaining part, i.e., the crystallization water, is lost at about 200 °C: this phenomenon is clearly highlighted by the end thermal peak observable in the DSC curves at 180 °C. The weight loss up to 180 °C is about 11–15%. At higher temperatures two exothermic effects are displayed in the DSC curves, namely at 400 and 550 °C, which are accompanied by a weight loss in the TG curves of about 25–30% and 8–10%, respectively.

These effects are related to the decomposition of carbonate and organo-phosphonates and to the formation of lithium metal(II) phosphate. Such experimental evidences are in good agreement with literature, according with the dehydration of some hydrate metal phosphates which proceeds by both anion disproportion and condensation. The X-ray powder diffraction patterns of $LiMn_{0.1}Fe_{0.9}PO_4$

prepared from either **P1** and Li_2CO_3 or **P2**, **P3**, and Li_2CO_3 precursors are very alike as well, as shown in Figure 4. Both belong to orthorhombic space group *Pnma* (olivine like structure) [13]. The similarity between these two samples is evident also from SEM (Scanning Electron Microscopy) images reported in Figures 5 and 6.

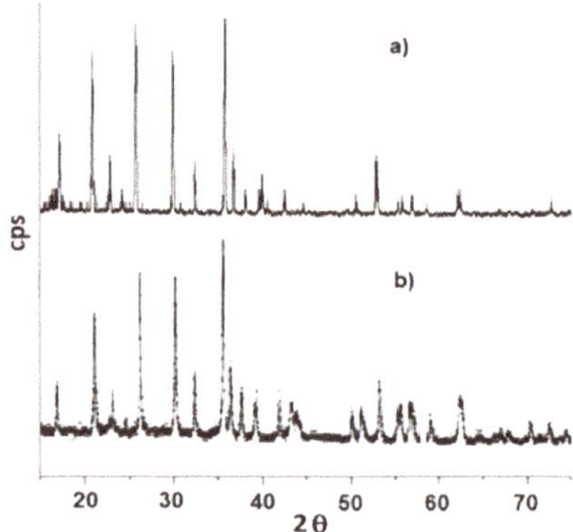

Figure 4. X-ray Diffraction pattern of $LiFe_{0.9}Mn_{0.1}PO_4$: (**a**) sample **S1**: **P1**/Li_2CO_3; and (**b**) sample **S2**: (**P2**:**P3** = 0.9:0.1)/Li_2CO_3.

Figure 5. SEM images of $LiFe_{0.9}Mn_{0.1}PO_4$: (**a**) sample **S1**: **P1**/Li_2CO_3; and (**b**) sample **S2**: (**P2**:**P3** = 0.9:0.1)/Li_2CO_3.

Figure 6. Morphology and size details at higher magnification.

The particles show a comparable morphology. In both samples the particles appear agglomerated and the presence of two phases can be noted: the former, likely carbon, is characterized by very small spheres, while the second one ($LiMn_{0.1}Fe_{0.9}PO_4$) is characterized by larger and less regular particles.

The particles are spheroidal, or in any case there is no dimension that prevails over the others during the growth, like for example in a needle structure; such experimental evidence suggests that during the thermal treatment strong nucleation with the formation of small nucleuses growing indifferently in all directions occurs. In Figure 6 a higher magnification highlights the formation of very small particles with nanometric size. Actually, the formation of carbon on the active material surface can inhibit the particle grow ensuring a tiny granulometry and possibly can provide good conductibility and electric contact between particles [13–28].

Electrochemical galvanostatic tests on both samples **S1** and **S2** are shown in Figure 7. The cathode electrodes have been charged and discharged with a current value of C/5 and a specific capacity of about 115–120 mAh/g has been obtained. The materials seem to show similar behaviour, even though sample **S2**, obtained by using **P2**, **P3** and Li_2CO_3 precursors, presents higher capacity and seems to be more stable upon cycling. On the other hand, the synthesis by **P1** precursor always produces a less performing material, even though it is not straightforward to give an explanation for such different behaviour. Several tests have been performed for each material, and the results are well reproducible.

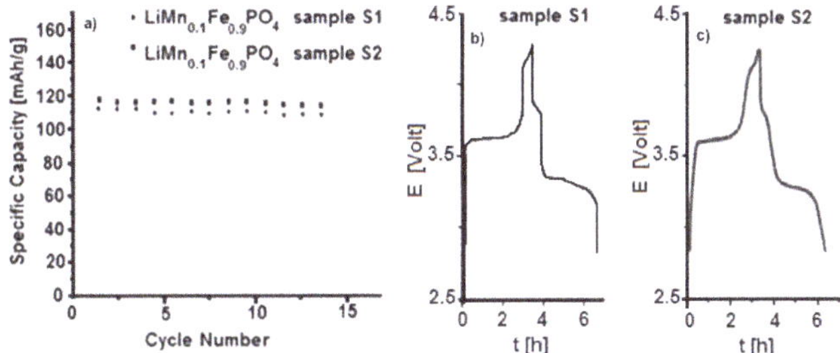

Figure 7. (a) Electrochemical galvanostatic tests at C/5 rate for samples **S1** and **S2**; and (b) voltage profile of sample **S1**; (c) voltage profile of Sample **S2**.

3.3. Characterization of $LiMn_xFe_{(1-x)}PO_4$ (with x = 0, 0.05, 0.1, 0.5, 0.9, 0.95, and 1)

Furthermore, taking in consideration this electrochemical results, a series of $LiMn_xFe_{(1-x)}PO_4$ (with x = 0, 0.05, 0.1, 0.5, 0.9, 0.95 and 1), obtained only from **P2** and/or **P3** and Li_2CO_3 precursors, have been prepared and characterized.

The powder X-ray diffraction spectra of $LiMn_xFe_{(1-x)}PO_4$ (x = 0; 0.05; 0.1; 0.5; 0.9; 0.95; 1) are shown in Figure 8. As it can be observed, substituting manganese in lithium manganese phosphate with iron(II) slightly moves all peaks to the right, although the similarity of the crystalline structure of the two lithium metal(II) phosphates is clear.

Moreover, in Table 3 the refinement results for cell parameters and crystallites size T, calculated by Scherrer equation (T = 0.9λ/Δ(2θ) cosθ), have been reported and it is clear that passing from $LiFePO_4$ to $LiMnPO_4$ the cell size slightly increases, while the crystallites size decreases.

Table 3. Cell parameters and crystallite size.

Compound	a (Å)	b (Å)	c (Å)	Crystallite Size T (Å)
$LiFePO_4$	10.330	6.010	4.690	393
$LiFe_{0.95}Mn_{0.05}PO_4$	10.335	6.011	4.693	370
$LiFe_{0.9}Mn_{0.1}PO_4$	10.347	6.020	4.699	368
$LiFe_{0.5}Mn_{0.5}PO_4$	10.372	6.054	4.706	365
$LiFe_{0.1}Mn_{0.9}PO_4$	10.448	6.103	4.743	282
$LiFe_{0.05}Mn_{0.95}PO_4$	10.448	6.104	4.743	280
$LiMnPO_4$	10.450	6.108	4.732	265

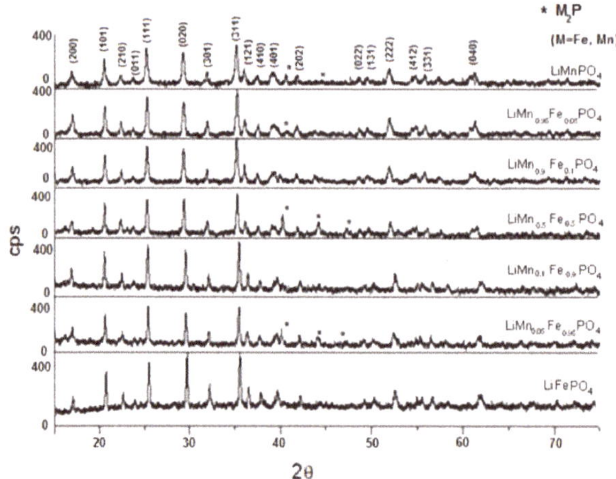

Figure 8. X-ray powder diffraction pattern of LiMn$_x$Fe$_{(1-x)}$PO$_4$ (x = 0; 0.05; 0.1; 0.5; 0.9; 0.95; 1).

In some samples (x = 0.05; 0.5; 0.95; 1) the presence of iron phosphide (Fe$_2$P) impurities have been detected. Replacement of the bivalent hetero-metal atom does not affect the crystalline structure of pure lithium iron and manganese phosphate. Indeed the crystalline structure (space group) is the same but substitution of iron with manganese (different atomic radius) in effect shifts the peak positions slightly and this is visible also in the reported XRD spectra. What is important here is that for all samples one, unique crystalline phase has been detected while the mechanic mixture clearly shows peak splitting due to the presence of two crystalline phases. On the other hand, when simply mixing together (0.5:0.5) LiMnPO$_4$ and LiFePO$_4$, the observed X-ray diffractogram shows a splitting of the peaks, as reported in Figure 9. Actually, in this case two similar crystalline structures presenting slightly different peak positions are present, therefore, two distinct phases and a splitting of peaks are evident. It is worth to note that in the case of manganese-iron phosphate synthesized from metal(II) phenylphosphonate mixtures as described before, even when Mn(II) and Fe(II) are present in equal ratio, as in LiMn$_{0.5}$Fe$_{0.5}$PO$_4$, formation of only one crystalline phase has been observed.

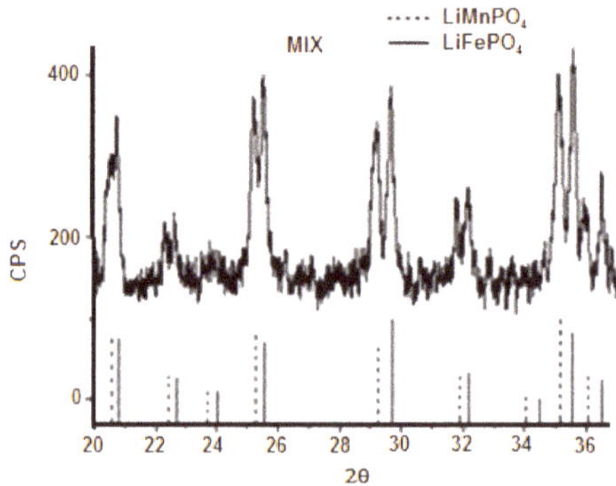

Figure 9. Comparison between X-ray Diffraction spectra of a binary mixture of LiFePO$_4$ and LiMnPO$_4$ and corresponding pure lithium metal phosphates.

The results of BET analysis for cathodic powders obtained are reported in Table 4 along with the carbon weight percentage determined by elemental analysis, which ranges from 10% to 13.7%.

This percentage value has been also confirmed by EDX analysis performed on some samples. The average specific surface is equal to about 115 $m^2 \cdot g^{-1}$.

Table 4. Specific surface of cathodic powder and carbon percentage.

Material (Symbol)	Specific Surface Area ($m^2 \cdot g^{-1}$)	Carbon Content (%)
$LiFePO_4$	105	10.0
$LiMn_{0.05}Fe_{0.95}PO_4$	100	10.2
$LiMn_{0.1}Fe_{0.9}PO_4$	130	12.5
$LiMn_{0.5}Fe_{0.5}PO_4$	115	10.3
$LiMn_{0.9}Fe_{0.1}PO_4$	110	11.3
$LiMn_{0.95}Fe_{0.05}PO_4$	151	13.7
$LiMnPO_4$	105	11.5

The SEM images in Figure 10 show similarity of both morphology and particle size of the various samples. Indeed, identical considerations already done for $LiMn_{0.1}Fe_{0.9}PO_4$ and no particular differences can be highlighted.

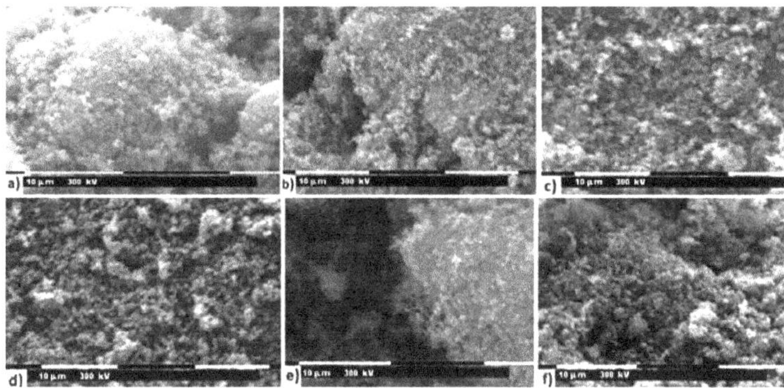

Figure 10. SEM images of: (**a**) $LiFePO_4$; (**b**) $LiMn_{0.05}Fe_{0.95}PO_4$; (**c**) $LiMn_{0.5}Fe_{0.5}PO_4$; (**d**) $LiMn_{0.9}Fe_{0.1}PO_4$; (**e**) $LiMn_{0.95}Fe_{0.05}PO_4$; and (**f**) $LiMnPO_4$.

The electrochemical tests indicate that pure lithium iron phosphate is the material with the highest specific capacity, i.e., about 150 $mAh \cdot g^{-1}$. Upon increasing of manganese content, the capacity gradually drops, reaching the significantly low value of 23 $mAh \cdot g^{-1}$ for $LiMn_{0.95}Fe_{0.05}PO_4$. Therefore, regarding the specific capacity, the presence of manganese does not seem to have particular advantages.

As it can be seen from Figure 11a, the only advantage shown by the presence of manganese is a higher insertion-deinsertion potential value. In fact, the potential value of the Mn^{3+}/Mn^{2+} redox couple is 4.15 V, while for the Fe^{3+}/Fe^{2+} redox couple it results to be 3.5 V, both vs. Li^0/Li^+.

The $LiMn_{0.5}Fe_{0.5}PO_4$ compound exhibits a poorly stability upon cycling, while the compounds with a higher percentage of manganese show a low capacities. Taking into consideration the two pure compounds, namely $LiFePO_4$ and $LiMnPO_4$, they both have capacity higher than the respective modified compounds. In particular, the presence of Mn(II) decreases the capacity of $LiFePO_4$ more than the substituition of manganese with iron in $LiMnPO_4$. Indeed the plateaus at 3.5 V for $LiMn_{0.9}Fe_{0.1}PO_4$ and $LiMn_{0.95}Fe_{0.05}PO_4$ are absent, while for $LiMn_{0.5}Fe_{0.5}PO_4$ the plateau at 3.5 V is shorter than the one at 4.15 V. Even if there is just one phase, as it is possible to note by XRD, lithium ion insertion into the structure induces either reduction of Fe^{+3} to Fe^{+2} at about 3.5 V or the reduction of Mn^{+3} to Mn^{+2} at about 4.1 V, producing in both cases an equilibrium between the oxidized and reduced form (namely, $MePO_4/LiMePO_4$, where Me is Fe or Mn), and then determining the stress inside the structure. Such stress, produced by hetero-atom reduction, is even enhanced when its amount is very low, since, in this case, around its position dissimilar atoms are present and the redox reaction could be inhibited. However, the presence of low quantity of manganese can help the lithium insertion into the iron-base structure, but not vice versa.

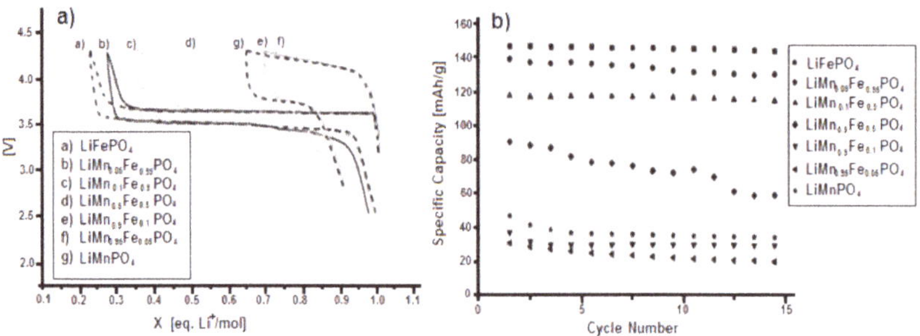

Figure 11. Results of tests at C/5 for LiMn$_x$Fe$_{(1-x)}$PO$_4$ (x = 0.05; 0.1; 0.5; 0.9; 0.95): (**a**) Charge-discharge curves; and (**b**) specific capacity.

Among the different compounds of the series, only those of lithium iron-phosphate with low manganese content have been taken into account for further investigation, since they have the highest capacities. To evaluate the performances of those samples, the reversibility degree of the insertion-deinsertion process, the pseudo-diffusion lithium coefficient and the charge transfer resistance, were calculate by PSV, PITT, and EIS experiments, respectively.

In Figure 12 potential step voltammetry is shown for LiFePO$_4$, LiMn$_{0.05}$Fe$_{0.95}$PO$_4$, and LiMn$_{0.1}$Fe$_{0.9}$PO$_4$. In this picture, the insertion-deinsertion process can be observed in correspondence of two peaks. The upward peaks correspond to the oxidation process at approximately 3.47 V for all samples (odd sweeps), while the downward peaks represent the reduction process at approximately 3.37, 3.40, and 3.42 V (even sweeps) for LiFePO$_4$, LiMn$_{0.05}$Fe$_{0.95}$PO$_4$, and LiMn$_{0.1}$Fe$_{0.9}$PO$_4$, respectively. In general terms, the shorter the distance between oxidation and reduction peaks, the higher the reversibility of the process. In the present case it is possible to recognize that manganese content in LiFePO$_4$ increases the process reversibility. The average value of pseudo-diffusion coefficient has been also evaluated determining the Cottrell region for the potential step voltammetry corresponding to the deinsertion process, by using the PITT technique [29–31], and assuming the average particle radius as the diffusion characteristic length, L, equal to about 0.1 × 10^{-4} cm. Indeed, as said before, the insertion of lithium takes place by means of several reaction fronts, and a pseudo-diffusivity coefficient should be more correctly defined, despite McKinnon and Hearing's assumption [32], who found that it is not possible to distinguish between two different diffusion models based on continuous (solid solution formation) or not continuous (two-phase formation) charging procedures. In Table 5 the pseudo-diffusion coefficient value for both pure LiFePO$_4$ and the materials with low manganese content has been reported. Increasing the manganese content, the value enhances very slightly, so that only few changes can be reached with manganese adding.

Impedance spectroscopy has been also performed on these three materials and in Figure 13 (Nyquist diagram) the real and imaginary parts of impedance have been reported.

Table 5. Pseudo-diffusion coefficients D (in cm^2·s^{-1}).

Material (Symbol)	D (cm^2·s^{-1})
LiFePO$_4$	2.0 × 10^{-14}
LiMn$_{0.05}$Fe$_{0.95}$PO$_4$	5.7 × 10^{-14}
LiMn$_{0.1}$Fe$_{0.9}$PO$_4$	7.7 × 10^{-14}

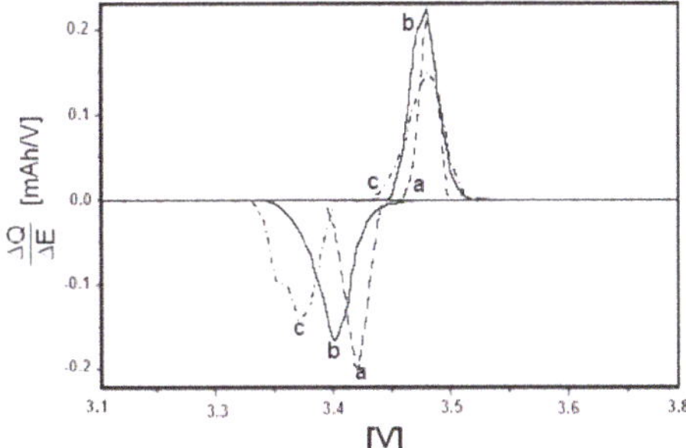

Figure 12. Potential spectroscopy of: (**a**) $LiMn_{0.1}Fe_{0.9}PO_4$; (**b**) $LiMn_{0.05}Fe_{0.95}PO_4$; and (**c**) $LiFePO_4$.

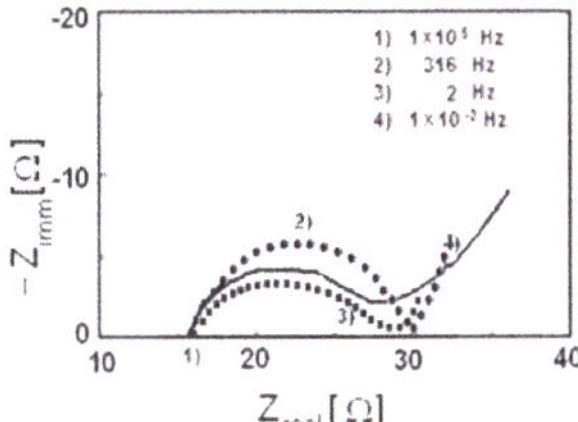

Figure 13. Impedance spectroscopy of: (**a**) • $LiMn_{0.1}Fe_{0.9}PO_4$; (**b**) − $LiMn_{0.05}Fe_{0.95}PO_4$; and (**c**) ■ $LiFePO_4$.

The electrolytic resistance R_{el} is the first intercept of the semicircle with the *x*-axis, while the second intercept minus the electrolytic resistance R_{el} represents the charge transfer resistance R_{ct}. Therefore, it is possible to state that for all cases R_{ct} is about 15 Ω. It is known by literature that the transfer charge resistance for pristine $LiFePO_4$ without in situ carbon formation is higher than 40–50 Ω [33,34], so that by using this synthetic method a decrease of the charge transfer resistance has been, overall, reached.

Finally, we compared the synthesis and the electrochemical performances of our materials with analogous $LiMn_xFe_{(1-x)}PO_4$ already described in the literature and reported in Table 6.

Table 6. Electrochemical performances of analogous $LiMn_xFe_{(1-x)}PO_4$ material described in the literature.

Compound	Method	C-Rate	Capacity (mAh/g)	Reference
$LiMn_xFe_{(1-x)}PO_4$/C (x = 0; 0.5; 1)	Solvothermal process	C/5	150; 65; 50	[35]
$LiMn_xFe_{(1-x)}PO_4$/C (x = 0; 0.5; 1)	freeze-dry process	C/20	140; 120; 95	[36]
$LiMn_xFe_{(1-x)}PO_4$/C (x = 0.7; 0.8; 0.9)	Solid state reaction	C/10	110; 120; 130	[37]
$LiMn_xFe_{(1-x)}PO_4$ (x = 0; 0.05; 0.1; 0.2; 0.4)	Hydrothermal process	C/10	140; 110; 95; 90; 78	[38]
$LiMn_xFe_{(1-x)}PO_4$ (x = 0; 0.1; 0.2; 0.3)	Mechano-activation synthesis	C/10	135; 108; 125; 80	[39]

Saravanan et al. [35] produced $LiMn_xFe_{(1-x)}PO_4$ /C (x = 0; 0.5; 1) with in situ carbon formation, by using the solvothermal method. They obtained a specific capacity equal to 150, 65, and 50 mAh/g for x equal to 0, 0.5 and 1, respectively, after 20 cycles at C/5. The same material obtained by Yoncheva et al. [36] at 500 °C starting from a phosphonate-formate precursor, freeze-drying an aqueous solution containing Li, Fe and Mn phosphate and formate ions, on the other hand, showed a capacity of 140, 120, and 95 mAh/g for x equal to 0, 0.5 and 1, respectively, a C/20. Zhang et al. [37] produced $LiMn_xFe_{(1-x)}PO_4$/C by solid state reaction with x = 0.7, 0.8, and 0.9, obtaining at C/10 a capacity ranging from 110 to 130 mAh/g as x decreases. Xu et al. [38] synthesized carbon free materials through a direct hydrothermal process a 170 °C achieving a capacity of 140, 110, 95, 90, and 78 mAh/g for x equal to 0.1, 0.2, 0.05, 0, and 0.4, respectively, at C/10. The lowest charge transfer resistance has been obtained for x = 0.1 and it is about 200 Ω, while, for x = 0.2 and x = 0, it is 450 and 1400 Ω, respectively.

Finally Wang et al. [39], which attained carbon-free $LiMn_xFe_{(1-x)}PO_4$ by mechano-activation assisted synthesis, reached, the best performance of 125 mAh/g for x equal to 0.2 at C/10.

On the basis of these considerations, our results are consistent and, in some cases, even better than those found in the literature.

4. Experimental

SEM analysis were obtained by the high-resolution microscope FE-SEM Auriga-Zeiss. The apparatus is also equipped with an EDX (Energy Dispersive X-Ray) detector (Bruker, Milan, Italy). Room temperature powder X-ray diffraction was performed by using Cu-Kα radiation λ = 0.15418 nm (Philips PW 1830 generator and Seifert XRD-3000 diffractometers). The data were collected with a step size of 0.02° and at count time of 4 s per step (0.3°·min^{-1}) over the range $15° \leqslant 2\theta \leqslant 80°$. The powder diffraction pattern was indexed by using a Rietveld profile analysis [40].

Thermogravimetric (DSC-TGA) data of the precursor mixtures were obtained in flowing dry nitrogen at a heating rate of 10 °C·min^{-1} on a TA Instruments SDT Q600 thermogravimetric analysis. The FT-IR absorption spectra were recorded on a Shimadzu Prestige 21 FT-IR spectrophotometer using KBr pellets. BET (Fisons Instruments) analyses have been performed at liquid nitrogen temperature and using gaseous N_2 to evaluate the specific surface of powders. Elemental analysis has been performed by the Servizio di Microanalisi del ISM-CNR, Monterotondo, Rome, Italy. Electrochemical characterization of samples was performed in T-shaped battery cells with lithium metal as counter (anode) and reference electrode. The cathode electrode contains about 10 mg of electroactive material with 10 wt % of Carbon Super S and 5 wt % of Teflon. The electrolyte is constituted by a glass wool separator filled with a 1 M solution of LiPF6 in 1/1 ethylene carbonate/diethyl carbonate. Potential step voltammetry (PSV) was carried out in a three-electrode cell configuration by using the following setting values: potential step: 0.02 V, relaxation time: 10 min, step duration: until I > Io/30 or 10 s if Io < 0.01 mA and in the range 3.2–3.7 V versus lithium. The same configuration was used for the potentiostatic–intermittent titration technique (PITT) experiments. The electrochemical impedance spectroscopy (EIS) has been performed in a frequency range from 10^5 to 10^{-2} Hz, with a voltage amplitude of 0.01 V applied on a cell voltage of 3.47 V. A frequency response analyzer (Solartron 1255 HF and Solartron 1286 models by EG and G) and a galvanostat–potentiostat (Mac-Pile II Biologic) were used for these experiments.

5. Conclusions

Hybrid organic-inorganic precursors based on metal(II) phenylphosphonates have been synthesized, characterized and used for the synthesis of different $LiFe_{(1-x)}Mn_xPO_4$ composites. First of all, $LiMn_{0.1}Fe_{0.9}PO_4$ has been prepared following two different synthetic routes, i.e., using as organic precursors either $Fe_{0.9}Mn_{0.1}(C_6H_5PO_3)\cdot H_2O$ (**P1** precursor) or a mixture of $Fe(C_6H_5PO_3)\cdot H_2O$ and $Mn(C_6H_5PO_3)\cdot H_2O$ (**P2** and **P3** precursors). The materials thus obtained show similar behaviour, even if the sample prepared by using a mixture of **P2** and **P3** precursors presents a slightly higher capacity and seems to be more stable upon cycling. Subsequently, a series of $LiMn_{(1-x)}Fe_xPO_4$ (with x = 0.05, 0.1, 0.5, 0.9 and 0.95) has been produced by using only **P2**, **P3**, and Li_2CO_3 precursors mixtures. Structural and morphological characterizations have been carried out analysing the effect of the reciprocal presence of iron and manganese on the electrochemical performances. Enhancing the manganese content, the capacity decreases remarkably and the only advantage is the presence of a second charge-discharge plateau with higher potential value. Moreover, the reversibility degree of the insertion-deinsertion process increases, the pseudo-diffusion lithium coefficient increases only slightly and the charge transfer resistance almost keeps constant, being in every cases quite lower than the corresponding values reported in literature for pristine $LiFePO_4$. This is due to the presence of carbon produced in situ during the synthesis, which seems to be the only component able to increase substantially the electrochemical performances of this cathode material.

Author Contributions: M.P. conceived and designed the experiments; A.D. and E.M.B. performed the experiments; A.D., E.M.B., S.V.C., F.A.S. and C.L. analysed the data; and A.D. and S.V.C. wrote the paper.

Conflicts of Interest: The authors declare no conflict of interest.

References

1. Lu, L.; Han, X.; Li, J.; Hua, J.; Ouyang, M. A review on the key issues for lithium-ion battery management in electric vehicles. *J. Power Sources* **2013**, *226*, 272–288. [CrossRef]
2. Mathew, M.; Kong, Q.H.; McGrory, J.; Fowler, M. Simulation of lithium ion battery replacement in a battery pack for application in electric vehicles. *J. Power Sources* **2017**, *349*, 94–104. [CrossRef]
3. Pelletier, S.; Jabali, O.; Laporte, G.; Veneroni, M. Battery degradation and behaviour for electric vehicles: Review and numerical analyses of several models. *Trans. Res. Part B Methodol.* **2017**, *103*, 158–187. [CrossRef]
4. Olivetti, E.A.; Ceder, G.; Gaustad, G.G.; Fu, X. Lithium-Ion Battery Supply Chain Considerations: Analysis of Potential Bottlenecks in Critical Metals. *Joule* **2017**, *1*, 229–243. [CrossRef]
5. Santiangeli, A.; Fiori, C.; Zuccari, F.; Dell'Era, A.; Orecchini, F.; D'Orazio, A. Experimental analysis of the auxiliaries consumption in the energy balance of a pre-series plug-in hybrid-electric vehicle. *Energy Procedia* **2014**, *45*, 779–78868. [CrossRef]
6. Orecchini, F.; Santiangeli, A.; Dell'Era, A. EVs and HEVs Using Lithium-Ion Batteries. In *Lithium-Ion Batteries: Advances and Applications*; Pistoia, G., Ed.; Elsevier B.V.: Amsterdam, The Netherlands, 2014; pp. 205–248. ISBN 978-044459513-3. [CrossRef]
7. Fabbri, G.; Mascioli, F.M.F.; Pasquali, M.; Mura, F.; Dell'Era, A. Automotive application of lithium-ion batteries: A new generation of electrode materials. In Proceedings of the IEEE 22nd International Symposium on Industrial Electronics (ISIE), Taipei, Taiwan, 28–31 May 2013. [CrossRef]
8. Dell'Era, A.; Pasquali, M.; Fabbri, G.; Pasquali, L.; Tarquini, G.; Santini, E. Automotive application of lithium-ion batteries: Control of commercial batteries in laboratory test. In Proceedings of the IEEE 23rd International Symposium on Industrial Electronics (ISIE), Istanbul, Turkey, 1–4 June 2014; pp. 1616–1621. [CrossRef]
9. Fergus, J. Recent Developments in Cathode Materials for Lithium Ion Batteries. *J. Power Sources* **2010**, *195*, 939–954. [CrossRef]
10. Scrosati, B.; Garche, J. Lithium Batteries: Status, Prospects and Future. *J. Power Sources* **2010**, *195*, 2419–2430. [CrossRef]
11. Li, Z.; Zhang, D.; Yang, F. Developments of Lithium Ion Batteries and Challenges of LiFePO$_4$ as One Promising Cathode Material. *J. Mater. Sci.* **2009**, *44*, 2435–2443. [CrossRef]
12. Padhi, A.K.; Nanjundaswamy, K.S.; Goodenough, J.B. Phospho_Olivines as Positive_Electrode Materials for Rechargeable Lithium Batteries. *J. Electrochem. Soc.* **1997**, *144*, 1188–1194. [CrossRef]
13. Bauer, E.M.; Bellitto, C.; Righini, G.; Pasquali, M.; Dell'Era, A.; Prosini, P.P. A versatile method of preparation of carbon-rich LiFePO$_4$: A promising cathode material for Li-ion batteries. *J. Power Sources* **2005**, *146*, 544–549. [CrossRef]
14. Pasquali, M.; Dell'Era, A.; Prosini, P.P. Fitting of the voltage-Li + insertion curve of LiFePO$_4$. *J. Solid State Electrochem.* **2009**, *13*, 1859–1865. [CrossRef]
15. Benoit, C.; Franger, S. Chemistry and Electrochemistry of Lithium Iron Phosphate. *J. Solid State Electrochem.* **2008**, *12*, 987–993. [CrossRef]
16. Dell'era, A.; Pasquali, M. Comparison between different ways to determine diffusion coefficient and by solving Fick's equation for spherical coordinates. *J. Solid State Electrochem.* **2009**, *13*, 849–859. [CrossRef]
17. Safronov, D.V.; PinusI, Y.; Profatilova, I.A.; Tarnopolskii, V.A.; Skundin, A.M.; Yaroslavtsev, A.B. Kinetics of Lithium Deintercalation from LiFePO$_4$. *Inorg. Mater.* **2011**, *47*, 303–307. [CrossRef]
18. Barker, J.; Saidi, M.Y.; Swoyer, J.L. Lithium Iron(II) Phospho-olivines Prepared by a Novel Carbothermal Reduction Method. *Electrochem. Solid-State Lett.* **2003**, *6*, A53–A55. [CrossRef]
19. Wang, D.; Li, H.; Shi, S.; Huang, X.; Chen, L. Improving the Rate Performance of LiFePO$_4$ by Fe_Site Doping. *Electrochim. Acta* **2005**, *50*, 2955–2958. [CrossRef]

20. Roberts, M.R.; Vitins, G.; Owen, J.R. High Throughput Studies of $Li_{1-x}Mg_{x/2}FePO_4$ and $LiFe_{1-y}Mg_yPO_4$ and the Effect of Carbon Coating. *J. Power Sources* **2008**, *179*, 754–762. [CrossRef]

21. Wang, G.X.; Bewlay, S.L.; Konstantinov, K.; Liu, H.K.; Dou, S.X.; Ahn, J.H. Physical and electrochemical properties of doped lithium iron phosphate electrodes. *Electrochim. Acta* **2004**, *50*, 443–447. [CrossRef]

22. Safronov, D.V.; Novikova, S.A.; Kulova, T.L.; Skundin, A.M.; Yaroslavtsev, A.B. Lithium Diffusion in Materials Based on $LiFePO_4$ Doped with Cobalt and Magnesium. *Inorg. Mater.* **2012**, *48*, 513–519. [CrossRef]

23. Nakamura, T.; Sakumoto, K.; Okamoto, M.; Seki, S.; Kobayashi, Y.; Takeuchi, T.; Tabuchi, M.; Yamada, Y. Electrochemical study on Mn^{2+} substitution in $LiFePO_4$ olivine compound. *J. Power Sources* **2007**, *174*, 435–441. [CrossRef]

24. Nakamura, T.; Miwa, Y.; Tabuchi, M.; Yamada, Y. Structural and Surface Modifications of $LiFePO_4$ Olivine Particles and Their Electrochemical Properties. *J. Electrochem. Soc.* **2006**, *153*, A1108–A1114. [CrossRef]

25. Lee, K.T.; Lee, K.S. Electrochemical properties of $LiFe0.9Mn0.1PO_4/Fe_2P$ cathode material by mechanical alloying. *J. Power Sources* **2009**, *189*, 435–439. [CrossRef]

26. Altomare, A.; Bellitto, C.; Ibrahim, S.A.; Mahmoud, M.R.; Rizzi, R. Iron(ii) phosphonates: A new series of molecule-based weak ferromagnets. *J. Chem. Soc. Dalton Trans.* **2000**, *21*, 3913–3919. [CrossRef]

27. Bellitto, C.; Federici, F.; Colapietro, M.; Portalone, G.; Caschera, D. X-ray single-crystal structure and magnetic properties of $Fe[CH_3PO_3)]\cdot H_2O$: A layered weak ferromagnet. *Inorg. Chem.* **2002**, *41*, 709–714. [CrossRef]

28. Bauer, E.M.; Bellitto, C.; Pasquali, M.; Prosini, P.P.; Righini, G. Versatile synthesis of carbon-Rich $LiFePO_4$ enhancing its Electrochemical Properties. *Electrochem. Solid-State Lett.* **2004**, *7*, A85–A87. [CrossRef]

29. Ho, C.; Raistrick, I.D.; Huggins, R.A. Application of A-C techniques to the study of lithium diffusion in tungsten trioxide thin films. *J. Electrochem. Soc.* **1980**, *127*, 343–350. [CrossRef]

30. Levi, M.D.; Levi, E.A.; Aurbach, D. The mechanism of lithium intercalation in graphite film electrodes in aprotic media Part 2 potentiostatic intermittent titration and in situ XRD studies of the solid-state ionic diffusion. *J. Electroanal. Chem.* **1997**, *421*, 89–97. [CrossRef]

31. Weppner, W.; Huggins, R.A. Determination of the kinetic parameter of mixed-conducting electrodes and application to the system Li3Sb. *J. Electrochem. Soc.* **1977**, 1569–1578. [CrossRef]

32. McKinnon, W.R.; Haering, R.R. *Modern Aspect in Electrochemistry*, 1st ed.; White, R.E., Bockris, J.O.M., Conway, B.E., Eds.; Plenum Press: New York, NY, USA; London, UK, 1983; ISBN 978-1-4615-7463-7. [CrossRef]

33. Wang, C.; Hong, J. Ionic/Electronic Conducting Characteristics of $LiFePO_4$ Cathode Materials. The Determining Factors for High Rate Performance. *Electrochem. Solid-State Lett.* **2007**, *10*, A65–A69. [CrossRef]

34. Park, C.Y.; Park, S.B.; Oh, S.H.; Jang, H.; Cho, W.I.; Ion, L. Diffusivity and Improved Electrochemical Performances of the Carbon Coated $LiFePO_4$. *Bull. Korean Chem. Soc.* **2011**, *32*, 836–840. [CrossRef]

35. Saravanan, K.; Vitta, J.J.; Reddy, M.V.; Chowdari, B.V.R.; Balaya, P. Storage performance of $LiFe_{1-x}Mn_xPO_4$ nanoplates (x = 0, 0.5, and 1). *J. Solid State Electrochem.* **2010**, *14*, 1755–1760. [CrossRef]

36. Yoncheva, M.; Koleva, V.; Mladenov, M.; Sendova-Vassileva, M.; Nikolaeva-Dimitrova, M.; Stoyanova, R.; Zhecheva, E. Carbon-coated nano-sized $LiFe_{1-x}Mn_xPO_4$ solid solutions ($0 \leqslant x \leqslant 1$) obtained from phosphate–formate precursors. *J. Mater. Sci.* **2011**, *46*, 7082–7089. [CrossRef]

37. Xu, J.; Chen, G.; Li, H.J.; Lv, Z.S. Direct-hydrothermal synthesis of $LiFe_{1-x}Mn_xPO_4$ cathode materials. *J. Appl. Electrochem.* **2010**, *40*, 575–580. [CrossRef]

38. Zhang, B.; Wang, X.; Li, H.; Huang, X. Electrochemical performances of $LiFe_{1-x}Mn_xPO_4$ with high Mn content. *J. Power Sources* **2011**, *196*, 6992–6996. [CrossRef]

39. Wang, Y.; Zhang, D.; Yu, X.; Cai, R.; Shao, Z.; Liao, X.Z.; Ma, Z.F. Mechanoactivation-assisted synthesis and electrochemical characterization of manganese lightly doped $LiFePO_4$. *J. Alloy. Compd.* **2010**, *492*, 675–680. [CrossRef]

40. Rietveld, H.M. A profile refinement method for nuclear and magnetic structures. *J. Appl. Crystallogr.* **1969**, *2*, 65–71. [CrossRef]

Article

Thermal Behavior and Structural Study of SiO$_2$/Poly(ε-caprolactone) Hybrids Synthesized via Sol-Gel Method

Stefano Vecchio Ciprioti [1,*], **Riccardo Tuffi** [2], **Alessandro Dell'Era** [1], **Francesco Dal Poggetto** [3] and **Flavia Bollino** [4,*]

[1] Department of Basic and Applied Science for Engineering (S.B.A.I.), Sapienza University of Rome, via del Castro Laurenziano 7, Roma, I-00161, Italy; alessandro.dellera@uniroma1.it

[2] Department of Sustainability, ENEA-Casaccia Research Center, Via Anguillarese 301, Rome, 00123, Italy; riccardo.tuffi@enea.it

[3] Ecoricerche S.r.l., Via Principi Normanni, Capua 81043, Italy; amm.ecoricerche@virgilio.it

[4] Department of Industrial and Information Engineering, University of Campania Luigi Vanvitelli, via Roma 29, Aversa, 81031, Italy

* Correspondence: stefano.vecchio@uniroma1.it (S.V.C.); flavia.bollino@unicampania.it (F.B.); Tel.: +39-064-976-6906 (S.V.C.); +39-081-501-0483 (F.B.)

Received: 15 January 2018; Accepted: 6 February 2018; Published: 10 February 2018

Abstract: SiO$_2$-based organic-inorganic hybrids (OIHs) are versatile materials whose properties may change significantly because of their thermal treatment. In fact, after their preparation at low temperature by the sol-gel method, they still have reactive silanol groups due to incomplete condensation reactions that can be removed by accelerating these processes upon heating them in controlled experimental conditions. In this study, the thermal behavior of pure SiO$_2$ and four SiO$_2$-based OIHs containing increasing amount (6, 12, 24 and 50 wt %) of poly(ε-caprolactone) (PCL) has been studied by simultaneous thermogravimetry (TG) and differential scanning calorimetry (DSC). The FTIR analysis of the gas mixture evolved at defined temperatures from the samples submitted to the TG experiments identified the mechanisms of thermally activated processes occurring upon heating. In particular, all samples already release ethanol at low temperature. Moreover, thermal degradation of PCL takes place in the richest-PCL sample, leading to 5-hexenoic acid, H$_2$O, CO$_2$, CO and ε-caprolactone. After the samples' treatment at 450, 600 and 1000 °C, the X-ray diffraction (XRD) spectra revealed that they were still amorphous, while the presence of cristobalite is found in the richest-PCL material.

Keywords: sol-gel method; SiO$_2$–based hybrids; poly(ε-caprolactone); TG-DSC; TG-FTIR; X-ray diffraction analysis

---

1. Introduction

In recent years, organic–inorganic hybrids (OIHs) have played a crucial role in the development of multifunctional nanostructured materials [1–7]. OIHs are not a simple physical mixture of organic and inorganic phases possessing properties that are the sum of those of both components; rather, they are intimately mixed, with average dimensions ranging from a few Å to several nanometers [8,9] These materials have been divided into two classes according to the nature of the bonds between them [10]. Class I consists of those forming weak hydrogen bonds or van der Waals forces, and Class II contains materials obtained by strong chemical bonds (covalent or ionic covalent bonds) between the components. As a matter of fact, research in this area is supported by the growing interest of all materials scientists who are looking to fully exploit this opportunity for creating smart materials that benefit from synergetic or complementary effects exerted by the two phases embedded in one [8,10].

A wide versatility in the design of OIHs may be achieved if they are synthesized at low temperature. In this regard, the sol-gel technique has several advantages over other synthesis processes: it is versatile, since glasses and ceramics may be produced at low temperatures. The transition of the system from a colloidal liquid ('sol') into a solid 'gel' occurs via hydrolysis of a metal alkoxide precursor and polycondensation reactions occurring in a water-alcohol solution [11,12]. After drying of the obtained 'wet gel', and depending on the heat treatment carried out on the 'dry gel' it is possible to obtain several products, such as xerogel, areogel or dense ceramics (by a sintering process of the xerogel or the aerogel). The starting low temperature condition allows the chemical homogeneity of the various elements to be controlled down to the atomic level, and thermolabile molecules (e.g., polymers and drugs) to be entrapped in the inorganic matrix, thus producing OIHs [8,13].

Recently, silica-based organic-inorganic hybrids have been attracting the growing interest of several research groups, leading to the development of functional materials for many application areas [14–18]. In the last two–three years, our group has also been involved in preparing (via the sol-gel method) and characterizing SiO_2-based glasses, ceramics [19–23], as well as SiO_2-based OIHs, with particular reference to SiO_2/polyethylene glycol (SiO_2/PEG) hybrids containing increasing percentages of PEG (from 6 to 70 wt %) [24,25]. More recently, SiO_2/PCL hybrids containing variable percentages of PCL (6, 12, 24 and 50 wt %) were synthesized via sol-gel and characterized by means of several instrumental techniques [26]. FT-IR, NMR, XRD and SEM analyses showed that the SiO_2/PCL materials were amorphous and homogeneous organic-inorganic hybrid materials in which the C=O groups in the PCL chains form H-bonds with the –OH groups of the silica matrix. Some studies reported in the literature proved that those hybrids were bioactive and biocompatible [27–29]. For this reason, their use was proposed in the biomedical field to prepare coatings able to enhance biological performances of metallic implants [30] or as drug delivery matrices [2].

Some studies reported in the literature showed that the biological properties of the sol-gel materials were affected by the heat treatment carried out after the gel formation [31–35]. When the transition from sol to gel occurs, indeed, both hydrolysis and condensation reactions are incomplete and, thus, reactive silanols are still present in the system. A heating treatment at relatively high temperatures (100–600 °C) is necessary to accelerate this phase, thus removing the organic species and leading to formation of covalent Si-O-Si bonds [11]. Therefore, heating temperature and rate can affect microstructure and crystallization degree of the final material, influencing, in turn, the ion release from the materials [36]. This property is a key factor in determining material biological characteristics, because it can cause modification in the material surface charge and, thus, in protein adsorption [37] and hydroxyapatite nucleation [38].

The aim of the present investigation has been, therefore, to examine and closely compare the thermal behavior of pure SiO_2 (denoted with S) with that of four SiO_2/PCL hybrids containing 6, 12, 24 and 50 wt % of PCL (with the symbols SP6, SP12, SP24 and SP50, respectively). The focus of the present study has been to determine the mechanisms of reactions occurring in these materials upon heating them under inert atmosphere by coupling TG and FTIR devices, similar to what has previously been done using TG and mass spectrometry [39]. Such information associated to specific biological studies could allow the rational fine tuning of biomaterials with properties (e.g., crystallization degree, ion release ability, protein adsorption ability, osseointegration ability, etc. [31,32,36–38]) adequate to specific applications. To this end, it is also useful to detect by XRD the modification of the solid phases induced upon heating and stables at these temperatures.

2. Results and Discussion

2.1. Thermal Behavior Study

The TG/DSC curves of pure SiO_2 (S) and of the SiO_2/PCL hybrids (SP6, SP12, SP24 and SP50) have been reported in Figure 1.

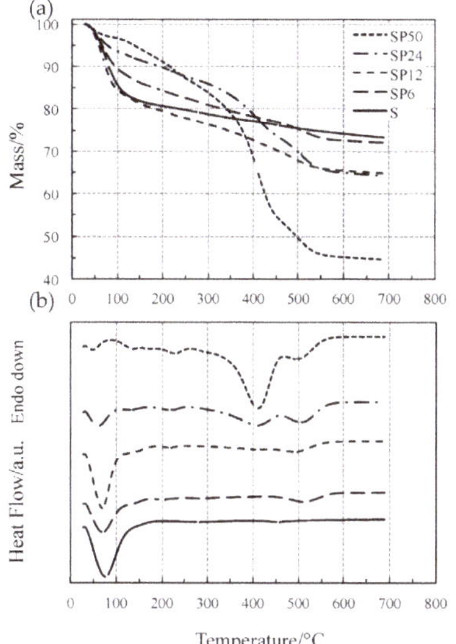

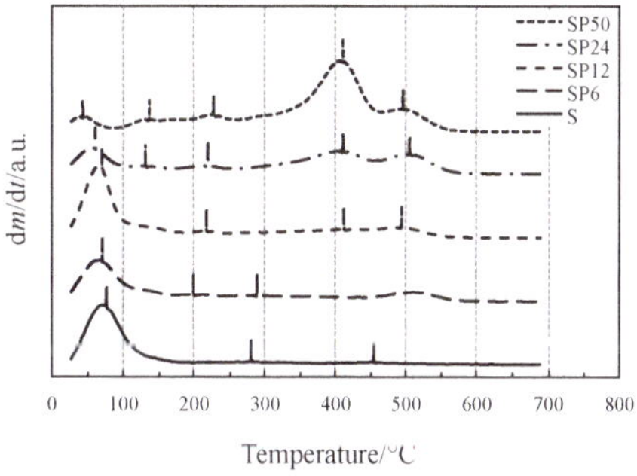

Figure 1. Simultaneous TG (**a**) and DSC (**b**) curves of all the materials tested at 10 °C·min^{-1} in flowing Ar atmosphere.

Initial and final temperatures of each process accompanied by a mass loss have been more clearly identified by the first-order derivative curves of TG (DTG) curves, displayed in Figure 2.

Figure 2. DTG curves of all the materials tested.

The TG/DSC curves of all the materials in Figure 1 showed an initial mass loss (corresponding to the first DTG peak) accompanied by an endothermic DSC peak, ascribed to the simultaneous loss of water and alcohol up to 140 °C, except for SP50 (short dotted lines) for which the process ends at around 85 °C. It is clearly evident from Figure 2 (low-temperature region) that the SP hybrid materials, except for SP24 and SP50, show the same thermal behavior as pure S. At temperature higher than 180 °C, dehydration is completed and S undergoes dehydroxylation, elimination of water due to condensation of the hydroxyl surface groups, with a slow and quite constant mass loss rate (linear portion of the TG curve up to 600 °C not detectable by the DTG curve), as found in previous studies [19–22,24,25]. SP materials (except for SP24 and SP50) show the same thermal behavior up to 300–400 °C, while at higher temperatures, a one- or two-step process took place up to 580–600 °C. This process is accompanied by an endothermic effect, and the intensity of the corresponding DSC peak was found to increase with the amount of PCL in the material, while the degradation temperature shifts towards lower values with an increase in the PCL content.

Similar to what has been observed in a previous study [40], this process is attributable to the thermal degradation of PLC, which usually takes place in two steps of mass loss. On the other hand, the thermal behavior of the PCL-richer materials (SP24 and SP50) is remarkably different from those of the other SP materials. When dehydration is completed at about 85 °C, SP50 undergoes a two-step process up to 260 °C, which can probably be ascribed to dehydroxylation, followed by the two-step thermal degradation of PLC between 300 and 600 °C. The DSC curve recorded two endothermic effects, expressed by two partially convoluted broad peaks: the first one intense up to 455 °C, followed by a second that is a shoulder.

2.2. FTIR Evolution Gas Analysis to Provide a Mechanistic Interpretation of the Thermally Stimulated Processes

Vertical bars displayed in Figure 2, close to the DTG peak temperatures where the reaction reaches the maximum rates, represent the temperatures at which the gas or gaseous mixture evolved from TG experiments was collected and sent to the FTIR device. The FTIR spectra of the mixtures collected from the TG/DSC experiments of all materials tested are shown in Figure 3.

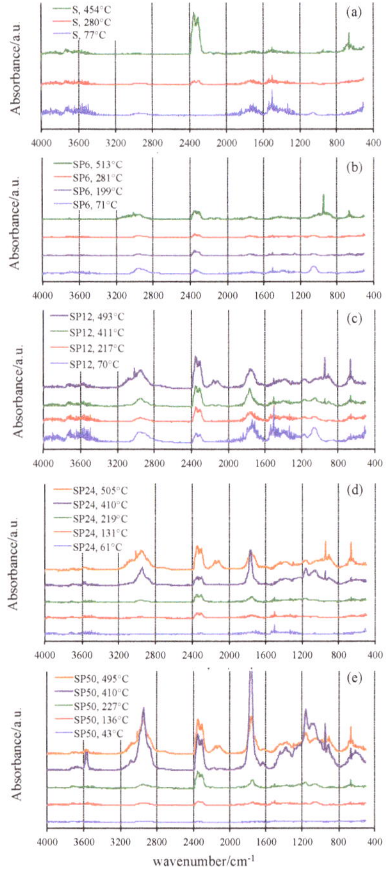

Figure 3. FTIR spectra of the gaseous mixture evolved at selected temperatures from TG experiments.

A confirmation of the mechanisms hypothesized was found by analyzing the FTIR spectrum of the gases evolved from the samples S during the TG experiment at low temperature (77 °C), showing the typical signals of water. Sharp peaks in the wavenumber regions 4000–3400 cm^{-1} and 2000–1200 cm^{-1} are visible due to the H–O–H stretching and bending vibrations. Moreover, the weak peak at about 1040 cm^{-1} suggests that ethanol [41], used as solvent in the synthesis process and also formed by the hydrolysis reaction that involves the alkoxide precursor tetraethyl orthosilicate (TEOS), is also released in this temperature range from the material in which it was previously embedded in the gel form. A higher release of ethanol was detected in SP6, revealed by the presence of the C–H stretching at 2955 cm^{-1}, as well as by the peaks related to the C–C and C–O bonds at 1373, 1249, 1040 and 875 cm^{-1}. This can be explained by a decrease of hydrolysis degree and condensation rate caused by the interaction of the –OH groups of the forming inorganic network with the polymer chains in the sol. Therefore, the –OH groups involved in the H–bonds with the C=O of the PCL [26] cannot react with other alkoxide precursors or other oligomers. As a consequence, a higher content of residual ethoxy group is retained in the gel. Moreover, the presence of water and CO_2 (duplet at 2345–2300 cm^{-1} [41]) is also observed.

Similarly, the amount of ethanol and water released even at low temperature in PCL-rich OIHs (SP12, SP24 and SP50) is higher, due to the higher amount of PCL and, thus, to the higher amount of –OH bonded with it. Moreover, the higher amount of ethanol leads to the formation of a higher amount of CO_2. The FTIR spectra of pure S and SP6 at 280 and 281 °C, respectively, show that a decrease of the bands attributed to water and ethanol, as well as the development of CO_2, were observed. SP12 revealed a similar thermal behavior (with respect to those of S and SP6) at low temperatures (69.5 and 217 °C).

At higher temperatures (513 °C for SP6), ethanol is completely degraded, thus leading to the formation of ethylene (as proved by the bands in the following regions: 3300–2900 and 1430 cm^{-1}, as well as the sharp band at 950 cm^{-1}), CO_2 and a low amount of CO. The higher amount of ethylene produced from the SP6 sample compared to that of S is due to the higher initial amount of ethanol developed from sample SP6.

FTIR spectra of SP12 at 411 °C showed new bands at 2940, 1770, 1150 and 1050 cm^{-1}, which can be ascribable to the formation of caproic acid and ε-caprolactone, both of which are produced from the thermal degradation of PCL, as affirmed by Persenaire and co-workers [40]. Moreover, the bands of CO_2, CO and the sharp one of ethylene are also visible, even if with low intensity.

SP24 and SP50 showed the same thermal behavior as SP12, but the presence of 5-hexenoic acid in the FTIR spectrum at 410 °C is more evident in the former, while that at 495 °C showed the least decrease; and in the gas phase, ε-caprolactone is mainly present. By increasing the temperature, the band at 3570 cm^{-1}, present only in the spectrum of 5-hexenoic acid, decreases. This finding is in agreement with the mechanism of degradation of PCL reported in the literature [40], which is reported to occur in two steps: in the first, the rupture of polyester chains via ester pyrolysis reactions is involved, leading to the formation of 5-hexenoic acid, H_2O, CO_2 and a low amount of CO. The second step is attributed to the formation of ε-caprolactone by an unzipping depolymerisation process. Therefore, the intensity of CO_2 and CO signals is higher in the spectra of those samples compared to those of S and SP6, because when the PCL degrades, CO_2 and CO also are produced [40,42].

Therefore, the obtained results suggest that in order to obtain OIHs free of internal toxic residual solvents, the materials should be heated at 400 °C.

2.3. XRD Analysis to Provide a Mechanistic Interpretation of the Thermally Stimulated Processes

Figure 4 shows the XRD spectra of both S and SP50 after their thermal treatment at 450 and 600 °C (plots (**a**) and (**b**), respectively). They are all practically amorphous, and only the broad characteristic peak of silica between 15 and 35° is observed [43].

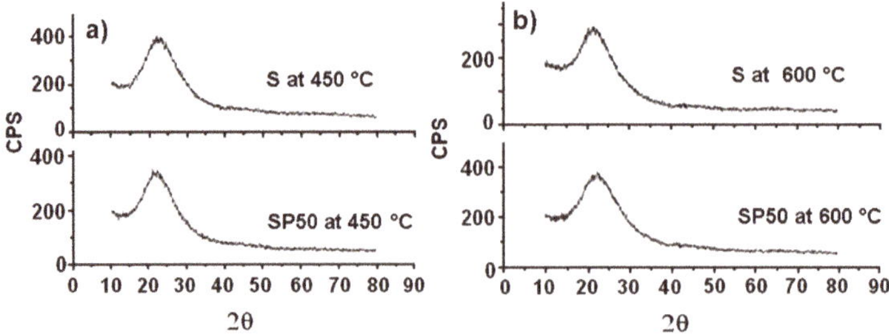

Figure 4. XRD spectra of S and SP50 materials after their treatment at 450 °C (**a**) and 600 °C (**b**).

Furthermore, the S and SP materials are revealed to be amorphous, even after their treatment at 1000 °C, as is clearly evident from the XRD spectra in Figure 5a.

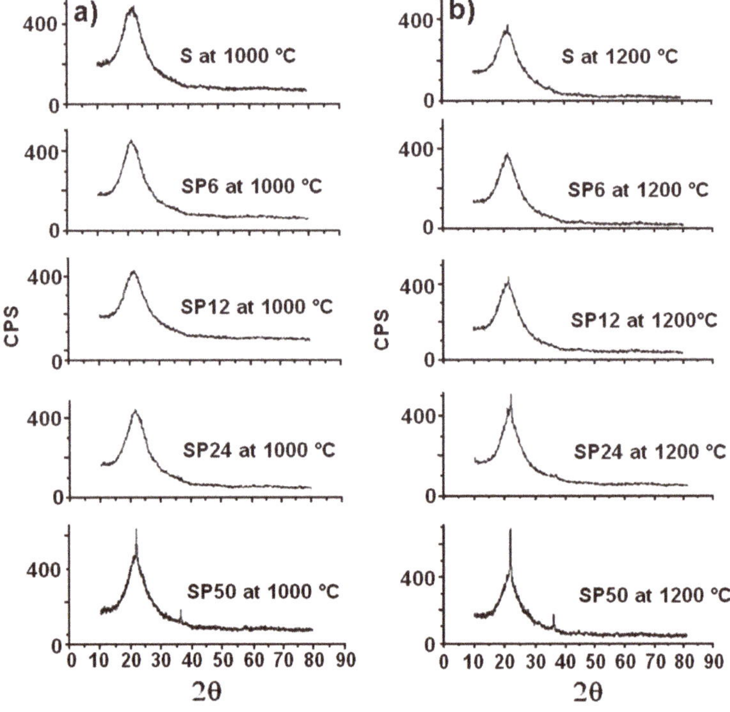

Figure 5. XRD spectra of all the materials tested after their treatment at 1000 °C (**a**) and 1200 °C (**b**).

SP50 shows an initial crystalline structure (that of β-cristobalite, a high-temperature stable polymorph of silica). Crystallization of β-cristobalite seems to occur more evidently (especially in the case of SP50) in all the materials (even in pure S) after their treatment at 1200 °C, as the XRD spectra in Figure 5b show clearly. This result partly confirmed what was obtained in a previous study [44], where amorphous silica crystallized in cristobalite at 1000 °C due to a local rearrangement of the amorphous material (similar to β-cristobalite). The explanation for this is that the instantaneous local atomic arrangement of amorphous SiO_2 is similar to that of β-cristobalite [45]. Usually, a phase change transformation from quartz to β-cristobalite only takes place when the temperature is about

1470 °C [38]. Then, at high temperature, it is easier to observe the crystallization of amorphous SiO_2 into β-cristobalite than the phase change from quartz.

3. Materials and Methods

3.1. Synthesis of the Hybrid Materials

The SiO_2/PCL hybrid materials were synthesized by means of the sol-gel method, according to a procedure reported in detail in Catauro et al. [26]. The PCL-free SiO_2 was obtained from a Silica sol prepared by adding drop by drop water to a solution of the alkoxide precursor tetraethyl orthosilicate (TEOS, sigma Aldrich, Milan, Italy) in pure ethanol (EtOH, 99.8%, Sigma Aldrich, Milan, Italy) and nitric acid (65%, Sigma Aldrich, Milan, Italy). The last was used as catalyst of the hydrolysis reaction, which involves the alkoxide precursor. In the sol, the following molar ratios between the reagents are achieved: H_2O/TEOS = 2; EtOH/TEOS = 6; TEOS/HNO_3 = 1.7.

To synthesize the hybrid materials, different amounts of PCL (Sigma Aldrich, Milan, Italy), with an average molar mass of 10.0 kg·mol^{-1}, were dissolved in chloroform (Sigma Aldrich, Milan, Italy) and then added to the silica sol. All reagents were used as received, without further purification. Finally, five materials were synthesized: pure SiO_2 and four SiO_2/PCL hybrids containing 6, 12, 24 and 50 wt % of PCL.

3.2. Instrumental Details to Study the Thermal Behavior of the Hybrid Materials

After their preparation, all the materials investigated were gently ground in an agate mortar for some minutes to reduce them into fine powders. Then, samples of the obtained powders were further characterized by coupled TG/Differential scanning calorimetry (TG/DSC), coupled thermogravimetry/Fourier transform infrared spectroscopy (TG/FTIR) and X-ray diffraction (XRD) analyses. The thermal behavior of the OIHs was studied under an inert nitrogen flowing atmosphere (60 mL·min^{-1}) up to 700 °C at a heating rate of 10 °C·min^{-1} using a simultaneous Mettler Toledo TG/DSC 2950 instrument (Mettler Toledo, Columbus, OH, USA), equipped with a STARe software (version 12.00, Mettler Toledo, Columbus, OH, USA). The instrument was equipped with two identical cylindrical crucibles, one for the reference filled with alumina in powder form and one for the sample, uniformly covered with about 20–25 mg of solid to uniformly cover the bottom surface area of the crucible. Calibration of the sample temperature was performed using very pure indium and zinc reference materials (purity higher than 99.998%), thus assuming a final average uncertainty $u(T) = \pm 1$ K was estimated over the whole temperature range.

In order to collect more information to support a reasonable mechanism associated to the thermally stimulated processes that take place in the OIHs submitted to the TG/DSC experiments, the TG/FTIR experiments were performed using a SETARAM 92 16.18 TG apparatus (SETARAM, Caluire, France) under a stream of argon of 40 mL min^{-1} in the temperature range between 25 and 700 °C at 10 °C min^{-1}. This instrument has been equipped with 250 μL alumina crucibles filled with about 100–150 mg of sample to obtain the minimum amount of gaseous species required for FTIR measurements. A preliminary blank experiment was performed using empty crucibles under identical experimental conditions of the samples tested. All the experimental data were collected and analyzed using the Calisto software (version 1.38, SETARAM, Caluire-et-Cuire, France). The vapors evolved during the TG experiments were conveyed to a Thermofisher Scientific Nicolet iS10 Spectrophotometer (Thermofisher, Waltham, MA, USA) linked through a heated transfer line kept at 200 °C. The instrument allows monitoring the actual reaction trend by collecting a spectrum each 11 s, being eight scans performed at 0.5 cm^{-1} intervals with a resolution of 4 cm^{-1}.

The thermal treatment of the sample powders was carried out in a muffle furnace for 2 h under argon purge gas atmosphere at 450, 600, and 1000 °C (temperatures selected on the basis of a careful examination of their thermal behavior shown by the TG experiments). In order to verify the occurrence of crystallization processes at higher temperatures a supplementary thermal treatment at 1200 °C was also performed in a tubular oven under argon flowing atmosphere for 90 min.

All the crystalline phases were identified by XRD analysis using a Philips diffractometer equipped with a PW 1830 generator (Philips, Amsterdam, The Netherlands), where the source of X-ray is given by a Cu-Kα radiation (λ = 0.15418 nm).

4. Conclusions

The thermal behavior of four SiO_2-PCL hybrid materials containing increasing amount of poly(ε-caprolactone) (PCL) has been studied by simultaneous TG/DSC and compared with that of pure SiO_2. The FTIR analysis of the gases evolved at defined temperatures from the samples submitted to TG experiments enabled identification of the mechanisms of dehydration, ethanol release (at low temperature) and PCL thermal degradation, occurring upon heating. In particular, the thermal degradation of PCL in SP50 leads to 5-hexenoic acid, H_2O, CO_2, CO and ε-caprolactone. After the samples' treatment at 450, 600 and 1000 °C, the X-ray diffraction (XRD) spectra revealed that they were still amorphous, while the presence of cristobalite was found in the richest-PCL material.

The knowledge of the thermal behavior of those bioactive and biocompatible sol-gel hybrid materials, as well as of the degradation processes that take place within them upon heating, has a key role in the development of new performing biomaterials. In fact, this information allows modulation of the microstructure of the obtained biomaterials and, thus, their biological properties. Therefore, the correlation between the obtained results and further information about both structural modifications induced by heating and biological performance as a function of the heat treatment could be useful in the near future for the design of biomaterials suitable for specific clinical applications.

Author Contributions: F.B. and F.D.P. conceived, designed and performed the sol-gel synthesis of all the materials and helped in the interpretation of FTIR spectra for the EGA analysis; R.T. performed the TG/DSC and TG/FTIR experiments; A.D. carried out the XRD experiments and interpreted the results obtained; S.V.C. gave his contribution in the interpretation of thermal analysis results and wrote the paper, although all the authors provided their own contributions as far as their expertise was concerned.

Conflicts of Interest: The authors declare no conflict of interest.

References

1. Joshua, D.Y.; Damron, M.; Tang, G.; Zheng, H.; Chu, C.-J.; Osborne, J.H. Inorganic/organic hybrid coatings for aircraft aluminum alloy substrates. *Prog. Org. Coat.* **2001**, *41*, 226–232. [CrossRef]
2. Catauro, M.; Bollino, F. Anti-inflammatory entrapment in polycaprolactone/silica hybrid material prepared by Sol-Gel route, characterization, bioactivity and in vitro release behavior. *J. Appl. Biomater. Funct. Mater.* **2013**, *11*, 172–179. [CrossRef] [PubMed]
3. Klukowska, A.; Posset, U.; Schottner, G.; Wis, M.L.; Salemi-Delvaux, C.; Malatesta, V. Photochromic hybrid Sol-Gel coatings: Preparation, properties, and applications. *Mater. Sci.* **2002**, *20*, 95–104.
4. Samuneva, B.; Djambaski, P.; Kashchieva, E.; Chernev, G.; Kabaivanova, L.; Emanuilova, E.; Salvado, I.M.M.; Fernandes, M.H.V.; Wu, A. Sol-Gel synthesis and structure of silica hybrid biomaterials. *J. Non-Cryst. Solids* **2008**, *354*, 733–740. [CrossRef]
5. Novak, B.M. Hybrid nanocomposite materials—Between inorganic glasses and organic polymers. *Adv. Mater.* **1993**, *5*, 422–433. [CrossRef]
6. Diaz, U.; Corma, A. Organic-inorganic hybrid materials: Multi-functional solids for multi-step reaction processes. *Chem. Eur. J.* **2018**, *2*, 1–16. [CrossRef] [PubMed]
7. You, N.; Liu, T.-H.; Fan, H.-T.; Shen, H. An efficient mercapto-functionalized organic-inorganic hybrid sorbent for selective separation and preconcentration of antimony(iii) in water samples. *RSC Adv.* **2018**, in press. [CrossRef]
8. Sanchez, C.; Ribot, F. Design of hybrid organic-inorganic materials synthesized via Sol-Gel chemistry. *New J. Chem.* **1994**, *18*, 1007–1047.
9. Wei, Y.; Jin, D.; Brennan, D.J.; Rivera, D.N.; Zhuang, Q.; DiNardo, N.J.; Qiu, K. Atomic force microscopy study of organic-inorganic hybrid materials. *Chem. Mater.* **1998**, *10*, 769–772. [CrossRef]

10. Judeinstein, P.; Sanchez, C. Hybrid organic-inorganic materials: A land of multidisciplinarity. *J. Mater. Chem.* **1996**, *6*, 511–525. [CrossRef]

11. Brinker, C.; Scherer, G. *Sol-Gel Science: The Physics and Chemistry of Sol-Gel Processing*; Academic Press: San Diego, CA, USA, 1989.

12. Catauro, M.; Bollino, F.; Papale, F. Synthesis of SiO_2 system via Sol-Gel process: Biocompatibility tests with a fibroblast strain and release kinetics. *J. Biomed. Mater. Res. Part A* **2014**, *102*, 1677–1680. [CrossRef] [PubMed]

13. Sanchez, C.; Julian, B.; Belleville, P.; Popall, M. Applications of hybrid organic-inorganic nanocomposites. *J. Mater. Chem.* **2005**, *15*, 3559–3592. [CrossRef]

14. Casarin, J.; Goncalves, A.C., Jr.; Segatelli, M.G.; Tarley, C.R.T. Poly(methacrylic acid)/SiO_2/AL_2O_3 based organic-inorganic hybrid adsorbent for adsorption of imazethapyr herbicide from aqueous medium. *React. Funct. Polym.* **2017**, *121*, 101–109. [CrossRef]

15. Niknahad, M.; Mannari, V. Corrosion protection of aluminum alloy substrate with nano-silica reinforced organic-inorganic hybrid coatings. *J. Coat. Technol. Res.* **2016**, *13*, 1035–1046. [CrossRef]

16. Qu, K.; Pu, Q.; Shan, G. Preparation and characterization of organic-inorganic hybrid flexible silica aerogel. *Huagong Xuebao Chin. Ed.* **2014**, *65*, 346–351.

17. Ren, Y.; Zhang, Y.; Gu, Y.; Zeng, Q. Flame retardant polyacrylonitrile fabrics prepared by organic-inorganic hybrid silica coating via Sol-Gel technique. *Prog. Org. Coat.* **2017**, *112*, 225–233. [CrossRef]

18. Zhang, X.; Lin, W.; Zheng, J.; Sun, Y.; Xia, B.; Yan, L.; Jiang, B. Insight into the organic-inorganic hybrid and microstructure tailor mechanism of SoL-GeL ORMOSIL antireflective coatings. *J. Phys. Chem. C* **2018**, *122*, 596–603. [CrossRef]

19. Catauro, M.; Bollino, F.; Dell'Era, A.; Ciprioti, S.V. Pure $AL_2O_3 \cdot 2SiO_2$ synthesized via a Sol-Gel technique as a raw material to replace metakaolin: Chemical and structural characterization and thermal behavior. *Ceram. Int.* **2016**, *42*, 16303–16309. [CrossRef]

20. Catauro, M.; Bollino, F.; Papale, F.; Vecchio Ciprioti, S. Investigation on bioactivity, biocompatibility, thermal behavior and antibacterial properties of calcium silicate glass coatings containing ag. *J. Non-Cryst. Solids* **2015**, *422*, 16–22. [CrossRef]

21. Catauro, M.; Dell'Era, A.; Vecchio Ciprioti, S. Synthesis, structural, spectroscopic and thermoanalytical study of Sol-Gel derived SiO_2-CaO-P_2O_5 gel and ceramic materials. *Thermochim. Acta* **2016**, *625*, 20–27. [CrossRef]

22. Vecchio Ciprioti, S.; Catauro, M. Synthesis, structural and thermal behavior study of four Ca-containing silicate gel-glasses: Activation energies of their dehydration and dehydroxylation processes. *J. Therm. Anal. Calorim.* **2016**, *123*, 2091–2101. [CrossRef]

23. Russo, V.; Masiello, D.; Trifuoggi, M.; Di Serio, M.; Tesser, R. Design of an adsorption column for methylene blue abatement over silica: From batch to continuous modeling. *Chem. Eng. J.* **2016**, *302*, 287–295. [CrossRef]

24. Catauro, M.; Renella, R.A.; Papale, F.; Vecchio Ciprioti, S. Investigation of bioactivity, biocompatibility and thermal behavior of Sol-Gel silica glass containing a high peg percentage. *Mater. Sci. Eng. C* **2016**, *61*, 51–55. [CrossRef] [PubMed]

25. Vecchio Ciprioti, S.; Catauro, M.; Bollino, F.; Tuffi, R. Thermal behavior and dehydration kinetic study of SiO_2/PEG hybrid gel glasses. *Polym. Eng. Sci.* **2017**, *57*, 606–612. [CrossRef]

26. Catauro, M.; Bollino, F.; Mozzati, M.C.; Ferrara, C.; Mustarelli, P. Structure and magnetic properties of SiO_2/PCL novel Sol-Gel organic-inorganic hybrid materials. *J. Solid State Chem.* **2013**, *203*, 92–99. [CrossRef]

27. Lee, E.-J.; Teng, S.-H.; Jang, T.-S.; Wang, P.; Yook, S.-W.; Kim, H.-E.; Koh, Y.-H. Nanostructured poly(ε-caprolactone)–silica xerogel fibrous membrane for guided bone regeneration. *Acta Biomater.* **2010**, *6*, 3557–3565. [CrossRef] [PubMed]

28. Rhee, S.-H.; Choi, J.-Y.; Kim, H.-M. Preparation of a bioactive and degradable poly(ε-caprolactone)/silica hybrid through a Sol−Gel method. *Biomaterials* **2002**, *23*, 4915–4921. [CrossRef]

29. Tian, D.; Dubois, P.; Grandfils, C.; Jérôme, R.; Viville, P.; Lazzaroni, R.; Brédas, J.-L.; Leprince, P. A novel biodegradable and biocompatible ceramer prepared by the Sol−Gel process. *Chem. Mater.* **1997**, *9*, 871–874. [CrossRef]

30. Catauro, M.; Bollino, F.; Papale, F. Surface modifications of titanium implants by coating with bioactive and biocompatible poly (ε-caprolactone)/SiO_2 hybrids synthesized via Sol-Gel. *Arab. J. Chem.* **2014**. [CrossRef]

31. Bollino, F.; Armenia, E.; Tranquillo, E. Zirconia/hydroxyapatite composites synthesized via Sol-Gel: Influence of hydroxyapatite content and heating on their biological properties. *Materials* **2017**, *10*, 757. [CrossRef] [PubMed]

32. Catauro, M.; Bollino, F.; Renella, R.A.; Papale, F. Sol-Gel synthesis of SiO_2-CaO-P_2O_5 glasses: Influence of the heat treatment on their bioactivity and biocompatibility. *Ceram. Int.* **2015**, *41*, 12578–12588. [CrossRef]

33. Chen, L.; Xu, L.-L. Research and preparing of SiO_2-CaO-P_2O_5 bioactive material. *Guisuanyan Tongbao* **2009**, *28*, 686–691.

34. Ma, J.; Chen, C.Z.; Wang, D.G.; Meng, X.G.; Shi, J.Z. Influence of the sintering temperature on the structural feature and bioactivity of Sol-Gel derived SiO_2-CaO-P_2O_5 bioglass. *Ceram. Int.* **2010**, *36*, 1911–1916. [CrossRef]

35. Voicu, G.; Popa, A.M.; Badanoiu, A.I.; Iordache, F. Influence of thermal treatment conditions on the properties of dental silicate cements. *Molecules* **2016**, *21*, 233. [CrossRef] [PubMed]

36. Kumar, R.; Münstedt, H. Polyamide/silver antimicrobials: Effect of crystallinity on the silver ion release. *Polym. Int.* **2005**, *54*, 1180–1186. [CrossRef]

37. Mavropoulos, E.; Costa, A.M.; Costa, L.T.; Achete, C.A.; Mello, A.; Granjeiro, J.M.; Rossi, A.M. Adsorption and bioactivity studies of albumin onto hydroxyapatite surface. *Colloids Surf. B Biointerfaces* **2011**, *83*, 1–9. [CrossRef] [PubMed]

38. Radev, L. Influence of thermal treatment on the structure and in vitro bioactivity of Sol-Gel prepared CaO-SiO_2-P_2O_5 glass-ceramics. *Process. Appl. Ceram.* **2014**, *8*, 155–166. [CrossRef]

39. Papadopoulos, C.; Kantiranis, N.; Vecchio, S.; Lalia-Kantouri, M. Lanthanide complexes of 3-methoxy-salicylaldehyde. *J. Therm. Anal. Calorim.* **2010**, *99*, 931–938. [CrossRef]

40. Persenaire, O.; Alexandre, M.; Degée, P.; Dubois, P. Mechanisms and kinetics of thermal degradation of poly(ε-caprolactone). *Biomacromolecules* **2001**, *2*, 288–294. [CrossRef] [PubMed]

41. Domán, A.; Madarász, J.; László, K. In situ evolved gas analysis assisted thermogravimetric (TG-FTIR and TG/DTA–MS) studies on non-activated copper benzene-1,3,5-tricarboxylate. *Thermochim. Acta* **2017**, *647*, 62–69. [CrossRef]

42. Vogel, C.; Siesler, H.W. Thermal degradation of poly(ε-caprolactone), poly(l-lactic acid) and their blends with poly(3-hydroxy-butyrate) studied by TGA/FT-IR spectroscopy. *Macromol. Symposia* **2008**, *265*, 183–194. [CrossRef]

43. Sarikaya, M.; Turan, M.D.; Aydogmus, R.; Yucel, A.; Kizilkaya, N.; Depci, T. Extraction of meso-pores amorphous SiO_2 from van pumice. *Curr. Phys. Chem.* **2017**, *7*, 301–304. [CrossRef]

44. Zhang, G.; Xu, Y.; Xu, D.; Wang, D.; Xue, Y.; Su, W. Pressure-induced crystallization of amorphous SiO2 with silicon–hydroxy group and the quick synthesis of coesite under lower temperature. *High Press. Res.* **2008**, *28*, 641–650. [CrossRef]

45. Keen, D.A.; Dove, M.T. Local structures of amorphous and crystalline phases of silica, SiO_2, by neutron total scattering. *J. Phys. Condens. Matter* **1999**, *11*, 9263. [CrossRef]

Article

Further Theoretical Insight into the Mechanical Properties of Polycaprolactone Loaded with Organic–Inorganic Hybrid Fillers

Saverio Maietta [1], Teresa Russo [2], Roberto De Santis [2], Dante Ronca [3], Filomena Riccardi [3], Michelina Catauro [4], Massimo Martorelli [1] and Antonio Gloria [2,*]

[1] Department of Industrial Engineering, Fraunhofer JL IDEAS–University of Naples Federico II, P.le Tecchio 80, 80125 Naples, Italy; smaietta@unina.it (S.M.); massimo.martorelli@unina.it (M.M.)
[2] Institute of Polymers, Composites and Biomaterials—National Research Council of Italy, V.le J.F. Kennedy 54–Mostra d'Oltremare Pad. 20, 80125 Naples, Italy; teresa.russo@unina.it (T.R.); rosantis@unina.it (R.D.S.)
[3] Institute of Orthopaedics and Traumathology, University of Campania "Luigi Vanvitelli", Via L. De Crecchio 2-4, 80138 Naples, Italy; dante.ronca@unicampania.it (D.R.); mena.riccardi@libero.it (F.R.)
[4] Department of Industrial and Information Engineering, University of Campania "Luigi Vanvitelli", Via Roma 29, 81031 Aversa, Italy; michelina.catauro@unicampania.it
* Correspondence: angloria@unina.it; Tel.: +39-081-242-5942.

Received: 29 December 2017; Accepted: 17 February 2018; Published: 21 February 2018

Abstract: Experimental/theoretical analyses have already been performed on poly(ε-caprolactone) (PCL) loaded with organic–inorganic fillers (PCL/TiO$_2$ and PCL/ZrO$_2$) to find a correlation between the results from the small punch test and Young's modulus of the materials. PCL loaded with Ti2 (PCL = 12, TiO$_2$ = 88 wt %) and Zr2 (PCL = 12, ZrO$_2$ = 88 wt %) hybrid fillers showed better performances than those obtained for the other particle composition. In this context, the aim of current research is to provide further insight into the mechanical properties of PCL loaded with sol–gel-synthesized organic–inorganic hybrid fillers for bone tissue engineering. For this reason, theoretical analyses were performed by the finite element method. The results from the small punch test and Young's modulus of the materials were newly correlated. The obtained values of Young's modulus (193 MPa for PCL, 378 MPa for PCL/Ti2 and 415 MPa for PCL/Zr2) were higher than those obtained from a previous theoretical modelling (144 MPa for PCL, 282 MPa for PCL/Ti2 and 310 MPa for PCL/Zr2). This correlation will be an important step for the evaluation of Young's modulus, starting from the small punch test data.

Keywords: computer-aided design (CAD); mechanical analysis; finite element analysis (FEA); composites; organic–inorganic hybrid materials; biomedical applications

---

1. Introduction

In the field of tissue engineering, the development of advanced substrates and scaffolds represents a great challenge. As reported in the literature, different kinds of polymers and polymer-based composite materials have been widely proposed together with several approaches to improve their biological and mechanical features [1–6].

Organic–inorganic hybrid materials have been synthesized and proposed as biomaterials with interesting properties [7–15].

Most hybrid organic–inorganic materials can be synthesized via the sol–gel method, benefiting from the combination of the best characteristics of polymers and inorganic materials.

The sol–gel chemistry is based on the hydrolysis and polycondensation of metal alkoxides (M(OR)$_x$ with M = Si, Sn, Zr, Ti, Al, Mo, V, W, Ce, and so forth [1,7,8,16,17]). The great advantage of the sol–gel method results in a process which can be performed at low temperature (i.e., room temperature) [1,4,17].

Hybrid materials with different physical properties and morphologies have been developed, introducing many polymers (the organic phase) into inorganic networks [1,7]. Regarding synthetic polymers, poly(ε-caprolactone) (PCL), which is a biodegradable aliphatic polyester [4,18,19], has been widely investigated for several biomedical applications (i.e., tissue engineering) [5,6,20–25].

On the other hand, the bioactivity of ZrO_2 and TiO_2 glasses was previously demonstrated, highlighting the formation of a bonelike apatite layer on the surfaces, hence the ability to bond to living bone [1].

In addition, previous works focused on the synthesis of PCL-based organic–inorganic materials via the sol–gel method, where PCL was incorporated into the network by means of hydrogen bonds between the carboxylic groups of the polymer and the hydroxyl groups of the inorganic phase, as well as on the analysis of the developed PCL/TiO_2 and PCL/ZrO_2.

The formation of a hydroxyapatite layer on the surfaces of samples soaked in a solution with a composition that simulated human blood plasma demonstrated the bioactivity of PCL/ZrO_2, PCL/TiO_2, and further organic–inorganic hybrid materials [4,26,27].

With regard to the design of substrates for hard-tissue engineering, mechanical performances play an important role, as hard tissues (i.e., bone) are stronger (higher strength) and stiffer (higher elastic modulus) compared to soft tissues (i.e., cartilage) [1,4].

Generally, polymers such as PCL do not match the required mechanical properties, as they are too flexible and weak, if hard-tissue engineering applications are considered. For this reason, to overcome these drawbacks, an alternative choice is represented by the use of polymer-based composites [1–6]. Benefiting from technical criteria and considerations in the design of composite materials, as well as from the concepts of the stress transfer mechanism and stress concentration, the amount of micro/nano-particles embedded in a polymer matrix may be properly optimized to avoid weakness in the structure [1–4,6]. The possibility to improve the properties of the neat PCL by embedding the bioactive PCL/TiO_2 or PCL/ZrO_2 organic–inorganic hybrid microfillers has already been demonstrated [1]. Specifically, it was found that both small punch tests and cell viability/proliferation analyses showed mechanical and biological performances for PCL reinforced with Ti2 (PCL = 12, TiO_2 = 88 wt %) and Zr2 (PCL = 12, ZrO_2 = 88 wt %) hybrid fillers, which were better than those obtained for the other particle composition [1].

In this context, starting from the optimization of 2D substrates, 3D additive manufactured composite scaffolds for hard-tissue engineering were developed and analyzed [4].

It is well known that the small punch test represents an interesting test method to evaluate the mechanical properties when a material is available only in small quantities [1,4]. However, this test method cannot be used for the determination of Young's modulus. A numerical simulation is also needed [4].

Thus, finite element (FE) analysis was performed on 2D substrates consisting of PCL loaded with sol–gel-synthesized PCL/TiO_2 or PCL/ZrO_2 hybrid fillers [4]. The results from the small punch test and Young's modulus of the materials were correlated [4].

As reported previously [4], the problem is axisymmetric and the model was previously meshed with a quadrilateral planar element in a representative axial section. In particular, six disk specimens with different values of Young's modulus (200 MPa, 500 MPa, 1000 MPa, 2000 MPa, 3500 MPa, and 5000 MPa) were already simulated, assuming a Poisson's ratio of 0.40. In any case, the obtained values were lower than the experimental Young's modulus.

Taking into account the results from previous experimental tests [1], as well as the approach already reported for the FE analysis [4], in the current research, further theoretical analyses were carried out on the 2D substrates. This consisted of a PCL matrix loaded with sol–gel-synthesized PCL/TiO_2 or PCL/ZrO_2 hybrid fillers, with the aim of finding a new correlation between the small punch test data and Young's modulus.

2. Results and Discussion

The starting point of the current research was a theoretical approach reported in the literature [4], which benefited from experimental results on PCL loaded with sol–gel-synthesized PCL/TiO$_2$ or PCL/ZrO$_2$ particles [1]. Specifically, the present study reports theoretical analyses on the 2D substrates consisting of a PCL matrix loaded with sol–gel-synthesized PCL/Ti2 or PCL/Zr2 hybrid fillers (PCL = 12 and TiO$_2$ = 88 wt % for Ti2; PCL = 12 and ZrO$_2$ = 88 wt % for Zr2), to provide further insight into the mechanical properties. All the results were compared to those obtained from a previous experimental/theoretical approach [4].

The important role of CAD, image and theoretical/experimental analyses has been widely reported for different kinds of applications [28–32].

With regard to experimental analysis, the small punch test is considered as a reproducible miniature specimen test method to evaluate the mechanical properties of ultra-high molecular weight polyethylene used in surgical implants and retrieved acrylic bone cements [1,4]. However, it was also employed to assess the punching properties of PCL/iron-doped hydroxyapatite nanocomposite substrates and PCL loaded with sol–gel-synthesized organic–inorganic hybrid fillers [1,2].

Previous experimental data from small punch tests on PCL and PCL loaded with sol–gel-synthesized organic–inorganic hybrid fillers showed load-displacement curves generally displaying an initial linear trend, a successive decrease in the curve slope until a maximum load was reached, and a final decrease in the load values until failure occurred. Furthermore, among the investigated composite substrate, better mechanical and biological features were found for PCL reinforced with Ti2 (PCL b = 12, TiO$_2$ = 8 wt %) and Zr2 (PCL = 12, ZrO$_2$ = 8 wt %) hybrid fillers [1,4].

In the present FE analysis, similar results were clearly achieved in terms of the displacement contour plot for all composites when different loads were applied according to Young's modulus of each disk specimen (from 200 to 5000 MPa) to obtain a final displacement value of 0.2 mm.

As an example, a displacement contour plot for a composite disk is shown in Figure 1.

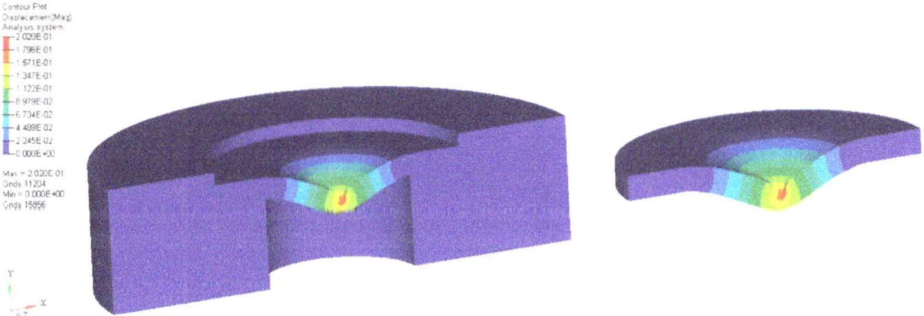

Figure 1. Results from finite element (FE) analysis: typical displacement contour plot for a composite disk.

Basically, the area of interest is represented by the disk placed in/on the support. As expected, the maximum displacement values (0.2 mm) were achieved in the center zone (red color) of the specimen.

Accordingly, FE analysis was used to calculate the force–displacement curves (Figure 2A) for all of the investigated materials, as well as a normalized force–displacement curve (Figure 2B) which was obtained by dividing the force value by Young's modulus for all materials.

Even though the problem was non-linear as a consequence of the friction, it was found that all the points of the normalized force–displacement curves collapsed into a single curve (Figure 2B).

The single curve was approximated according to the following equation:

$$\frac{F}{E} = A \cdot \delta \tag{1}$$

where F is the applied force, E represents Young's modulus of the material, A is a constant equal to 45.1 N·GPa^{-1}·mm^{-1}, and δ is the displacement of the punch.

Computational results and Equation (1) may represent interesting findings, providing the possibility to assess Young's modulus of materials subjected to the small punch test, once the value of the Poisson's ratio (close to 0.40) and the coefficient of friction (around 0.20) are considered.

Thus, according to Equation (1), the values of Young's modulus (which better describe the experimental results from small punch tests performed on PCL, PCL/Ti2 and PCL/Zr2 [1,4]) are shown in Table 1.

Even though Young's modulus evaluated for PCL (193 MPa) was higher than that already assessed by a previous theoretical analysis (144 MPa) [4], it is still lower if compared to the neat PCL (i.e., 343.9–571.5 MPa) [4,33,34]. Such lower value should probably be ascribed to the intrinsic porosity of the substrate as a direct consequence of the solvent evaporation during the preparation via moulding and solvent casting techniques [1,4].

In addition, Young's modulus evaluated for PCL/Ti2 (378 MPa) and PCL/Zr2 (415 MPa) was also higher than those previously computed in theoretical modelling (282 MPa and 310 MPa) [4].

The higher values of Young's modulus obtained for PCL, PCL/Ti2 and PCL/Zr2 would seem to suggest a better approximation of the experimental values.

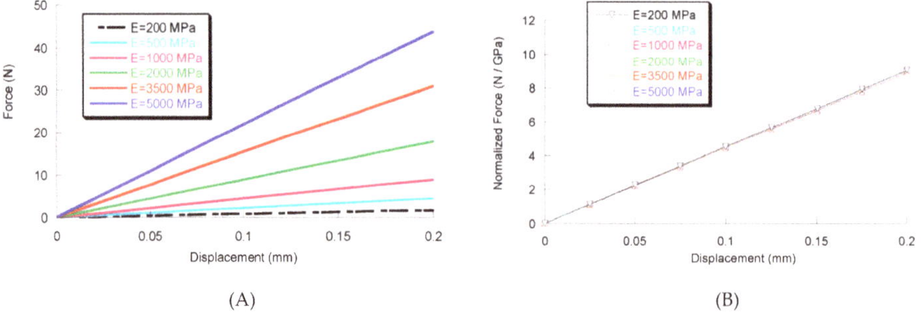

Figure 2. Results from FE analysis: force–displacement curves (**A**) and normalized force–displacement curves (**B**).

Table 1. Results obtained from theoretical analysis (Young's modulus) benefiting from experimental small punch tests (displacement and force). The experimental results were adapted from [1,4].

Materials	Displacement (mm)	Force (N)	Young's Modulus (MPa)
PCL	0.195	1.70	193
PCL/Ti2	0.191	3.26	378
PCL/Zr2	0.192	3.59	415

In a previous work on advanced composites for hard-tissue engineering (based on PCL/organic–inorganic hybrid fillers to design 3D additive manufactured scaffolds starting from 2D substrates [4]), results from compression tests on 3D porous structures showed mean values of compressive modulus which were 90 MPa and 79 MPa for 3D composite (PCL/Ti2, PCL/Zr2) and PCL scaffolds, respectively. Such values were lower than Young's moduli obtained in the current research (Table 1) and those computed in the previous theoretical modelling [4], as a consequence of the controlled morphology and macro-porosity of the 3D additive manufactured scaffolds, where "apparent" stress and strain were considered.

A potential limitation of the current study was the lack of correlation between the obtained results and the preparation method (i.e., moulding and solvent casting techniques), which clearly affects the morphology, surface topography, and, consequently, the punching performances. In any case, the present work should be considered as further insight into the mechanical properties of the analyzed microcomposites, as well as an initial step towards future research with the aim of finding a complex correlation, which will also involve the effect of the preparation method.

3. Materials and Methods

3.1. Materials

Class I PCL/TiO$_2$ and PCL/ZrO$_2$ organic–inorganic hybrid particles were synthesized via the sol–gel method, and composites consisting of poly(ε-caprolactone) (PCL) loaded with the hybrid particles were developed, as described elsewhere [1]. Briefly, in designing PCL-based composite substrates, PCL pellets (weight-average molecular weight M_w = 65,000, Aldrich, St. Louis, MO, USA) were dissolved in tetrahydrofuran (THF, Aldrich, St. Louis, MO, USA) with stirring at room temperature [1].

Organic–inorganic particles with a diameter lower than 38 µm were dispersed in the polymer solution (ultrasonic bath, Branson 1510 MT, Danbury, CT, USA). Moulding and solvent casting techniques were used to manufacture disk-shaped specimens (6.4 mm in diameter and 0.5 mm in thickness) [1,2].

3.2. Generation of Solid Model and Numerical Simulation

The 3D CAD (computer-aided design) model was obtained using SolidWorks® software (2016, Waltham, MA, USA). All the elements were considered as volumes and then connected. The assembly function was suitably employed and the model was built up.

The geometric model related to the small punch test is schematically reported in Figure 3. The disk specimen is pictured in blue, whereas the steel punch and support are in red and green, respectively.

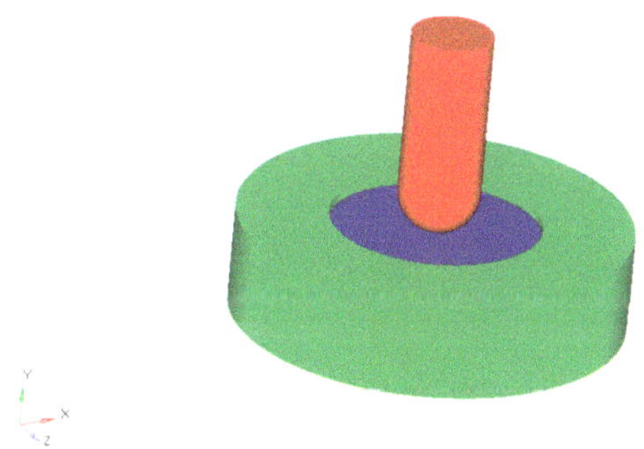

Figure 3. Schematic representation of the geometric model.

The geometric model of the small punch test was imported into the HyperWork® 14.0 (Altair Engineering Inc., Troy, MI, USA) environment using the STEP (Standard for the Exchange of Product) format, and the finite element analysis (FEA) was performed. In the model, all the materials were assumed to be linearly elastic and isotropic. Each component of the model was defined in terms of mechanical properties (i.e., Young's modulus and Poisson's ratio).

The values of Young's modulus and Poisson's ratio considered for the different materials are listed in Table 2. In particular, according to a previous work [4], with regard to the steel, a modulus of 210 GPa and a Poisson's ratio of 0.35 were assumed, and six disk specimens with different values of Young's modulus (200 MPa, 500 MPa, 1000 MPa, 2000 MPa, 3500 MPa, and 5000 MPa) were simulated. In addition, a Poisson's ratio of 0.40 was assumed for all the disk specimens [4]. Contact elements were used to model the contact between disk and punch as well as between disk and support, assuming a friction coefficient of 0.20 for all the mating surfaces [4].

Table 2. Mechanical properties of materials: Young's modulus and Poisson's ratio. All values were adapted from [4].

Material	Young's Modulus (MPa)	Poisson's Ratio
Disk 1	200	0.40
Disk 2	500	0.40
Disk 3	1000	0.40
Disk 4	2000	0.40
Disk 5	3500	0.40
Disk 6	5000	0.40

In a previous study [4], the geometry of the test was reproduced using the commercial finite element software ANSYS 11 (Ansys Inc., Canonsburg, PA, USA) and the model was meshed only with quadrilateral planar element (PLANE82) in a representative axial section (as the problem is axisymmetric). In the present research, a 3D mesh was created and each model was divided into 3D four-sided solid elements (CTETRA). With the aim to minimize the mesh-dependent results, an appropriate mesh size was employed for all components of the model.

Table 3 shows the number of grids, elements, contact elements, and degrees of freedom for the analyzed model.

Table 3. Number of grids, elements, contact elements and degrees of freedom for the analyzed model.

Total # of Grids	Total # of Elements	Total # of Contact Elements	Total # of Degrees of Freedom
55,559	267,342	1461	158,939

The results previously obtained from experimental analysis (small punch test-ASTM F2183) [1] and theoretical modelling [4] on 2D disk specimens (6.4 mm in diameter and 0.5 mm in thickness), consisting of PCL loaded with sol–gel-synthesized organic–inorganic hybrid fillers, were also considered.

4. Conclusions

The following conclusions were drawn:

1. The results from the small punch test and Young's modulus of PCL loaded with sol–gel-synthesized organic–inorganic hybrid fillers were newly correlated.
2. The data were compared to those obtained from a previous theoretical modelling, suggesting a better approximation of the experimental Young's modulus.
3. The found correlation will be an important step in the evaluation of Young's modulus starting from the small punch test data.

Acknowledgments: Mr. Rodolfo Morra (Institute of Polymers, Composites and Biomaterials – National Research Council of Italy) is acknowledged for providing information on the standard test method for small punch testing.

Author Contributions: S.M. and A.G. wrote the paper and performed FEA; R.D.S. and T.R. provided information on the experimental/theoretical mechanical data; D.R. and F.R. provided contributions and interpretations related to bone structure and tissue engineering; M.C. performed the synthesis of organic–inorganic hybrid particles in a previous experimental work, provided contributions and interpretations related to sol–gel synthesis; M.M. performed the optimization of geometric features and CAD design; A.G. conceived and designed the research.

Conflicts of Interest: The authors declare no conflict of interest.

References

1. Russo, T.; Gloria, A.; D'Antò, V.; D'Amora, U.; Ametrano, G.; Bollino, F.; de Santis, R.; Ausanio, G.; Catauro, M.; Rengo, S. Poly(ε-caprolactone) reinforced with sol-gel synthesized organic-inorganic hybrid fillers as composite substrates for tissue engineering. *J. Appl. Biomater. Biomech.* **2010**, *8*, 146. [PubMed]
2. Gloria, A.; Russo, T.; D'Amora, U.; Zeppetelli, S.; D'Alessandro, T.; Sandri, M.; Banobre-Lopez, M.; Pineiro-Redondo, Y.; Uhlarz, M.; Tampieri, A.; et al. Magnetic poly(ε-caprolactone)/iron-doped hydroxyapatite nanocomposite substrates for advanced bone tissue engineering. *J. R. Soc. Interface* **2013**, *10*, 20120833. [CrossRef] [PubMed]
3. Borzacchiello, A.; Gloria, A.; Mayol, L.; Dickinson, S.; Miot, S.; Martin, I.; Ambrosio, L. Natural/synthetic porous scaffold designs and properties for fibro-cartilaginous tissue engineering. *J. Bioact. Compati. Polym.* **2011**, *26*, 437–451. [CrossRef]
4. Santis, R.; Gloria, A.; Russo, T.; D'Amora, U.; D'Antò, V.; Bollino, F.; Catauro, M.; Mollica, F.; Rengo, S.; Ambrosio, L. Advanced composites for hard-tissue engineering based on PCL/organic-inorganic hybrid fillers: From the design of 2D substrates to 3D rapid prototyped scaffolds. *Polym. Compos.* **2013**, *34*, 1413–1417. [CrossRef]
5. De Santis, R.; Russo, A.; Gloria, A.; D'Amora, U.; Russo, T.; Panseri, S.; Sandri, M.; Tampieri, A.; Marcacci, M.; Dediu, V.A. Towards the design of 3D fiber-deposited poly(ε-caprolactone)/iron-doped hydroxyapatite nanocomposite magnetic scaffolds for bone regeneration. *J. Biomed. Nanotechnol.* **2015**, *11*, 1236–1246. [CrossRef] [PubMed]
6. Domingos, M.; Gloria, A.; Coelho, J.; Bartolo, P.; Ciurana, J. Three-dimensional printed bone scaffolds: The role of nano/micro-hydroxyapatite particles on the adhesion and differentiation of human mesenchymal stem cells. *Proc. Inst. Mech. Eng. H* **2017**, *231*, 555–564. [CrossRef] [PubMed]
7. Catauro, M.; Raucci, M.G.; de Marco, D.; Ambrosio, L. Release kinetics of ampicillin, characterization and bioactivity of TiO$_2$/PCL hybrid materials synthesized by sol-gel processing. *J. Biomed. Mater. Res. A* **2006**, *77*, 340–350. [CrossRef] [PubMed]
8. Catauro, M.; Raucci, M.; Ausanio, G. Sol-gel processing of drug delivery zirconia/polycaprolactone hybrid materials. *J. Mater. Sci. Mater. Med.* **2008**, *19*, 531–540. [CrossRef] [PubMed]
9. Catauro, M.; Pacifico, S. Synthesis of bioactive chlorogenic acid-silica hybrid materials via the sol–gel route and evaluation of their biocompatibility. *Materials* **2017**, *10*, 840. [CrossRef] [PubMed]
10. Ciprioti, S.V.; Bollino, F.; Tranquillo, E.; Catauro, M. Synthesis, thermal behavior and physicochemical characterization of ZrO$_2$/PEG inorganic/organic hybrid materials via sol–gel technique. *J. Therm. Anal. Calorim.* **2017**, *130*, 535–540. [CrossRef]
11. Catauro, M.; Bollino, F.; Giovanardi, R.; Veronesi, P. Modification of Ti6Al4V implant surfaces by biocompatible TiO$_2$/PCL hybrid layers prepared via sol-gel dip coating: Structural characterization, mechanical and corrosion behavior. *Mater. Sci. Eng. C Mater. Biol. Appl.* **2017**, *74*, 501–507. [CrossRef] [PubMed]
12. Catauro, M.; Bollino, F.; Papale, F. Response of SAOS-2 cells to simulated microgravity and effect of biocompatible sol–gel hybrid coatings. *Acta Astronaut.* **2016**, *122*, 237–242. [CrossRef]
13. Catauro, M.; Bollino, F.; Papale, F.; Piccolella, S.; Pacifico, S. Sol–gel synthesis and characterization of SiO$_2$/PCL hybrid materials containing quercetin as new materials for antioxidant implants. *Mater. Sci. Eng. C* **2016**, *58*, 945–952. [CrossRef] [PubMed]
14. Catauro, M.; Bollino, F.; Papale, F.; Marciano, S.; Pacifico, S. TiO$_2$/PCL hybrid materials synthesized via sol–gel technique for biomedical applications. *Mater. Sci. Eng. C* **2015**, *47*, 135–141. [CrossRef] [PubMed]
15. Catauro, M.; Bollino, F.; Mozzati, M.C.; Ferrara, C.; Mustarelli, P. Structure and magnetic properties of SiO$_2$/PCL novel sol–gel organic–inorganic hybrid materials. *J. Solid State Chem.* **2013**, *203*, 92–99. [CrossRef]
16. Catauro, M.; Verardi, D.; Melisi, D.; Belotti, F.; Mustarelli, P. Novel sol-gel organic-inorganic hybrid materials for drug delivery. *J. Appl. Biomater. Biomech.* **2010**, *8*, 42–51. [PubMed]
17. Catauro, M.; Tranquillo, E.; Illiano, M.; Sapio, L.; Spina, A.; Naviglio, S. The Influence of the Polymer Amount on the Biological Properties of PCL/ZrO$_2$ Hybrid Materials Synthesized via Sol-Gel Technique. *Materials* **2017**, *10*, 1186. [CrossRef] [PubMed]
18. David, I.; Scherer, G. An organic/inorganic single-phase composite. *Chem. Mater.* **1995**, *7*, 1957–1967. [CrossRef]

19. Choi, E.-J.; Kim, C.-H.; Park, J.-K. Synthesis and characterization of starch-*g*-polycaprolactone copolymer. *Macromolecules* **1999**, *32*, 7402–7408. [CrossRef]

20. Causa, F.; Battista, E.; Della Moglie, R.; Guarnieri, D.; Iannone, M.; Netti, P.A. Surface investigation on biomimetic materials to control cell adhesion: The case of RGDconjugation on PCL. *Langmuir* **2010**, *26*, 9875–9884. [CrossRef] [PubMed]

21. Zhong, Z.; Sun, X.S. Properties of soy protein isolate/polycaprolactone blends compatibilized by methylene diphenyl diisocyanate. *Polymer* **2001**, *42*, 6961–6969. [CrossRef]

22. Hutmacher, D.W.; Schantz, T.; Zein, I.; Ng, K.W.; Teoh, S.H.; Tan, K.C. Mechanical properties and cell cultural response of polycaprolactone scaffolds designed and fabricated via fused deposition modeling. *J. Biomed. Mater. Res. A* **2001**, *55*, 203–216. [CrossRef]

23. Patrício, T.; Domingos, M.; Gloria, A.; D'Amora, U.; Coelho, J.; Bártolo, P. Fabrication and characterisation of PCL and PCL/PLA scaffolds for tissue engineering. *Rapid Prototyp. J.* **2014**, *20*, 145–156. [CrossRef]

24. Huang, R.; Li, W.; Lv, X.; Lei, Z.; Bian, Y.; Deng, H.; Wang, H.; Li, J.; Li, X. Biomimetic LBL structured nanofibrous matrices assembled by chitosan/collagen for promoting wound healing. *Biomaterials* **2015**, *53*, 58. [CrossRef] [PubMed]

25. Xin, S.; Zeng, Z.; Zhou, X.; Luo, W.; Shi, X.; Wang, Q.; Deng, H.; Du, Y. Recyclable Saccharomyces cerevisiae loaded nanofibrous mats with sandwich structure constructing via bio-electrospraying for heavy metal removal. *J. Hazard. Mater.* **2017**, *324*, 365–372. [CrossRef] [PubMed]

26. Ohtsuki, C.; Kokubo, T.; Yamamuro, T. Mechanism of apatite formation on CaO-SiO$_2$-P$_2$O$_5$ glasses in a simulated body fluid. *J. Non. Cryst. Solids* **1992**, *143*, 84–92. [CrossRef]

27. Kokubo, T.; Kim, H.-M.; Kawashita, M. Novel bioactive materials with different mechanical properties. *Biomaterials* **2003**, *24*, 2161–2175. [CrossRef]

28. Vezzetti, E.; Speranza, D.; Marcolin, F.; Fracastoro, G. Diagnosing cleft lip pathology in 3D ultrasound: A landmarking-based approach. *Image Anal. Stereol.* **2015**, *35*, 53–65. [CrossRef]

29. Clemente, C.; Esposito, L.; Speranza, D.; Bonora, N. Firecracker eye exposure: Experimental study and simulation. *Biomech. Modeling Mechanobiol.* **2017**, *16*, 1401–1411. [CrossRef] [PubMed]

30. Moos, S.; Marcolin, F.; Tornincasa, S.; Vezzetti, E.; Violante, M.G.; Fracastoro, G.; Speranza, D.; Padula, F. Cleft lip pathology diagnosis and foetal landmark extraction via 3D geometrical analysis. *Int. J. Interac. Des. Manuf.* **2017**, *11*, 1–18. [CrossRef]

31. Martorelli, M.; Maietta, S.; Gloria, A.; de Santis, R.; Pei, E.; Lanzotti, A. Design and analysis of 3D customized models of a human mandible. *Procedia CIRP* **2016**, *49*, 199–202. [CrossRef]

32. Caputo, F.; De Luca, A.; Greco, A.; Maietta, S.; Marro, A.; Apicella, A. Investigation on the static and dynamic structural behaviours of a regional aircraft main landing gear by a new numerical methodology. *Frattura Integr. Strutt.* **2018**, *12*, 191–204.

33. Eshraghi, S.; Das, S. Mechanical and microstructural properties of polycaprolactone scaffolds with one-dimensional, two-dimensional, and three-dimensional orthogonally oriented porous architectures produced by selective laser sintering. *Acta Biomater.* **2010**, *6*, 2467–2476. [CrossRef] [PubMed]

34. De Santis, R.; Gloria, A.; Russo, T.; D'Amora, U.; Zeppetelli, S.; Dionigi, C.; Sytcheva, A.; Herrmannsdörfer, T.; Dediu, V.; Ambrosio, L. A basic approach toward the development of nanocomposite magnetic scaffolds for advanced bone tissue engineering. *J. Appl. Polym. Sci.* **2011**, *122*, 3599–3605. [CrossRef]

Article

Theoretical Design of Multilayer Dental Posts Using CAD-Based Approach and Sol-Gel Chemistry

Saverio Maietta [1], **Roberto De Santis** [2], **Michelina Catauro** [3], **Massimo Martorelli** [1] and **Antonio Gloria** [2,*]

[1] Department of Industrial Engineering, Fraunhofer JL IDEAS—University of Naples Federico II, P.le Tecchio 80, 80125 Naples, Italy; smaietta@unina.it (S.M.); massimo.martorelli@unina.it (M.M.)
[2] Institute of Polymers, Composites and Biomaterials—National Research Council of Italy, V.le J.F. Kennedy 54—Mostra d'Oltremare Pad. 20, 80125 Naples, Italy; rosantis@unina.it
[3] Department of Industrial and Information Engineering, University of Campania "Luigi Vanvitelli", Via Roma 29, 81031 Aversa, Italy; michelina.catauro@unicampania.it
* Correspondence: angloria@unina.it; Tel.: +39-081-242-5942

Received: 10 March 2018; Accepted: 4 May 2018; Published: 7 May 2018

Abstract: A computer-aided design (CAD)-based approach and sol-gel chemistry were used to design a multilayer dental post with a compositional gradient and a Young's modulus varying from 12.4 to 2.3 GPa in the coronal-apical direction. Specifically, we propose a theoretical multilayer post design, consisting of titanium dioxide (TiO_2) and TiO_2/poly(ε-caprolactone) (PCL) hybrid materials containing PCL up to 24% by weight obtained using the sol-gel method. The current study aimed to analyze the effect of the designed multilayer dental post in endodontically treated anterior teeth. Stress distribution was investigated along and between the post and the surrounding structures. In comparison to a metal post, the most uniform distributions with lower stress values and no significant stress concentration were found when using the multilayer post.

Keywords: computer-aided design (CAD); mechanical analysis; finite element analysis (FEA); composites; hybrid materials; biomedical applications

1. Introduction

The role of computer-aided design (CAD) for theoretical and experimental analyses has been widely used for different applications [1–6]. Such methods have been used to develop several kinds of polymeric and composite devices and have received considerable attention in biomedical applications [7–10].

The restoration of endodontically treated teeth represents a challenge as it generally involves the use of both metals and non-metallic materials [11]. In this field, many dental post-core systems have been used [11,12]. Initially, metal posts were chosen due to their long-term safety. As a consequence of mismatch between the elastic modulus of metal alloys and the surrounding structures, stress concentration generally occurs, often leading to catastrophic root fracture [12]. For this reason, studies have been devoted to the development of different shapes, sizes, and materials for the post [12].

Considering the results of previous studies, the use of materials with a lower elastic modulus, such as fiberglass-reinforced composites, may provide more favorable stress distribution. However, these composite posts have an elastic modulus, often ranging from 45.7 to 53.8 GPa [12,13], that is lower than that of metal posts, e.g., 95 GPa for gold and 110 GPa for titanium [12,14], but is still higher than those of natural tissues, which is 18.6 GPa for dentin [12,15].

Studies on endodontically treated canine teeth showed interesting results in terms of stress distribution, focusing on the ferrule effect and on the role of the specific material-shape combination of the post [16].

The mechanical behavior of a restored tooth is negatively affected by a dental post created using a high modulus material [11]. A dental post should stabilize the core without weakening the root [11]. As reported in the literature [11], stress concentration generally occurs at the apical and cervical regions of the tooth. Thus, an ideal post would possess a stiffness that decreases from the coronal part to the apical end, optimizing the stress transfer mechanism [11]. Given this context, functionally graded materials have also been considered for the development of dental posts with tailored properties, to overcome the drawbacks related to the use of both flexible and rigid posts [11].

Titanium [17], poly(ε-caprolactone) (PCL) [9,10,18–21], and several organic-inorganic hybrid materials obtained via sol-gel method [22–31] have been proposed for different biomedical applications. For example, titanium dioxide (TiO_2) and TiO_2/PCL hybrid materials containing PCL up to 24% by weight were obtained using the sol-gel method. In this case, heat and pressure were applied for powder compaction. The effects of the processing conditions and the amount of polymer on the performance of the materials were properly evaluated [17].

In this study, we theoretically design a multilayer dental post with a stiffness decreasing from the coronal part to the apical end using a CAD-based approach and sol-gel chemistry. In particular, a multilayer post with a compositional gradient of sol-gel synthesized materials and a Young's modulus ranging from 12.4 to 2.3 GPa in the coronal-apical direction was designed according to the values experimentally obtained [17] for TiO_2/PCL 94/6 (12.4 GPa), TiO_2/PCL 88/12 (9.2 GPa), TiO_2 (4.1 GPa), and TiO_2/PCL 76/24 (2.3 GPa). In endodontically treated canine teeth, the stress distribution along the multilayer post and at the interface between the post and the surrounding structure was assessed and compared to that of a titanium post. The null hypothesis was that the proposed multilayer post with a compositional gradient and a Young's modulus varying in the coronal-apical direction would not affect the stress distribution.

2. Materials and Methods

2.1. Materials and Post

A titanium post (post A) was used as the control. TiO_2 and TiO_2/PCL hybrid materials containing PCL up to 24% by weight were obtained via sol-gel method as described in a previous study [17].

As the experimentally-obtained values of the Young's modulus and Poisson's ratio for these materials (12.4 GPa and 0.27 for TiO_2/PCL 94/6; 9.2 GPa and 0.30 for TiO_2/PCL 88/12; 4.1 GPa and 0.27 for TiO_2; and 2.3 GPa and 0.30 for TiO_2/PCL 76/24) [17], a multilayer post with a compositional gradient and a modulus varying from 12.4 to 2.3 GPa in the coronal-apical direction (post B) was designed, analyzed, and compared with a titanium post (post A) hypothetically having the same geometrical characteristics.

The geometrical characteristics of the posts are reported in Table 1.

Table 1. Geometrical characteristics of the posts: length of coronal part, length of conicity part, coronal diameters, and apical diameters.

Total Length (mm)	Length of Coronal Part (Cylindrical) (mm)	Length on Conicity Part (mm)	Coronal Diameter	Apical Diameter
15 mm	7 mm	8 mm	Ø 1.05–Ø 1.25–Ø 1.45	Ø 0.55–Ø 0.75–Ø 0.95

2.2. Generation of the Tooth Solid Model

An upper canine was analyzed using a micro-CT scanner system (Bruker microCT, Kontich, Belgium). Micro-CT scan images were obtained and the three-dimensional (3D) CAD model of the tooth was generated as in a previous study [16], where a total of 951 slices were collected (1024 × 1024 pixels) and 252 slices were used. To process the image data sets, ScanIP® (3.2, Simpleware Ltd., Exeter, U.K.) was used. A previously adopted approach was used to generate the 3D model [16]. Briefly, procedures related to image segmentation and filtering were used, and the 3D tessellated model was created [16].

Blending operations were performed via converting cross sections of the tessellated models into surfaces. ScanTo3D® (SolidWorks® 2017, Dassault Systemes, Paris, France) was used to manage the tessellated geometry. Specific procedures were used to create lofting surfaces and to ensure the congruence of interfacial boundaries of tooth tissues [16]. The system of coordinates, the geometrical model, and features were previously reported [16].

Two different geometric models of the restored tooth were analyzed. Specifically, two posts were considered: posts A and B (15 mm in length) with a conical-tapered shape. A 0.1 mm thick cement layer was added between the abutment and crown. In the canal, the cement was added between the post and the root. In addition, the periodontal ligament with a thickness of 0.25 mm was modelled around the root [16].

2.3. Numerical Simulation

The geometric models of endodontically treated anterior teeth were imported into HyperMesh® (HyperWorks®-14.0, Altair Engineering Inc., Troy, MI, U.S.).

Finite element (FE) analyses were performed on two models: (1) Model A (a tooth with Post A) and (2) Model B (a tooth with Post B). The values of Young's modulus and Poisson's ratio for the components of the tooth model are reported in Table 2.

Table 2. Young's modulus and Poisson's ratio for the components of the tooth model [16]. * The values varied from the coronal to the apical part of the part according to the different regions [17] of the proposed multilayer post.

Component	Young's Modulus (GPa)	Poisson's Ratio
Lithium disilicate crown	70	0.30
Crown cement	8.2	0.30
Abutment	12	0.30
Post A	110	0.35
Post B	12.4–2.3 *	0.27–0.30 *
Post cement	8.2	0.30
Root	18.6	0.31
Periodontal ligament	$0.15\ (\times 10^{-3})$	0.45
Food (apple pulp)	$3.41\ (\times 10^{-3})$	0.10

As previously reported [16], a 3D mesh was created and 3D solid CTETRA elements with four grid points were considered for the models. Consistent with a previous methodology [16], the study focused on the closing phase of the chewing cycle and solid food acting on the crown surface, using apple pulp with a Poisson's ratio and Young's modulus of 0.10 and 3.41 MPa, respectively (Figure 1). Slide-type contact elements were considered between the food and tooth surface. For the contact condition between each part of the post restoration, the "freeze" type was used.

Briefly, convergence and mesh independence studies were also performed to obtain accurate results. Mesh convergence was performed to determine the number of elements needed in the model to ensure that the results were not affected by varying the mesh size. The complexity of the model vs. response (i.e., stress) was recorded. Following convergence, mesh refinement was performed. Thus, further technical features of the analyzed models included the total number of grids (structural) (51,552), elements excluding contact (213,361), node-to-surface contact elements (14,094), and degrees of freedom (188,127).

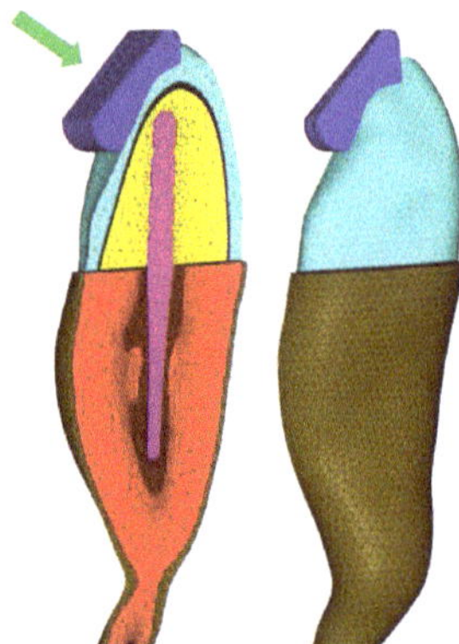

Figure 1. Finite element (FE) model according to the components in the geometric model.

With regard to nodal displacements, the FE models of the restored tooth constraints were applied in all the directions on the surface of the periodontal ligament. On the surface of the crown, a load of 50 N was applied at 45° to the longitudinal axis of the tooth [16]. As a linear elastic behavior was assumed for all the components, a non-failure condition was considered and linear static analyses were performed. The maximum principal stress and von Mises stress distributions were evaluated along the post and at the interface between the post and the surrounding structure.

3. Results

The maximum principal stress and von Mises stress distributions were observed in the abutment, post, post cement, root, and periodontal ligament. The considered cross sections are depicted in Figures 2 and 3.

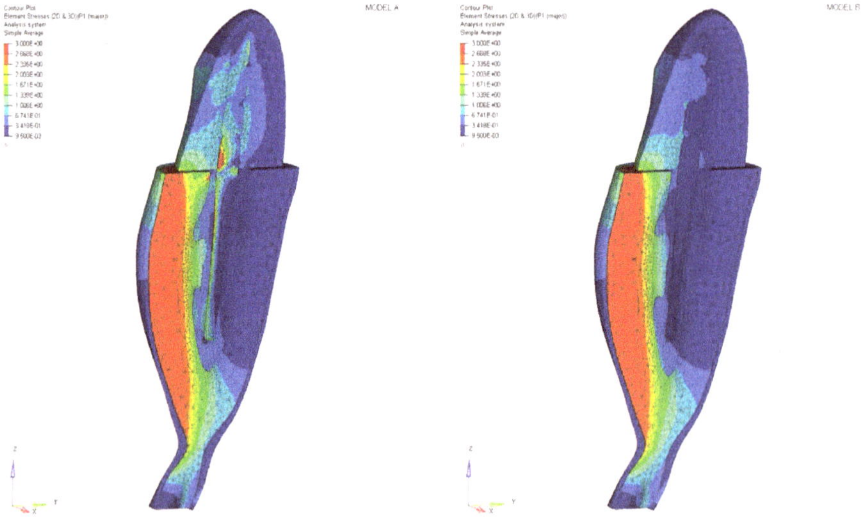

Figure 2. Maximum principal stress distribution (MPa): model A and model B.

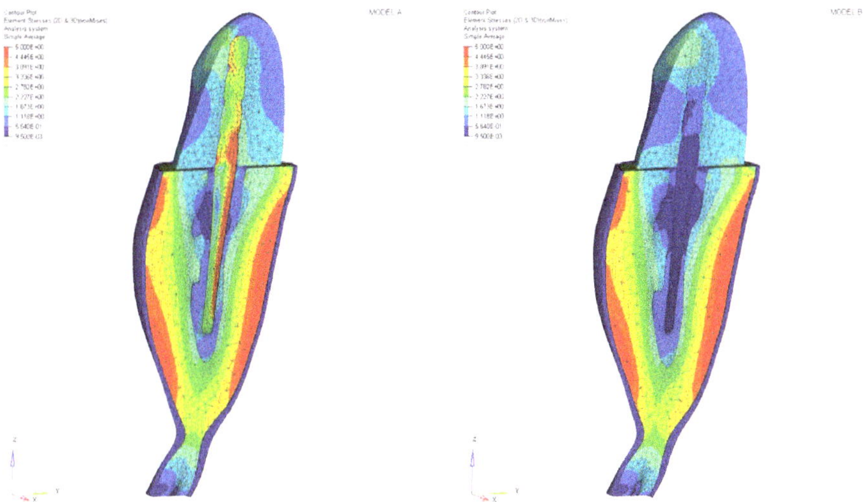

Figure 3. Von Mises stress distribution (MPa): model A and model B.

Differences were found between the two models in terms of the maximum principal stress and von Mises stress distributions. If compared to model B, higher stress regions were observed for model A along the post near the cervical margin of the tooth. Comparing the analyzed models, the most uniform stress distribution was achieved for post B. The maximum principal stress and von Mises stress distributions along the post are displayed in Figures 4–7.

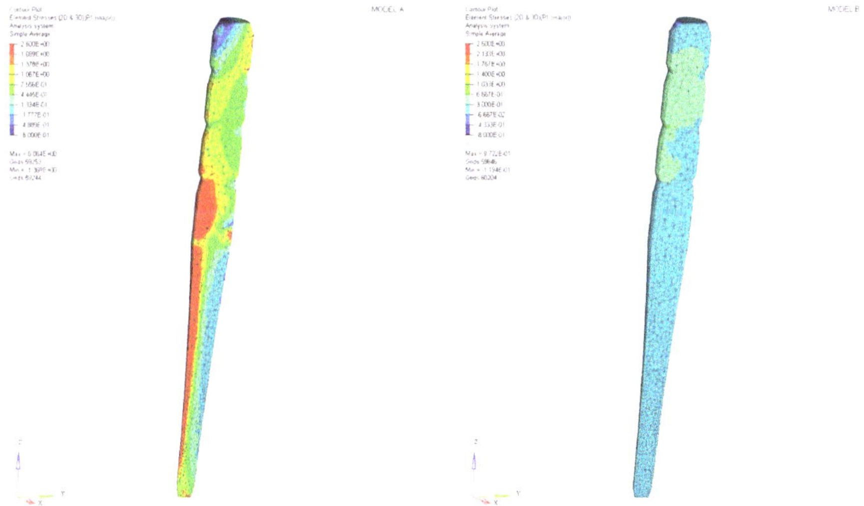

Figure 4. Maximum principal stress distribution (MPa): post A and post B.

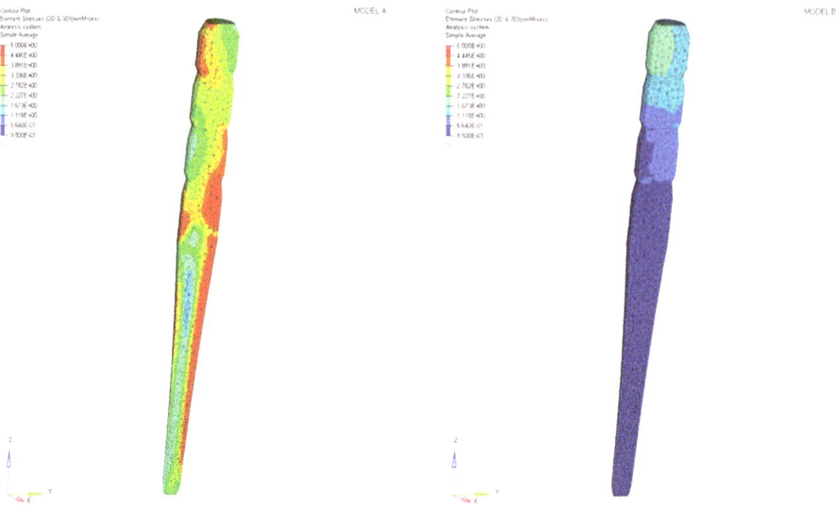

Figure 5. Von Mises stress distribution (MPa): post A and post B.

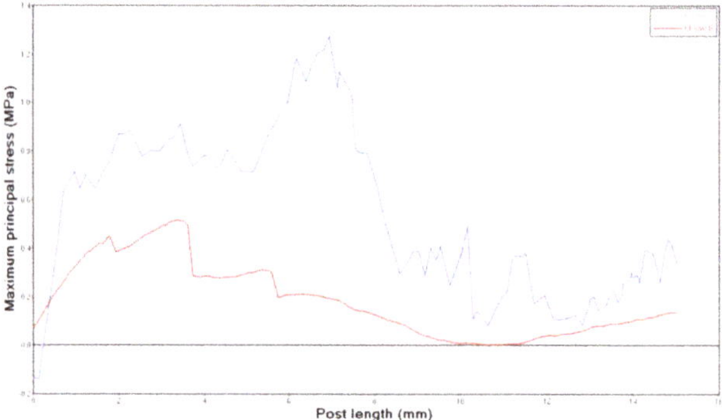

Figure 6. Maximum principal stress distribution along the center of the post from the coronal to the apical part.

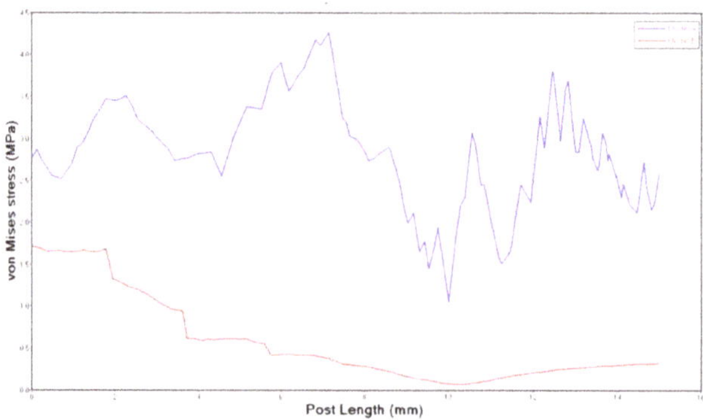

Figure 7. Von Mises stress distribution along the center of the post from the coronal to the apical part.

Stress concentrations were observed along the post in model A, whereas lower stress values were evident for model B. In addition, with regard to the stress distribution at the interface between the post and the surrounding structures (Figures 8–11), for model A, high stress gradients were found as well as fluctuations and changes up to the apical part, which were much more marked than in model B.

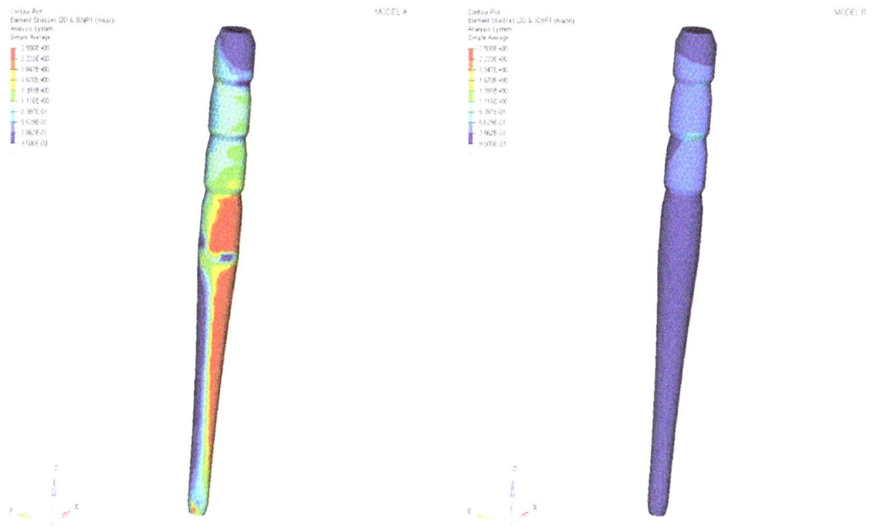

Figure 8. Maximum principal stress distribution (MPa) at the interface between the post and surrounding structures.

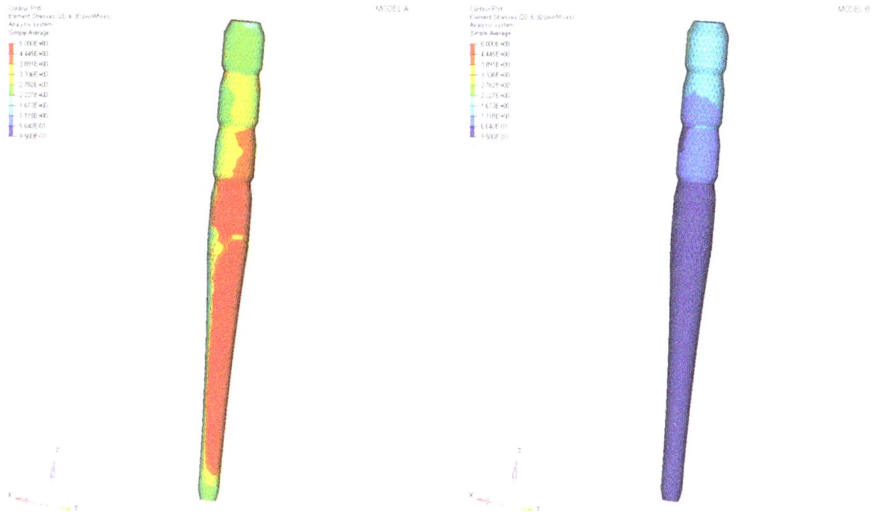

Figure 9. Von Mises stress distribution (MPa) at the interface between the post and surrounding structures.

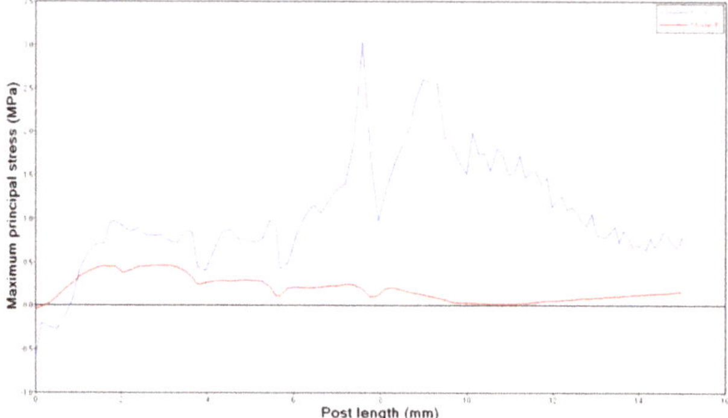

Figure 10. Maximum principal stress distribution at the interface between the post and surrounding structures from the coronal to the apical part.

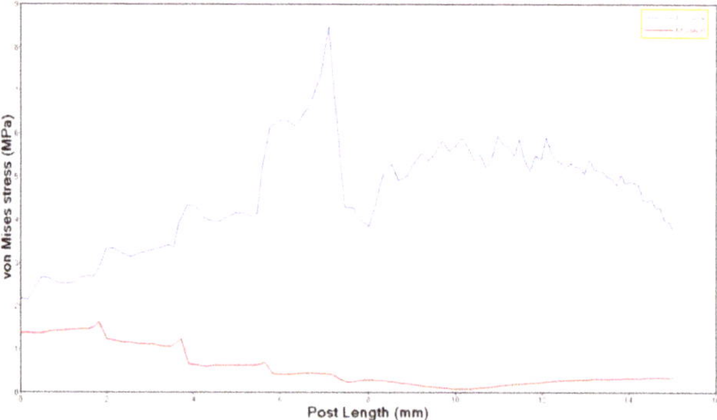

Figure 11. Von Mises stress distribution at the interface between the post and surrounding structures from the coronal to the apical part.

In comparison to model A, model B showed gradual changes and lower stress values (Figures 8–11). The differences between a cross section at the cervical margin of the tooth of the two models were compared. The results are displayed in Figures 12–14.

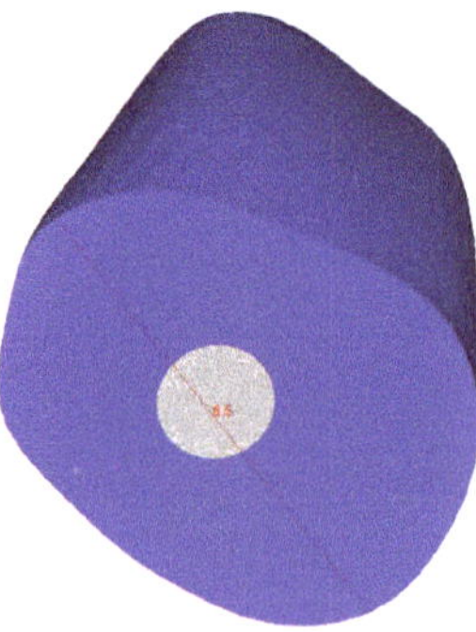

Figure 12. Cross section at the cervical margin of the tooth that was further analyzed.

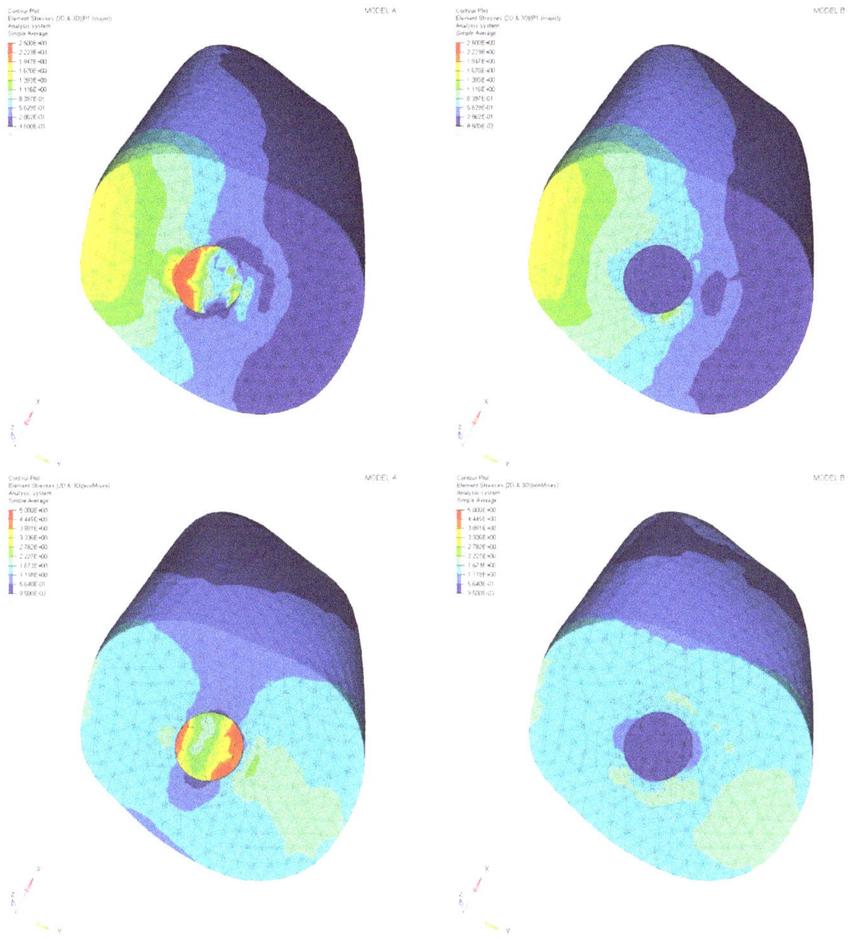

Figure 13. Maximum principal stress and von Mises stress distributions (MPa) in the cross-section at the cervical margin—scheme of Figure 12.

In particular, Figures 13 and 14 report the maximum principal stress and von Mises stress distributions in the cross-section at the cervical margin along the direction indicated by the red line in Figure 12. The obtained results demonstrate high stress gradients for model A at the interface between the surrounding structure and the post.

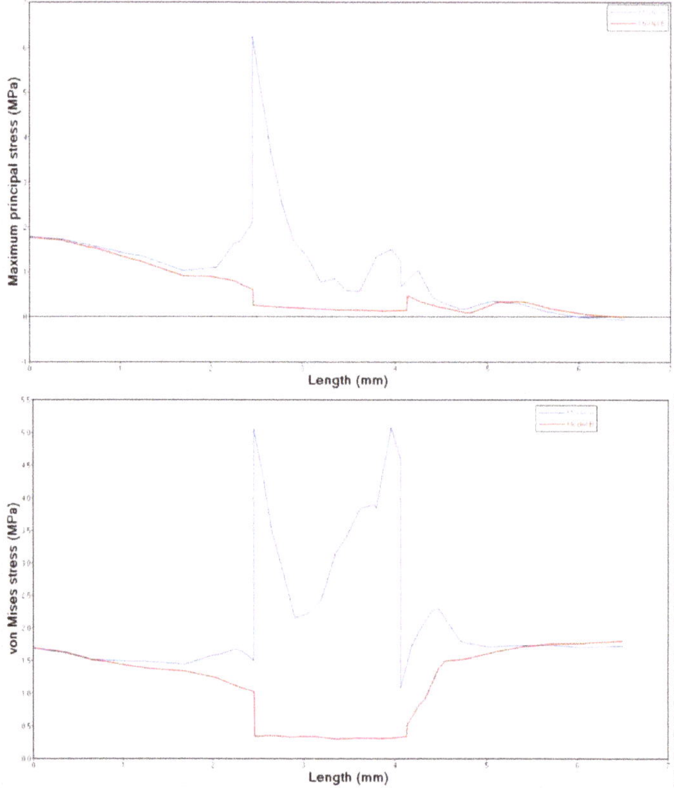

Figure 14. Maximum principal stress and von Mises stress distributions in the cross-section at the cervical margin, along the direction indicated by the red line shown in Figure 12.

4. Discussion

A dental post designed using a high-modulus material clearly alters the mechanical behavior of the restored tooth [11,32]. To prevent catastrophic root fracture, fiberglass posts and resin cores are currently used as post-core systems [12]. The performance of post-and-core systems have been widely investigated [33]. As many efforts have been made to develop composite posts using different shapes and kinds of fibers, such as carbon, glass, and quartz, clinical procedures have been continuously modified [34–36]. Although many experimental and theoretical analyses and clinical studies have been completed, no precise recommendations have been made [35]. A general procedure includes selection of the post, the preparation of the root canal, the use of adhesive resin cements or self-adhesive cements to bond the post, which must suitably extend to retain the core, and the placement of a crown [35]. However, with regard to devices, materials, and clinical procedures, contradictory opinions still remain [35,37]. The performance of the fiber posts depends on the manufacturing process, matrix, fiber properties, distribution and amount of fibers [37]. Several clinical studies have also been performed on patients with teeth restored using posts fabricated from carbon fiber-, quartz fiber-, or glass fiber-reinforced composites [37–42].

During loading, a high stress concentration normally occurs at the apical part of the post [11,43]. When the tooth structure is compromised, an increase in flexure may cause stress concentration at the cervical region. Furthermore, stress concentration should be ascribed to the tapering of the root canal at the apical region as well as to the characteristics of the post [11,44]. High stress concentrations arise from the stiffness mismatch between the post and surrounding structures [11,45]. An ideal post should possess a stiffness decreasing from the coronal to apical part to optimize stress distribution.

As many technical features related to the development of fiber-reinforced composite posts have been widely discussed in the literature, a CAD-based approach and sol-gel chemistry were considered in the current research to theoretically design a multilayer post with a stiffness decreasing from the coronal part to the apical end.

Sol-gel chemistry has been proposed as a method to develop organic-inorganic hybrid materials with specific properties for biomedical applications [22–31]. Thus, benefiting from previous experimental results [17], TiO_2 and TiO_2/PCL hybrid materials containing PCL up to 24% by weight obtained using the sol-gel method were used to design a multilayer dental post with tailored properties. In particular, with regard to endodontically treated anterior teeth, the effect of a multilayer post with a compositional gradient of sol-gel synthesized materials and a Young's modulus ranging from 12.4 to 2.3 GPa in the coronal-apical direction was evaluated in this study.

As a result of the multilayer structural design for post B, the performed analyses evidenced that higher values of maximum principal and von Mises stresses were found along the post near the cervical margin of the tooth for model A compared with model B, which showed no stress concentration (Figures 2 and 3). The multilayer structure, having different mechanical properties, allowed us to tailor the performance in the coronal-apical direction and avoid stress concentration, thus providing a better stress distribution in the restored tooth. Figures 6 and 7 confirm that the designed multilayer post (post B) provided better stress distribution along the center of the post from the coronal to the apical part, if compared to the titanium post (post A).

At the interface between the surrounding structures and the post, the maximum principal stress and von Mises stress distributions proved the important role of the designed post (Figures 10 and 11). In the case of the titanium post, the stress transfer mechanism involved higher values of stress as well as much more marked fluctuations and changes that were evident up to the apical part (Figures 10 and 11). Consistently, the analysis results of a cross section at the cervical margin of the tooth showed stress gradients for model A that were higher than those observed for model B (Figures 13 and 14). Finally, the null hypothesis that the proposed multilayer post with a compositional gradient and a Young's modulus varying in the coronal-apical direction in the restored model would not affect the stress distribution was rejected.

Potential limitations include the linear static analyses performed considering a non-failure condition and the approach used to design of the multilayer post, which was based on the results obtained in a previous work [17]. Regardless of these shortcomings, the current study should be considered as a first work toward the theoretical design of a multilayer dental post consisting of TiO_2 and TiO_2/PCL hybrid materials obtained using sol-gel method, with a compositional gradient and a Young's modulus varying in the coronal-apical direction.

5. Conclusions

Within the limitations of the present study, the following conclusions were drawn: (1) A theoretical design of a multilayer dental post was reported using CAD-based approach and sol-gel chemistry; (2) a model of an anterior tooth restored with a multilayer post, consisting of TiO_2 and TiO_2/PCL hybrid materials obtained via sol-gel method, was analyzed; and (3) in comparison to a titanium post, the most uniform stress distribution with no significant stress concentrations was found in the proposed multilayer dental post with a compositional gradient and a Young's modulus varying in the coronal-apical direction.

Author Contributions: S.M. and A.G. wrote the paper and performed FE analysis; R.D.S. and M.M. provided information on the experimental/theoretical mechanical data; M.C. synthesized the hybrid particles in a previous experimental work, and provided contributions and interpretations related to sol-gel synthesis; M.M. and A.G. performed the optimization of geometric features and CAD design; A.G. conceived and designed the research.

Acknowledgments: Rodolfo Morra (Institute of Polymers, Composites and Biomaterials—National Research Council of Italy) is acknowledged for providing information on the mechanical test methods employed in the previous research related to hybrid materials.

Conflicts of Interest: The authors declare no conflict of interest.

References

1. Martorelli, M.; Maietta, S.; Gloria, A.; De Santis, R.; Pei, E.; Lanzotti, A. Design and analysis of 3D customized models of a human mandible. *Procedia CIRP* **2016**, *49*, 199–202. [CrossRef]
2. Ausiello, P.; Ciaramella, S.; Garcia-Godoy, F.; Gloria, A.; Lanzotti, A.; Maietta, S.; Martorelli, M. The effects of cavity-margin-angles and bolus stiffness on the mechanical behavior of indirect resin composite class II restorations. *Dent. Mater.* **2017**, *33*, e39–e47. [CrossRef] [PubMed]
3. Caputo, F.; De Luca, A.; Greco, A.; Maietta, S.; Marro, A.; Apicella, A. Investigation on the static and dynamic structural behaviours of a regional aircraft main landing gear by a new numerical methodology. *Frattura Integr. Strutt.* **2018**, *12*, 191–204.
4. De Santis, R.; Gloria, A.; Maietta, M.; Martorelli, M.; De Luca, A.; Spagnuolo, G.; Riccitiello, F.; Rengo, S. Mechanical and Thermal Properties of Dental Composites Cured with CAD/CAM Assisted Solid-State Laser. *Materials* **2018**, *11*, 504. [CrossRef] [PubMed]
5. Caputo, F.; De Luca, A.; Greco, A.; Maietta, S.; Bellucci, M. FE simulation of a SHM system for a large radio-telescope. *Int. Rev. Model. Simul.* **2018**, *11*. [CrossRef]
6. Maietta, S.; Russo, T.; De Santis, R.; Ronca, D.; Riccardi, F.; Catauro, M.; Martorelli, M.; Gloria, A. Further Theoretical Insight into the Mechanical Properties of Polycaprolactone Loaded with Organic-Inorganic Hybrid Fillers. *Materials* **2018**, *11*, 312. [CrossRef] [PubMed]
7. Borzacchiello, A.; Gloria, A.; Mayol, L.; Dickinson, S.; Miot, S.; Martin, I.; Ambrosio, L. Natural/synthetic porous scaffold designs and properties for fibro-cartilaginous tissue engineering. *J. Bioact. Compat. Polym.* **2011**, *26*, 437–451. [CrossRef]
8. Giordano, C.; Albani, D.; Gloria, A.; Tunesi, M.; Batelli, S.; Russo, T.; Forloni, G.; Ambrosio, L.; Cigada, A. Multidisciplinary perspectives for Alzheimer's and Parkinson's diseases: Hydrogels for protein delivery and cell-based drug delivery as therapeutic strategies. *Int. J. Artif. Organs* **2009**, *32*, 836–850. [CrossRef] [PubMed]
9. Reitmaier, S.; Shirazi-Adl, A.; Bashkuev, M.; Wilke, H.J.; Gloria, A.; Schmidt, H. In vitro and in silico investigations of disc nucleus replacement. *J. R. Soc. Interface* **2012**, *9*, 1869–1879. [CrossRef] [PubMed]
10. Domingos, M.; Gloria, A.; Coelho, J.; Bartolo, P.; Ciurana, J. Three-dimensional printed bone scaffolds: The role of nano/micro-hydroxyapatite particles on the adhesion and differentiation of human mesenchymal stem cells. *Proc. Inst. Mech. Eng. Part H J. Eng. Med.* **2017**, *231*, 555–564. [CrossRef] [PubMed]
11. Abu Kasim, N.H.; Madfa, A.A.; Hamdi, M.; Rahbari, G.R. 3D-FE analysis of functionally graded structured dental posts. *Dent. Mater.* **2011**, *30*, 869–880. [CrossRef] [PubMed]
12. Lee, K.-S.; Shin, J.-H.; Kim, J.-E.; Kim, J.-H.; Lee, W.-C.; Shin, S.-W.; Lee, J.-Y. Biomechanical evaluation of a tooth restored with high performance polymer PEKK post-core system: A 3D finite element analysis. *BioMed Res. Int.* **2017**. [CrossRef] [PubMed]
13. Cheleux, N.; Sharrock, P.J. Mechanical properties of glass fiber-reinforced endodontic posts. *Acta Biomater.* **2009**, *5*, 3224–3230. [CrossRef] [PubMed]
14. Sakaguchi, R.L.; Powers, J.M. *Craig's Restorative Dental Materials-e-Book*; Elsevier Health Sciences: New York, NY, USA, 2012.
15. Craig, R.; Peyton, F. Elastic and mechanical properties of human dentin. *J. Dent. Res.* **1958**, *37*, 710–718. [CrossRef] [PubMed]
16. Ausiello, P.; Ciaramella, S.; Martorelli, M.; Lanzotti, A.; Zarone, F.; Watts, D.C.; Gloria, A. Mechanical behavior of endodontically restored canine teeth: Effects of ferrule, post material and shape. *Dent. Mater.* **2017**, *33*, 1466–1472. [CrossRef] [PubMed]
17. De Santis, R.; Catauro, M.; Di Silvio, L.; Manto, L.; Raucci, M.G.; Ambrosio, L.; Nicolais, L. Effect of polymer amount and processing conditions on the in vitro behaviour of hybrid titanium dioxide/polycaprolactone composites. *Biomaterials* **2007**, *28*, 2801–2809. [CrossRef] [PubMed]

18. Causa, F.; Battista, E.; Della Moglie, R.; Guarnieri, D.; Iannone, M.; Netti, P.A. Surface investigation on biomimetic materials to control cell adhesion: The case of RGDconjugation on PCL. *Langmuir* **2010**, *26*, 9875–9884. [CrossRef] [PubMed]

19. Zhong, Z.; Sun, X.S. Properties of soy protein isolate/polycaprolactone blends compatibilized by methylene diphenyl diisocyanate. *Polymer* **2001**, *42*, 6961–6969. [CrossRef]

20. Hutmacher, D.W.; Schantz, T.; Zein, I.; Ng, K.W.; Teoh, S.H.; Tan, K.C. Mechanical properties and cell cultural response of polycaprolactone scaffolds designed and fabricated via fused deposition modeling. *J. Biomed. Mater. Res. Part A* **2001**, *55*, 203–216. [CrossRef]

21. Patrício, T.; Domingos, M.; Gloria, A.; D'Amora, U.; Coelho, J.; Bártolo, P. Fabrication and characterisation of PCL and PCL/PLA scaffolds for tissue engineering. *Rapid Prototyp. J.* **2014**, *20*, 145–156. [CrossRef]

22. Catauro, M.; Raucci, M.G.; De Marco, D.; Ambrosio, L. Release kinetics of ampicillin, characterization and bioactivity of TiO_2/PCL hybrid materials synthesized by sol-gel processing. *J. Biomed. Mater. Res. A* **2006**, *77*, 340–350. [CrossRef] [PubMed]

23. Catauro, M.; Raucci, M.; Ausanio, G. Sol-gel processing of drug delivery zirconia/polycaprolactone hybrid materials. *J. Mater. Sci. Mater. Med.* **2008**, *19*, 531–540. [CrossRef] [PubMed]

24. Catauro, M.; Pacifico, S. Synthesis of bioactive chlorogenic acid-silica hybrid materials via the sol–gel route and evaluation of their biocompatibility. *Materials* **2017**, *10*, 840. [CrossRef] [PubMed]

25. Ciprioti, S.V.; Bollino, F.; Tranquillo, E.; Catauro, M. Synthesis, thermal behavior and physicochemical characterization of ZrO_2/PEG inorganic/organic hybrid materials via sol–gel technique. *J. Therm. Anal. Calorim.* **2017**, *130*, 535–540. [CrossRef]

26. Catauro, M.; Bollino, F.; Giovanardi, R.; Veronesi, P. Modification of ti6al4v implant surfaces by biocompatible TiO_2/PCL hybrid layers prepared via sol-gel dip coating: Structural characterization, mechanical and corrosion behavior. *Mater. Sci. Eng. C Mater. Biol. Appl.* **2017**, *74*, 501–507. [CrossRef] [PubMed]

27. Catauro, M.; Bollino, F.; Papale, F. Response of saos-2 cells to simulated microgravity and effect of biocompatible sol-gel hybrid coatings. *Acta Astronaut.* **2016**, *122*, 237–242. [CrossRef]

28. Catauro, M.; Bollino, F.; Papale, F.; Piccolella, S.; Pacifico, S. Sol-gel synthesis and characterization of SiO_2/PCL hybrid materials containing quercetin as new materials for antioxidant implants. *Mater. Sci. Eng. C* **2016**, *58*, 945–952. [CrossRef] [PubMed]

29. Catauro, M.; Bollino, F.; Papale, F.; Marciano, S.; Pacifico, S. TiO_2/PCL hybrid materials synthesized via sol–gel technique for biomedical applications. *Mater. Sci. Eng. C* **2015**, *47*, 135–141. [CrossRef] [PubMed]

30. Catauro, M.; Bollino, F.; Mozzati, M.C.; Ferrara, C.; Mustarelli, P. Structure and magnetic properties of SiO_2/PCL novel sol-gel organic-inorganic hybrid materials. *J. Solid State Chem.* **2013**, *203*, 92–99. [CrossRef]

31. Catauro, M.; Verardi, D.; Melisi, D.; Belotti, F.; Mustarelli, P. Novel sol-gel organic-inorganic hybrid materials for drug delivery. *J. Appl. Biomater. Biomech.* **2010**, *8*, 42–51. [PubMed]

32. Mahmoudi, M.; Saidi, A.R.; Amini, P.; Hashemipour, M.A. Influence of inhomogeneous dental posts on stress distribution in tooth root and interfaces: Three-dimensional finite element analysis. *J. Prosthet. Dent.* **2017**, *118*, 742–751. [CrossRef] [PubMed]

33. Cagidiaco, M.C.; Radovic, I.; Simonetti, M.; Tay, F.; Ferrari, M. Clinical performance of fiber post restorations in endodontically treated teeth: 2-year results. *Int. J. Prosthodont.* **2007**, *20*, 293–298. [PubMed]

34. Grandini, S.; Sapio, S.; Simonetti, M. Use of anatomic post and core for reconstructing an endodontically treated tooth: A case report. *J. Adhes. Dent.* **2003**, *5*, 243–247. [PubMed]

35. Wilson, P.D.; Wilson, N.; Dunne, S. *Manual of Clinical Procedures in Dentistry*; John Wiley & Sons: Hoboken, NJ, USA, 2018.

36. Manhart, J. Fiberglass reinforced composite endodontic posts. *Endod. Pract.* **2009**, 16–20. Available online: http://www.voco.com/en/product/rebilda_post/AB_Rebilda_Post__Rebilda_DC_Dr__Manhart_Endodontic_Practice_September_2009.pdf (accessed on 7 May 2018).

37. Faria, A.C.; Rodrigues, R.C.; de Almeida Antunes, R.P.; de Mattos Mda, G.; Ribeiro, R.F. Endodontically treated teeth: Characteristics and considerations to restore them. *J. Prosthodont. Res.* **2011**, *55*, 69–74. [CrossRef] [PubMed]

38. Fredriksson, M.; Astbäck, J.; Pamenius, M.; Arvidson, K. A retrospective study of 236 patients with teeth restored by carbon fiber-reinforced epoxy resin posts. *J. Prosthet. Dent.* **1998**, *80*, 151–157. [CrossRef]

39. Martelli, R. Fourth-generation intraradicular posts for the aesthetic restoration of anterior teeth. *Pract. Periodontics Aesthet. Dent.* **2000**, *12*, 579–584. [PubMed]

40. Malferrari, S.; Monaco, C.; Scotti, R. Clinical evaluation of teeth restored with quartz fiber-reinforced epoxy resin posts. *Int. J. Prosthodont.* **2003**, *16*, 39–44. [PubMed]

41. Göhring, T.N.; Peters, O.A. Restoration of endodontically treated teeth without posts. *Am. J. Dent.* **2003**, *16*, 313–317. [PubMed]

42. Grandini, S.; Goracci, C.; Tay, F.R.; Grandini, R.; Ferrari, M. Clinical evaluation of the use of fiber posts and direct resin restorations for endodontically treated teeth. *Int. J. Prosthodont.* **2005**, *18*, 399–404. [PubMed]

43. Kishen, A.; Asundi, A. Photomechanical investigations on postendodontically rehabilitated teeth. *J. Biomed. Opt.* **2002**, *7*, 262–270. [CrossRef] [PubMed]

44. Kishen, A. Mechanisms and risk factors for fracture predilection in endodontically treated teeth. *Endod. Top.* **2006**, *13*, 57–83. [CrossRef]

45. Genovese, K.; Lamberti, L.; Pappalettere, C. Finite element analysis of a new customized composite post system for endodontically treated teeth. *J. Biomech.* **2005**, *38*, 2375–2389. [CrossRef] [PubMed]

 *materials*

Article

Au Nanoparticle Sub-Monolayers Sandwiched between Sol-Gel Oxide Thin Films

Enrico Della Gaspera [1], Enrico Menin [2], Gianluigi Maggioni [3], Cinzia Sada [4]
and Alessandro Martucci [2,5,*]

[1] School of Science, RMIT University, Melbourne 3000, Australia; enrico.dellagaspera@rmit.edu.au
[2] Department of Industrial Engineering, University of Padova, via Marzolo 9, Padova 35131, Italy;
     enrico.menin@unipd.it
[3] Materials and Detectors Division, INFN, Legnaro National Laboratories, Viale dell'Università,
     Legnaro 35020, Italy; Gianluigi.Maggioni@lnl.infn.it
[4] Department of Physiscs and Astronomy, University of Padova, via Marzolo 8, Padova 35131, Italy;
     cinzia.sada@unipd.it
[5] National Research Council of Italy, Institute for Photonics and Nanotechnologies, Padova, via Trasea 7,
     Padova 35131, Italy
*   Correspondence: alex.martucci@unipd.it; Tel.: +39-049-8275506

Received: 22 February 2018; Accepted: 14 March 2018; Published: 14 March 2018

Abstract: Sub-monolayers of monodisperse Au colloids with different surface coverage have been embedded in between two different metal oxide thin films, combining sol-gel depositions and proper substrates functionalization processes. The synthetized films were TiO_2, ZnO, and NiO. X-ray diffraction shows the crystallinity of all the oxides and verifies the nominal surface coverage of Au colloids. The surface plasmon resonance (SPR) of the metal nanoparticles is affected by both bottom and top oxides: in fact, the SPR peak of Au that is sandwiched between two different oxides is centered between the SPR frequencies of Au sub-monolayers covered with only one oxide, suggesting that Au colloids effectively lay in between the two oxide layers. The desired organization of Au nanoparticles and the morphological structure of the prepared multi-layered structures has been confirmed by Rutherford backscattering spectrometry (RBS), Secondary Ion Mass Spectrometry (SIMS), and Scanning Electron Microscopy (SEM) analyses that show a high quality sandwich structure. The multi-layered structures have been also tested as optical gas sensors.

Keywords: metal oxides; multi-layer; surface plasmon resonance; optical sensors

---

1. Introduction

There is a growing need for nanostructured materials with tailored optical and electrical properties, however the material itself does not always provide the required properties: for this reason, a combination of different materials with accurately controlled organization is sometimes necessary in order to enhance the device performances and/or to acquire new properties. In this regard, the combination of semiconducting oxides and noble metals has been extensively investigated for applications in several fields, including photocatalysis, sensing, optoelectronics, and energy conversion [1–3]. The presence of noble metals on the surface of metal oxides enables efficient charge separation and electron transfer in optoelectronics devices, but also enhanced optical properties if the noble metals show localized Surface Plasmon Resonance (SPR) peaks in the spectral range of interest. This is usually the case for gold and silver, which have found use in many oxide-based nanocomposites, for example, for enhanced photocatalysis and solar fuel generation [4,5].

The discovery of the strong SPR coupling of close packed Au and Ag nanoparticles (NPs) [6], which leads to an increase of the intensity of the local electromagnetic field in the immediate surroundings of the metal particles, has driven an additional research effort that is devoted to the precise assembly of plasmonic NPs and their integration within optoelectronic devices. Several reports have been published discussing the distinctive optical and electrical properties of two-dimensional arrays of Au NPs, which can be exploited for different applications for example in Surface Enhanced Raman Scattering (SERS), sensing, and catalysis [7–11]. The combination of these ordered assemblies of Au NPs with catalytically and/or electrically active materials, such as semiconducting metal oxides, can generate a synergistic effect between the two components, enhancing the overall nanocomposite properties, for example, in optical recognition of reducing gases and Volatile Organic Compounds (VOCs) [9,12]. This nano-engineering of precisely ordered metal nanostructures and oxide surfaces can be achieved with a variety of experimental techniques, including lithography, sputtering, Chemical Vapor Deposition (CVD), and ion implantation. However, all of these techniques require either complex synthetic procedures or expensive equipment, and sometimes both. In this work, we present a simple and straightforward approach to synthesize high quality oxide/metal nanocomposites where plasmonic nanoparticles are assembled in a close-packed fashion, and interfaced two different metal oxides. We expand on our previous study on Au colloids deposited on properly functionalized substrates, and then covered with metal oxides [10], and by using only wet-chemistry techniques, we fabricate sub-monolayer of Au NPs that are sandwiched between two metal oxide layers. In detail, layers of monodisperse Au NPs are deposited over a semiconductive sol-gel film (NiO, TiO_2) and are then covered with a different sol-gel layer (TiO_2, NiO, ZnO). Within these structures, the Au NPs layer faces one material on one side and a different material on the other side, with potentially exciting electrical and optical properties that can find applications in several fields, including optoelectronic devices [13], sensors [14], and photovoltaics [15]. In addition to the simplicity of the presented method, such a synthetic procedure can be easily extended for many other metal oxides coatings, and to more complex multi-layered structures with different metal NPs that are embedded in between different semiconducting layers.

2. Materials and Methods

Spherical Au NPs of about 13 nm in diameter were synthesized with the Turkevich method by reducing Au ions in water at 100 °C with sodium citrate. The whole synthetic and purification protocol has been described previously [10].

To deposit a TiO_2 layer, a solution of Ethanol (0.413 mL), titanium butoxide (0.447 mL), and acetylacetone (0.216 mL) was prepared under vigorous stirring at room temperature. After 10 min, 0.1 mL Milli-Q water were added and were let stir for additional 20 min. Then, 1.83 mL ethanol was added, the total solution was let stir for 5 more minutes, and then it was used for films deposition.

To deposit the NiO layer, 300 mg of Nickel Acetate tetrahydrate were dissolved in 2 mL methanol, and subsequently 0.18 mL diethanolamine were added under stirring. After 40 min, 1.4 mL ethanol were added, and after additional 5 min, the solution was used for films deposition.

To deposit the ZnO layer, 200 mg of Zinc acetate dehydrate were dissolved in 0.9 mL ethanol, and subsequently 0.066 mL monoethanolamine (MEA) were added under stirring. After 30 min, 0.35 mL ethanol are added and the solution was used for film depositions after five more minutes of stirring.

The bottom oxide coating was deposited on either Si or SiO_2 (fused silica) substrates by spin coating with rotating speed ranging from 2000 rpm to 3000 rpm for 30 s, and then the sample was annealed directly at 500 °C for 10 min. The spinning rate was calibrated and adjusted in order to obtain films of about 45 nm after the 500 °C annealing for all of the three oxides used. The accuracy and reproducibility of the spinning procedure was tested after repeated depositions and gave a ±5 nm error on the sample thickness. To promote Au NPs bonding, the outer oxide surface is functionalized with aminopropyltrimethoxsilane (APS), after re-activation of the surface to promote formation of hydroxyl groups, which are necessary for the reaction with APS molecules (as a consequence of the thermal annealing, all of the hydroxyl groups were removed). The optimized activating procedure for NiO and TiO_2 films consisted in dipping the samples into a 4% H_2O_2 aqueous solution at room temperature for 1 min, followed by a thorough rinsing with deionized water. After this procedure, the previously reported protocols of substrate functionalization and Au NPs layer deposition were performed [10], followed by the deposition of the top oxide layer using the sol-gel recipes described earlier; eventually, the samples were thermally treated at 500 °C for one hour. The surface coverage of Au NPs was tailored simply diluting the Au colloidal solution: in this study we prepared samples with three different Au surface coverages, hereafter indicated as low (L, 6%), medium (M, 19%), and high (H, 35%). The surface coverage was estimated from SEM images, as reported in reference [10]. ZnO was not used as bottom layer because the hydroxylation protocol caused etching of the porous ZnO films, even if it was performed in milder conditions.

The films were characterized by XRD using a Philips PW1710 diffractometer (Amsterdam, The Netherlands) equipped with glancing-incidence X-ray optics. The analysis was performed at 0.5° incidence, using CuKα Ni filtered radiation at 30 kV and 40 mA. Optical absorption spectra of samples that were deposited on fused silica substrates were measured in the 300–2000 nm range using a Jasco V-570 spectrophotometer (Japan) Ellipsometry measurements were carried out on a J.A. Woollam V-VASE Spectroscopic Ellipsometer (Lincoln, NE, USA) in vertical configuration, in the 300–1500 nm range at three different angles of incidence (65°, 70°, 75°). The nanocomposites were modeled with Cauchy dispersions for the non-absorbing region, while Gaussian or Tauc-Lorentz oscillators were used for the UV absorption onset fitting. Rutherford backscattering spectrometry (RBS) was performed with an electrostatic accelerator, Van de Graaff type, using single-charged alpha particles (^{4}He$^+$) at 2.0 MeV and 20 nA. RBS analysis was performed on samples deposited on Si substrates. The incident beam was perpendicular to the sample, while the scattering angle was 160°. The surface and cross-sectional structure of the nanocomposite films were investigated with a xT Nova NanoLab Scanning Electron Microscopy (SEM). Secondary Ion Mass Spectrometry (SIMS) was exploited to measure the elemental in-depth profiles of chemical species in the deposited film. SIMS measurements were carried out by means of an IMS 4f mass spectrometer (Cameca, Padova, Italy), using a 14.5 KeV Cs$^+$ primary beam and by negative secondary ion detection. The charge build up while profiling the insulating samples was compensated by an electron gun without any need to cover the surface with a metal film. The SIMS spectra were carried out at different primary beam intensity (20 nA, stability 0.2%) rastering over a 150×150 μm^2 area and detecting secondary ions from a sub region close to 7×7 μm^2 to avoid crater effects. The primary beam was chosen in order to optimise the depth resolution and the multilayer interface determination. The signals were detected in beam blanking mode (i.e., interrupting the sputtering process during magnet stabilization periods) in order to improve the in-depth resolution. Moreover, the dependence of the erosion speed on the matrix composition was taken into account by measuring the erosion speed at various depths for each sample. The erosion speed was then evaluated by measuring the depth of the erosion crater at the end of each analysis by means of a Tencor Alpha Step profilometer with a maximum uncertainty of few nanometers (final value given by the average on 8 measures). The measurements were performed in High Mass Resolution configuration to avoid mass interference artifacts. The film thickness was determined by analysing the element signal dynamics. The error of the film thickness contains, therefore, contributes of the element inter-diffusion, of the film roughness, and finally of the technique artefacts. Optical gas sensing tests were performed by making optical absorption measurements in the 350–1500 nm wavelength range on films deposited on SiO$_2$ glass substrates using a Harrick gas flow cell (with an optical path length of 5.5 cm), coupled with a Jasco V-570 spectrophotometer. The operating temperature (OT) was set at 300 °C and gases at concentrations of 1 vol% for H$_2$ and of 1 vol% for CO in dry air at a flow rate of 0.4 L/min were used. The incident spectrophotometer beam was set normal to the film surface and illuminated an area of ~13 mm^2.

3. Results and Discussion

As described in the experimental section, the different samples that were prepared consist on a bottom layer (TiO$_2$, NiO), an intermediate layer of Au NPs with different surface coverage, and a top layer (TiO$_2$, NiO, ZnO). The Au colloids surface coverage can be easily tuned by changing the NPs concentration in the spin coating solution (more concentrated solutions lead to greater surface coverages) or by modifying the spinning speed (increasing the rotational speed leads to lower surface coverages).

Optical spectroscopy is a powerful tool to investigate Au NPs amount and organization in between the two oxide layers. Figure 1 shows the absorption spectra of some of the prepared nanostructures: for all of the NPs-containing samples the SPR peak appears in the visible-near IR range (Figure 1a,b). As can be seen in Figure 1a, an increase in intensity of the SPR peak component at higher wavelengths with increasing Au NPs surface coverage is observed, as already reported previously for bare Au NPs layers [8,9], and for Au NPs layers that were covered with metal oxides [10]. This effect is due to the reduced mutual distance between close-packed Au NPs, which leads to a stronger coupling of the plasmon resonances [6]. The broad absorption feature of the Au-free sandwich structure is due to optical interference because of the high refractive index of the oxide films (see below).

By comparing a sandwich structure that was prepared in this study with Au NPs layers that were deposited on glass with the same particles density and covered with only one metal oxide (Figure 1b), it can be noticed that the optical features of Au NPs that are embedded in between two metal oxides (in this case NiO as bottom layer and ZnO as top layer) are effectively in between the properties of the Au-NiO and Au-ZnO systems: in fact, the SPR peak of Au NPs deposited on glass and covered with ZnO is registered at about 605 nm, while when the Au colloids deposited on glass are covered with NiO, the SPR frequency is 690 nm. The plasmon peak of the Au NPs embedded in between the two oxides is definitely blue shifted compared to Au-NiO films, while due to the low frequency component related to plasmon coupling of neighboring NPs, it is difficult to appreciate the red shift compared to Au-ZnO. Nevertheless, the low frequency component of the SPR band is definitely red shifted in the sandwich structure compared to the Au-ZnO composite.

The blue or red shift observed in the SPR band is related to the difference in refractive index between the two oxides: NiO has a higher refractive index value when compared to ZnO, as will be discussed later on along with the ellipsometry measurements. So, according to Mie theory [16], the greater the refractive index value of the matrix in which the Au NPs are embedded, the greater the SPR wavelength arising from the metal NPs.

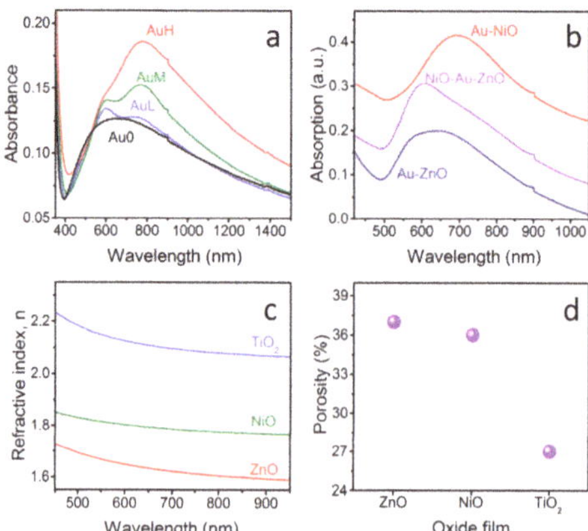

Figure 1. (**a**) Optical absorption spectra of TiO$_2$-NiO double layer with different Au nanoparticles (NPs) amount embedded in between. Au surface coverages: L = 6%, M = 19% and H = 35%; (**b**) Optical absorption spectra of Au-ZnO, Au-NiO and NiO-Au-ZnO nanocomposites deposited on fused silica susbtrate; (**c**) Refractive index dispersion curves for the three oxide films after the 500 °C annealing; and, (**d**) Porosity of the three oxide films evaluated with the Bruggeman model from the ellipsometric measurements.

The actual refractive index n and the thickness of the oxide layers that were deposited on glass substrates have been measured using spectroscopic ellipsometry, and the results are presented in Figure 1c. As can be seen, the refractive index values for the three oxides that are used are rather different between each other, but also they differ from the bulk values for the respective oxides. This difference is ascribed to the residual porosity of the thin films, which is a well-known effect for oxides films that are prepared from sol-gel solutions and annealed at relatively low temperatures, outside the sintering range [17–19]. For this reason, the oxide layer is modeled as an effective medium that is composed of dense matrix and pores, and through effective medium approximation (EMA) models, it is possible to evaluate the porosity amount. Using the bulk refractive index values at ~600 nm for the three oxides (n_{ZnO} = 2.01 [20]; n_{NiO} = 2.33 [21]; n_{TiO2} = 2.51 [20]), the pores volume fraction evaluated with the Bruggeman [22] relationship are 37%, 36%, and 27% for ZnO, NiO, and TiO_2, respectively. As a consequence, as can be also visualized in Figure 1c,d, the actual refractive index of the prepared samples follows the order of the bulk and dense materials, but the porosity of the TiO_2 layer is lower when compared to the other two oxides. In fact, anatase layers are more compact and smooth as compared to NiO and ZnO, as will be clarified later along with SEM characterization.

X-ray diffraction analysis gives a confirmation of the different Au amount according to the concentration of the solutions that are used for the Au layer deposition, and also verifies the crystallinity of the three oxides: all of these results are reported in Figure 2. Typical diffraction patterns for anatase TiO_2 (ICDD No. 86-1157, highlighted with ●), bunsenite NiO (ICDD No. 47-1049, highlighted with ■), wurtzite ZnO (ICDD No. 36-1451, highlighted with ▼) and cubic Au (ICDD No.04-0784, highlighted with ▲) can be easily identified in the prepared samples, according to their respective composition. Analyzing the oxide diffraction peaks, they do not undergo any relevant change from one sample to another, nor in the intensity or broadening (the full width at half maximum, FWHM, is related to the crystallite size, according to the Scherrer equation), validating the reproducibility of the different sol-gel recipes adopted. However, it has to be said that such a comparison, especially for the intensity of the diffraction peaks, is merely qualitative. In fact, although all of the samples had approximately the same films thickness and the same substrate size, the difference in XRD peaks intensity is strongly related to the thickness of the samples, the X-ray beam spot size, the careful alignment of the sample stage, because the measurements have been performed at glancing angle (0.5°), and so a quantitative comparison would be rather speculative. As far as Au diffraction peaks are concerned, few differences can be observed among the different samples: by increasing the Au NPs amount (from Low, to Medium, to High), a clear progressive increase in Au peaks intensity is detected, confirming the different surface coverage.

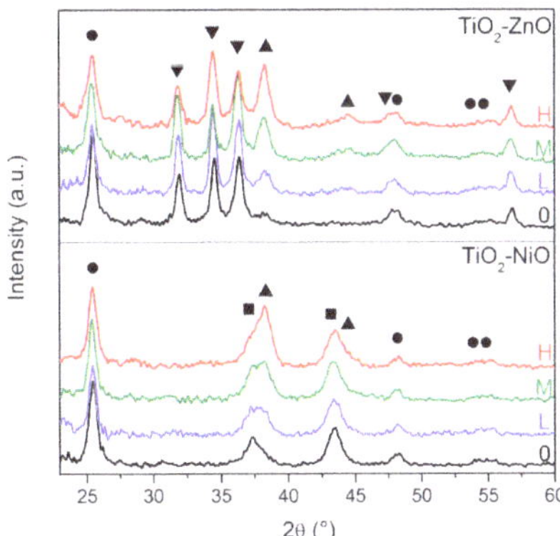

Figure 2. XRD patterns of TiO_2-NiO and TiO_2-ZnO series; Au NPs surface coverage in the different samples is highlighted with H (high), M (medium), L (low) and 0 (zero). Predicted diffraction peaks positions for TiO_2 (●), NiO (■), ZnO (▼) and Au (▲) according to the respective ICDD cards are also reported.

XRD has also been adopted to evaluate the effect of the activation of the bottom oxide layer before performing the APS functionalization process: as described in the experimental section, NiO and TiO$_2$ films were immersed in a hydrogen peroxide dilute solution in order to create –OH surface bonds. XRD measurements performed before and after the etching treatment (not reported) do not show any modification of the oxide diffraction peaks, nor in intensity or FWHM, excluding any change in the morphology and chemical composition of the oxide layers.

NiO-TiO$_2$ sandwich structures—with NiO as bottom layer and TiO$_2$ as top layer—with and without Au NPs, have been characterized with Rutherford Backscattering Spectrometry (RBS): This technique is useful to gain information about thickness, composition, and spatial distribution over thickness of the different components. The spectrum of the Au-free sample (Figure 3) shows two distinct peaks, which are centered at about 1.44 MeV and 1.5 MeV, which can be ascribed to Ti and Ni signals, respectively. The predicted energy positions for Ti and Ni (with the experimental setup used) are 1.44 MeV and 1.53 MeV, respectively: Ni signal is found at lower energies because the NiO layer is slightly far from the surface, so it is probed after the TiO$_2$ film. A simulation has been performed when considering a simple sandwich structure composed of a bottom layer of NiO and a top layer of TiO$_2$, letting the thickness vary: the best fit was obtained with a TiO$_2$ layer of 38 nm and a NiO film of 40 nm. The two values are extremely close to each other, confirming the correct choice of the deposition parameters in order to get similar thicknesses, even if the thickness values are slightly lower when compared to the expected ones (about 45 nm), as measured by ellipsometry and SEM analyses (see below): this is because both SEM and ellipsometry take into account the porosity of the films, while the RBS technique is based on nominal density for the different materials, and measuring the atoms/cm^2 values, the apparent thickness evaluated with RBS is reliable only if measuring fully dense materials. Nonetheless, having obtained similar thicknesses for both oxide coatings is a further proof of the accuracy of the experimental procedure.

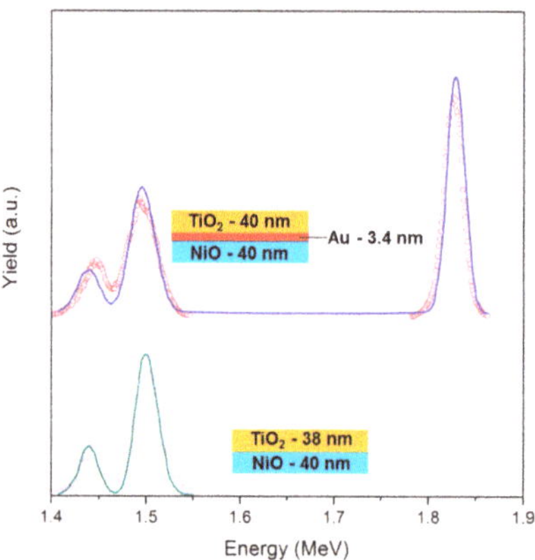

Figure 3. RBS spectra (open circles) and respective simulations (straight lines) of NiO-TiO$_2$ sample with and without an interlayer of Au NPs. The schematic representation of the layer stacks used for the fitting is also shown.

The same sample architecture, but with a layer of Au NPs in between the two oxides (Figure 3), shows the same two peaks at 1.44 MeV and 1.5 MeV, and an additional peak at 1.82 MeV, due to metallic Au. These experimental data have been modeled using a bottom layer of fully dense NiO of 40 nm thickness, an intermediate layer of Au (3.4 nm) and a top layer of dense TiO$_2$ of 40 nm. Again, the two oxide films are of the same thickness, slightly lower than the expectations due to the porosity effect described before. Since the software that is used for the fitting procedure does not take into account the possibility of having a layer composed of NPs, the simulation has been carried out with a bulk gold layer, obtaining a thickness of 3.4 nm as best fit. Using the integral of the Au peak, the dose of Au atoms can be estimated, being it 2×10^{16} at/cm^2; knowing the actual size of the Au NPs (13 nm), and using simple mathematics it is possible to estimate the Au NPs surface coverage, being it about 2.9×10^{11} NPs/cm^2. When considering the area of a single Au NP having a diameter of 13 nm, the estimated surface coverage is about 38%. This value, although being affected by a considerable error due to the simple calculations that for example do not take into account NPs size dispersity and crystalline structure, is quite close to the surface coverage value that was evaluated from a bare Au NPs layer deposited using the same experimental parameters (34%). So, the surface coverage of Au NPs is thereby qualitatively confirmed.

SEM characterization has been carried out performing the measurements in top view and in cross section, in order to evaluate the presence of the two layers, their thickness and morphology, and to examine the Au NPs distribution across the samples; all of the results are reported in Figure 4. Figure 4a–c shows the Au NPs layer embedded between TiO$_2$ and NiO films: anatase film is the bottom layer in Figure 4a, and the top layer in Figure 4b,c. The morphology of the two oxides is clearly different: TiO$_2$ films are more compact and smooth, while NiO layers have a more structured morphology, with the crystalline grains being clearly identifiable. Moreover, from the morphological difference between the two oxides, it seems that the NiO film has a higher porosity when compared to the TiO$_2$ layer, and effectively this has been confirmed by the ellipsometric evaluation discussed before. The thicknesses evaluated from the SEM images is in good agreement with the predicted values: In Figure 4a, TiO$_2$ and NiO films have been measured to be around 45 nm and 47 nm thick, respectively, while in Figure 4b, samples the evaluated thickness is 43 nm and 46 nm, respectively. Therefore, the target thickness of ~45 nm is confirmed. Au NPs can be seen as brighter spots, but since the difference in contrast with NiO crystals is quite low, it is sometimes difficult to distinguish them. Nevertheless, especially in Figure 4b, few bright particles in between the two oxide films can be recognized. From the low magnification image (Figure 4c), the high quality of the sandwich structure over few microns can be appreciated, and also Au NPs as brighter spots can be seen throughout the whole image, in between the two oxides.

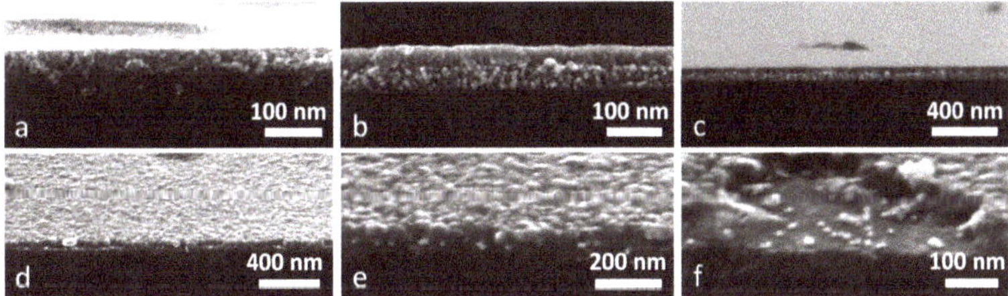

Figure 4. (**a**–**c**) SEM images of Au NPs layers embedded between TiO$_2$ and NiOfilms: TiO$_2$ is the bottom layer in (**a**) and the top layer in (**b**,**c**); (**d**–**f**) SEM images of Au NPs layers embedded between TiO$_2$ (bottom) and ZnO (top) at different magnifications. Brighter spots correspond to Au NPs.

Figure 4d–f show some images of a sample composed of Au NPs that are embedded between a bottom TiO$_2$ layer and a top ZnO layer. Again, the difference in morphology between the two oxides can be seen, being ZnO rougher and TiO$_2$ smoother, but also, Au NPs can be clearly seen due to the higher contrast difference. Bright circular spots exactly at the ZnO-TiO$_2$ interface are seen in all three images, and their size has been estimated in the 10 nm–15 nm range, as consistent with the value of the as-synthesized particles. We previously reported that Au NPs deposited on glass tend to sinter upon thermal treatment, but when the Au NPs are covered with a metal oxide film, this provides a physical barrier that strongly reduces the temperature-driven sintering [10]. Figure 4f shows a picture of the double layer where a portion of the ZnO layer is missing (probably as a consequence of the sample cutting and handling): few Au NPs that are deposited over the TiO$_2$ film can be easily seen, giving another proof of the metal NPs presence at the interface between the two oxides. Moreover, the thickness of TiO_2 and ZnO layers has been evaluated as well, being 46 nm and 51 nm, respectively; again, the predicted thickness is hereby confirmed, even if the ZnO film is slightly thicker than expected, possibly because its high surface roughness makes the precise evaluation of the thickness quite challenging.

SIMS provides another confirmation of the actual structure of the layered films (Figure 5): we evaluated the compositional depth profiles for two TiO_2-ZnO films that were deposited on silicon substrates, with (b) and without (a) an Au NPs layer that is embedded in between. The total samples thickness has been estimated to be around 80 nm (based on Si and O signals), in good agreement with previous characterizations. The Zn signal is detected in both samples at the surface, while the Ti signal is centered around 40–60 nm far from the surface. The sample containing Au NPs (Figure 5b) shows Au signal that is centered about 20 nm from the surface, not exactly in between the two oxides, but slightly closer to the surface. This is understandable when considering the structure and morphology of the sample (Figure 4d–f), where it can be seen that Au NPs are laying on top of the anatase layer, and they are surrounded and submerged by the top ZnO coating. The schematic presented in Figure 5c shows the sample structure, highlighting the position of Au NPs. Moreover, with such thin layered structures the thickness estimation using SIMS is affected by a substantial error. However, a further confirmation of the results that were presented in the previous characterizations has been obtained.

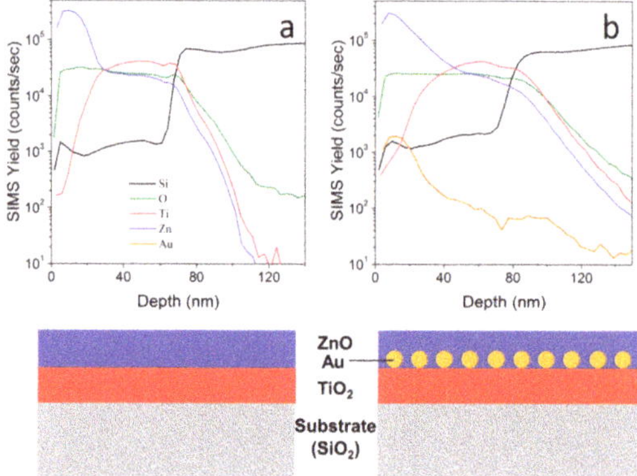

Figure 5. Secondary Ion Mass Spectrometry (SIMS) data showing depth profiles of different ions for (**a**) TiO_2-ZnO and (**b**) TiO_2-Au-ZnO thin films. A schematic representation of the respective sample structure is shown at the bottom.

We already studied the optical gas sensing properties of NiO, ZnO, and TiO_2 film containing Au NPs [23,24], showing how the interaction of the target gas with the metal oxide matrix can be monitored by looking at the SPR of the Au NPs. Here, we embedded the Au NPs between two different metal oxide layers for studying the effect, if any, of their coupling on the optical gas sensing properties. For this preliminary study, H_2 and CO have been tested as target gas because they were also used in our previous study on single metal oxides.

Among the different synthetized multilayer structures, the NiO-Au-TiO_2 (NAT) and TiO_2-Au-NiO (TAN) multilayers have been selected for the gas sensing measurements, because in our previous studies, the NiO-Au and TiO_2-Au films showed good optical gas sensing properties toward H_2 and CO [23,24].

Figure 6 shows the absorption spectra of the two multi-layer structures and their Optical Absorbance Change (OAC) parameter, defined as the difference between absorbance during gas exposure and absorbance in air (OAC = Abs_{Gas} − Abs_{Air}). The two samples respond rather differently: outside the 600–900 nm range, a decrease in absorption when exposed to both gases is seen for both samples, because of the interaction of the reducing gas with the NiO film [23]. Inside the 600–900 nm range, the NAT sample shows a sharp and strongly wavelength dependent signal, which is more intense for H_2 when compared to CO, while for the TAN sample, only a weak modulation of the OAC curve is observed. This large difference can be related to the optical absorption spectra of the two samples (Figure 6a): NAT sample shows a narrow and sharp Au SPR peak, possibly due to a partial detachment of Au NPs when depositing the top TiO_2 layer, while TAN optical spectrum presents a much broader, weaker, and red shifted plasmon peak. As a consequence, the difference from the spectra collected during gas exposure and the spectra collected in air, i.e., the OAC parameter, is strongly affected, being much higher when the optical spectrum has steep features, and much lower when the optical spectrum has plainer features.

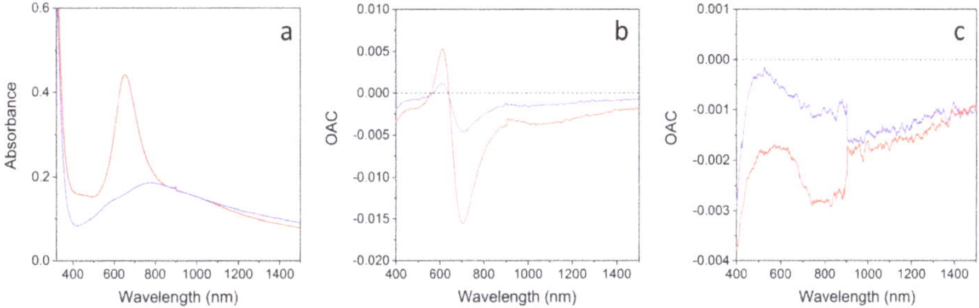

Figure 6. (**a**) Absorbance spectra of NiO-Au-TiO_2 (NAT) (red line) and TiO_2-Au-NiO (TAN) (blue line). Optical Absorbance Change (OAC) curves for (**b**) NAT sample and (**c**) TAN sample after exposure to 1% CO (blue lines) and 1% H_2 (red lines) at 300 °C OT.

In any case, for both of the samples, a reversible response for both gases is observed and some distinctive wavelengths corresponding to maximum and minimum (or null) response can be identified, theoretically permitting selective gas recognition through an appropriate choice of the analysis wavelength [25,26].

Figure 7 shows time-resolved tests at a fixed wavelength for multiple air-gas-air cycles. The wavelengths have been selected for obtaining a very high signal for H_2 and the smallest signal for CO, in order to demonstrate the selectivity of the sensor. An easily detectable signal for both of the gases is observed, which is much higher for hydrogen when compared to CO, as predicted from OAC curves (see Figure 6b,c), with relatively fast response times (between 30 s and 60 s) and acceptable recovery times (between 60 s and 90 s).

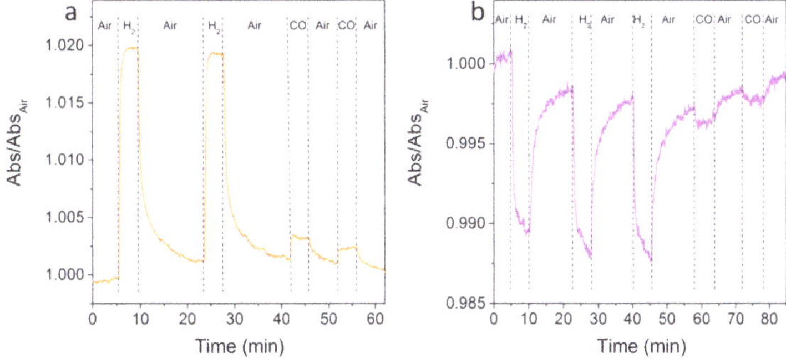

Figure 7. Dynamic tests after several air-gas-air cycles for: (**a**) NAT sample at 609 nm; (**b**) TAN sample at 580 nm.

4. Conclusions

Gold NPs have been successfully embedded at the interface between two different semiconducting oxides with an easy and straightforward procedure: first, monodisperse Au nanocrystals are synthesized with standard colloidal techniques, purified, and then deposited over a pre-functionalized sol-gel based metal oxide thin film; eventually, this structure is covered with a second oxide layer. The surface coverage of Au colloids can be easily tuned, and optical spectroscopy measurements that are coupled to morphological characterizations confirm the successful embedding of the metal spheres in between the two oxides, with the predicted surface coverage. In this nano-architecture, Au NPs are facing two different materials, with possible new interesting properties due to the multiple noble metal/metal oxide interfaces. These multilayered structures represent a high level of materials engineering, providing accurate control on NPs morphology, organization, and proper interface with the desired semiconducting material. Moreover, this process can be easily extended to a great variety of multilayered structures, which can find applications in several fields, including optical sensors, catalysts, and optoelectronic devices in general.

Acknowledgments: This work has been supported through Progetto Strategico PLATFORMS of Padova University. Enrico Della Gaspera thanks RMIT University for funding through the Vice Chancellor's Fellowship scheme. Alessandro Martucci acknowledges support from the RMIT Philanthropic Fund through an International Visiting Fellowship.

Author Contributions: Enrico Della Gaspera and Alessandro Martucci were involved with all aspects of the study including conceiving, designing, data interpretation and writing the manuscript. Enrico Menin prepared the samples and performed the gas sensing tests. Gianluigi Maggioni performed the RBS experiments; Cinzia Sada performed the SIMS experiments; all the authors read and approved the manuscript.

Conflicts of Interest: The authors declare no conflict of interest.

References

1. Ng, C.; Cadusch, J.J.; Dligatch, S.; Roberts, A.; Davis, T.J.; Mulvaney, P.; Gómez, D.E. Hot Carrier Extraction with Plasmonic Broadband Absorbers. *ACS Nano* **2016**, *10*, 4704–4711. [CrossRef] [PubMed]
2. Venditti, I. Gold nanoparticles in photonic crystals applications: A review. *Materials* **2017**, *10*, 97. [CrossRef] [PubMed]
3. Della Gaspera, E.; Martucci, A. Sol-gel thin films for plasmonic gas sensors. *Sensors* **2015**, *15*, 16910–16928. [CrossRef] [PubMed]
4. Wu, M.; Chen, W.-J.; Shen, Y.-H.; Huang, F.-Z.; Li, C.-H.; Li, S.-K. In situ growth of matchlike ZnO/Au plasmonic heterostructure for enhanced photoelectrochemical water splitting. *ACS Appl. Mater. Interfaces* **2014**, *6*, 15052–15060. [CrossRef] [PubMed]
5. Valdés, A.; Brillet, J.; Grätzel, M.; Gudmundsdóttir, H.; Hansen, H.A.; Jónsson, H.; Klüpfel, P.; Kroes, G.-J.; Le Formal, F.; Man, I.C.; et al. Solar hydrogen production with semiconductor metal oxides: New directions in experiment and theory. *Phys. Chem. Chem. Phys.* **2012**, *14*, 49–70. [CrossRef] [PubMed]
6. Freeman, R.G.; Grabar, K.C.; Allison, K.J.; Bright, R.M.; Davis, J.A.; Guthrie, A.P.; Hommer, M.B.; Jackson, M.A.; Smith, P.C.; Walter, D.G.; et al. Self-assembled metal colloid monolayers: An approach to SERS substrates. *Science* **1995**, *267*, 1629–1632. [CrossRef] [PubMed]
7. Haes, A.J.; Van Duyne, R.P. A nanoscale optical biosensor: sensitivity and selectivity of an approach based on the localized surface plasmon resonance spectroscopy of triangular silver nanoparticles. *J. Am. Chem. Soc.* **2002**, *124*, 10596–10604. [CrossRef] [PubMed]
8. Gao, S.; Koshizaki, N.; Tokuhisa, H.; Koyama, E.; Sasaki, T.; Kim, J.K.; Ryu, J.; Kim, D.S.; Shimizu, Y. Highly stable Au nanoparticles with tunable spacing and their potential application in surface plasmon resonance biosensors. *Adv. Funct. Mater.* **2010**, *20*, 78–86. [CrossRef]
9. Buso, D.; Palmer, L.; Bello, V.; Mattei, G.; Post, M.; Mulvaney, P.; Martucci, A. Self-assembled gold nanoparticle monolayers in sol-gel matrices: Synthesis and gas sensing applications. *J. Mater. Chem.* **2009**, *19*, 2051–2057. [CrossRef]

10. Della Gaspera, E.; Karg, M.; Baldauf, J.; Jasieniak, J.; Maggioni, G.; Martucci, A. Au Nanoparticle Monolayers Covered with Sol–Gel Oxide Thin Films: Optical and Morphological Study. *Langmuir* **2011**, *27*, 13739–13747. [CrossRef] [PubMed]

11. Larsson, E.M.; Langhammer, C.; Zorić, I.; Kasemo, B. Nanoplasmonic probes of catalytic reactions. *Science* **2009**, *326*, 1091–1094. [CrossRef] [PubMed]

12. Brigo, L.; Cittadini, M.; Artiglia, L.; Rizzi, G.A.; Granozzi, G.; Guglielmi, M.; Martucci, A.; Brusatin, G. Xylene sensing properties of aryl-bridged polysilsesquioxane thin films coupled to gold nanoparticles. *J. Mater. Chem. C* **2013**, *1*, 4252–4260. [CrossRef]

13. Ung, T.; Liz-Marzan, L.M.; Mulvaney, P. Optical properties of thin films of Au@SiO$_2$ particles. *J. Phys. Chem. B* **2001**, *105*, 3441–3452. [CrossRef]

14. Ruiz, A.M.; Cornet, A.; Shimanoe, K.; Morante, J.R.; Yamazoe, N. Effects of various metal additives on the gas sensing performances of TiO$_2$ nanocrystals obtained from hydrothermal treatments. *Sens. Actuators B* **2005**, *108*, 34–39. [CrossRef]

15. Hagfeldt, A.; Gratzel, M. Molecular photovoltaics. *Acc. Chem. Res.* **2000**, *33*, 269–277. [CrossRef] [PubMed]

16. Kreibig, U.; Vollmer, M. *Optical Properties of Metal Cluster*; Springer: Berlin/Heidelberg, Germany, 1995; pp. 41–53.

17. Kołodziejczak-Radzimska, A.; Jesionowski, T. Zinc oxide-from synthesis to application: A review. *Materials* **2014**, *7*, 2833–2881. [CrossRef] [PubMed]

18. Bandyopadhyay, S.; Paul, G.K.; Sen, S.K. Study of optical properties of some sol–gel derived films of ZnO. *Sol. Energy Mater. Sol. Cells* **2002**, *71*, 103–113. [CrossRef]

19. Brezesinski, T.; Wang, J.; Polleux, J.; Dunn, B.; Tolbert, S.H. Templated nanocrystal-based porous TiO$_2$ films for next-generation electrochemical capacitors. *J. Am. Chem. Soc.* **2009**, *131*, 1802–1809. [CrossRef] [PubMed]

20. Lide, D.R. *CRC Handbook of Chemistry and Physic*, 87th ed.; Taylor and Francis: Boca Raton, FL, USA, 2007; pp. 10–248.

21. Powell, R.J.; Spicer, W.E. Optical properties of NiO and CoO. *Phys. Rev. B* **1970**, *2*, 2182–2193. [CrossRef]

22. Bruggeman, D.A.G. Berechnung verschiedener physikalischer Konstanten von heterogenen Substanzen. I. Dielektrizitätskonstanten und Leitfähigkeiten der Mischkörper aus isotropen Substanzen. *Ann. Phys.* **1935**, *24*, 636–664. [CrossRef]

23. Della Gaspera, E.; Guglielmi, M.; Martucci, A.; Giancaterini, L.; Cantalini, C. Enhanced optical and electrical gas sensing response of sol-gel based NiO-Au and ZnO-Au nanostructured thin films. *Sens. Actuators B* **2012**, *164*, 54–63. [CrossRef]

24. Della Gaspera, E.; Mura, A.; Menin, E.; Guglielmi, M.; Martucci, A. Reducing gases and VOCs optical sensing using surface plasmon spectroscopy of porous TiO$_2$-Au colloidal films. *Sens. Actuators B* **2013**, *187*, 363–370. [CrossRef]

25. Della Gaspera, E.; Buso, D.; Guglielmi, M.; Martucci, A.; Bello, V.; Mattei, G.; Post, M.; Cantalini, C.; Agnoli, S.; Granozzi, G.; et al. Comparison study of conductometric, optical and SAW gas sensors based on porous sol-gel silica films doped with NiO and Au nanocrystals. *Sens. Actuators B* **2010**, *143*, 567–573. [CrossRef]

26. Della Gaspera, E.; Guglielmi, M.; Agnoli, S.; Granozzi, G.; Post, M.L.; Bello, V.; Mattei, G.; Martucci, A. Au Nanoparticles in Nanocrystalline TiO$_2$-NiO Films for SPR-Based, Selective H$_2$S Gas Sensing. *Chem. Mater.* **2010**, *22*, 3407–3417. [CrossRef]

Article

Sol-Gel Derived Active Material for Yb Thin-Disk Lasers

Rui M. Almeida *, Tiago Ribeiro and Luís F. Santos *

Centro de Química Estrutural/DEQ, Instituto Superior Técnico, Universidade de Lisboa, Av. Rovisco Pais, 1049-001 Lisboa, Portugal; tiago.velez.ribeiro@tecnico.ulisboa.pt
* Correspondence: rui.almeida@tecnico.ulisboa.pt (R.M.A.); luis.santos@tecnico.ulisboa.pt (L.F.S.)

Received: 27 July 2017; Accepted: 28 August 2017; Published: 2 September 2017

Abstract: A ytterbium doped active material for thin-disk laser was developed based on aluminosilicate and phosphosilicate glass matrices containing up to 30 mol% $YbO_{1.5}$. Thick films and bulk samples were prepared by sol-gel processing. The structural nature of the base material was assessed by X-ray diffraction and Raman spectroscopy and the film morphology was evidenced by scanning electron microscopy. The photoluminescence (PL) properties of different compositions, including emission spectra and lifetimes, were also studied. Er^{3+} was used as an internal reference to compare the intensities of the Yb^{3+} PL peaks at ~1020 nm. The Yb^{3+} PL lifetimes were found to vary between 1.0 and 0.5 ms when the Yb concentration increased from 3 to 30 mol%. Based on a figure of merit, the best active material selected was the aluminosilicate glass composition 71 SiO_2-14 $AlO_{1.5}$-15 $YbO_{1.5}$ (in mol%). An active disk, ~36 µm thick, consisting of a Bragg mirror, an aluminosilicate layer doped with 15 mol% Yb and an anti-reflective coating, was fabricated.

Keywords: sol-gel; thin-disk laser; Yb-doped glasses; aluminosilicate glasses; photoluminescence

---

1. Introduction

The use of laser technology in industrial and scientific applications is widely spread, given its added value in tasks such as welding, cutting or marking, where the speed and quality of the laser beam are determinant, thus influencing the competitiveness of the industry. Among the various types of lasers currently available, dielectric solid state lasers play an important role due mainly to the beam quality. They normally use a crystalline material as the lasing medium and, contrarily to semiconductor diode lasers, they are usually optically pumped. Fiber and disk lasers, in particular Yb^{3+}-doped ones, such as Yb:YAG, present high average output power, excellent beam quality and high efficiency [1,2]. Due to their unique features, thin-disk lasers are one of the best laser solutions for material micro-processing in automotive, aerospace, and heavy industry, with laser powers of up to 16 kW and beam qualities ≥2 mm·mrad [3], or for military applications as recently demonstrated by Boeing, where more than 30 kW power was attained [4].

The thin-disk laser concept was developed by Adolf Giesen et al. [5], who used a thin laser crystal disc with one face mounted on a water cooled block. The use of a very small thickness of active material (~100–200 µm), corresponding to a high surface area to volume ratio, leads not only to superior beam quality, but also to highly efficient cooling, resulting in almost negligible thermal gradients. Moreover, the residual gradient is symmetrical, which also contributes to an optimal beam quality. However, the small thickness of the active material results in an insufficient absorption length that must be compensated by a multi-pass pumping scheme [6]. The gain medium is typically a thin disk of Yb:YAG crystal grown by the Czochralski method, or another ytterbium-doped medium such as $Yb:Lu_3Al_5O_{12}$ or $Yb:Lu_2O_3$ [7,8], with a highly reflective coating on one side and an anti-reflective coating on the other side. The advantages of Yb-doped active material include the simple electronic structure of Yb^{3+}, with only one excited state level ($^2F_{5/2}$) with a lifetime of the order of 1–2 ms and ~10,000 cm^{-1} above the ground state ($^2F_{7/2}$), therefore with low non-radiative decay rates in most solid matrices and the possibility of achieving high doping levels without excessive luminescence quenching [9]. However, cutting such a small crystal thickness presents problems that could be overcome with a different approach: the active medium could be prepared by sol-gel (SG) processing. This is a relatively cheap wet chemistry technique which involves the hydrolysis and condensation of precursor species like organometallic compounds and nitrates in the presence of a catalyst, forming a porous gel structure that can be densified by subsequent heat treatment. Thick films can be doped with different concentrations of an active species like Yb and deposited on an appropriate substrate, e.g., a single crystal Si wafer. Moreover, anti-reflective and highly reflective coatings can be obtained as 1-D Photonic Bandgap Structures (PBGs), or photonic crystals (PCs), which are multilayered structures at the optical nanoscale that control the propagation of light, including the inhibition or enhancement of spontaneous emission (SE) of light at certain wavelengths [10,11]. Such structures may be deposited with high optical quality by the SG process [12].

Recently, our research group obtained polycrystalline Yb:YAG ceramic films as well as Yb-doped silicate glass films by the sol-gel process, with potential application as the active material in thin-disk lasers [13,14]. In the present work, multilayered structures based on Yb-doped SiO_2-Al_2O_3 or SiO_2-P_2O_5 oxide glass matrices were investigated and active films prepared; the purpose of the addition of Al or P to the silica matrix was to create non-bridging oxygen species to help disperse the Yb^{3+} ions homogeneously in the glass network, reducing possible concentration quenching phenomena.

2. Results

2.1. X-Ray Diffraction

In order to avoid the presence of scattering centers, it is important to determine whether the heat treatments of the bulk samples (up to ~1050 °C) or the films (up to ~1100 °C) may cause any crystallization. The difference in the temperatures used was due to the fact that multilayer films are treated sequentially, with fast heating from room temperature to the final heat treatment temperature, while bulk samples are monolithic and slowly heated until the final temperature, in order to allow complete combustion of the organic residues. For this purpose, both films and powdered bulk samples were analyzed by XRD. Figure 1 includes the XRD patterns of four aluminosilicate bulk samples, showing the amorphous nature of the aluminosilicates up to 15 mol% $YbO_{1.5}$, whereas higher Yb concentrations led to partial crystallization, namely of the $Yb_2Si_2O_7$ phase. The same behavior had already been observed for phosphosilicate glass films, where partial crystallization occurred for compositions with more than 5 mol% $YbO_{1.5}$, as reported in [14]. Figure 2 shows the X-ray diffraction (XRD) patterns of Yb-doped aluminosilicate films plus a silicon wafer reference, where partial crystallization is also observed for the 20 mol% $YbO_{1.5}$ film, while the amorphous character of a film with 15 mol% $YbO_{1.5}$ is confirmed. The patterns are truncated at $2\theta = 68°$, due to the main Si peak at 69.2°, corresponding to the {400} planes.

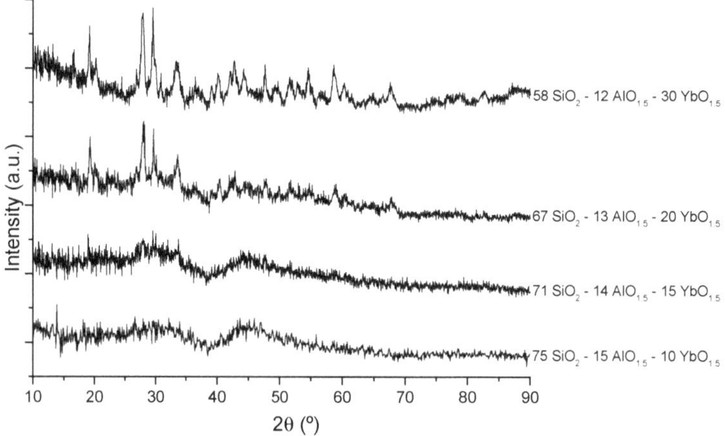

Figure 1. X-ray diffraction patterns of sol-gel derived Yb-doped aluminosilicate bulk samples, heat treated at 1050 °C.

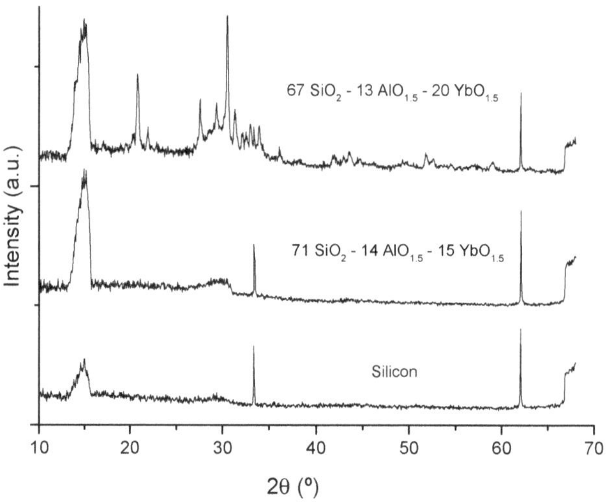

Figure 2. X-ray diffraction patterns of silicon and two aluminosilicate films with 15 and 20 mol% YbO$_{1.5}$, heat treated at 1100 °C.

2.2. Raman Spectroscopy

Figure 3 shows the Raman spectra of the heat treated Yb-doped aluminosilicate bulk glass samples. The Raman spectra, up to 15 mol% ytterbia, are relatively similar to that of pure vitreous silica, with the growth of a peak near 940 cm^{-1} with increasing Yb concentrations, probably due to the formation of Si-O$^-$ non-bridging oxygen bonds with the Yb^{3+} ions, rather than Si-O-Al sequences as previously assigned [14], since there is no correlation with the alumina content, which remains practically constant. For heat treated aluminosilicate compositions with higher Yb contents, the appearance of several sharp peaks suggests the formation of one or more crystalline phases, in agreement with the XRD results. However, this behavior depends also on the heat treatment temperature, as observed for the 71 SiO$_2$-14

AlO$_{1.5}$-15 YbO$_{1.5}$ samples, where an increase of 50 °C is sufficient to promote significant crystallization for this composition.

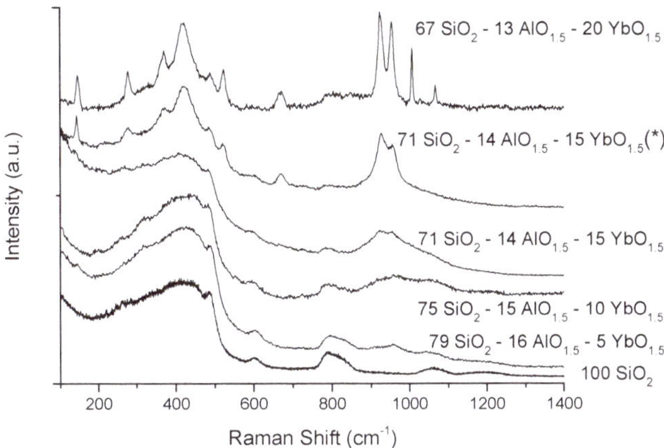

Figure 3. Raman spectra of Yb-doped aluminosilicate bulk glasses. The samples were heat treated at 1050 °C, except for the sample labelled (*) which was heat-treated at 1100 °C.

2.3. Fourier Transform Infrared (FTIR) Spectroscopy

Yb^{3+} ion absorption was measured in the near infrared (NIR), for both aluminosilicate and phosphosilicate monolithic gels. The Yb^{3+} ion is known to absorb at wavelengths near ~975 nm, due to the $^2F_{7/2} \rightarrow {}^2F_{5/2}$ transition starting from the ground state, with a shoulder near ~940 nm [14]. Figure 4 shows the Yb^{3+} absorption spectrum for the 79 SiO$_2$-16 AlO$_{1.5}$-5 YbO$_{1.5}$ bulk gel sample, before heat treatment, when it was still monolithic and transparent.

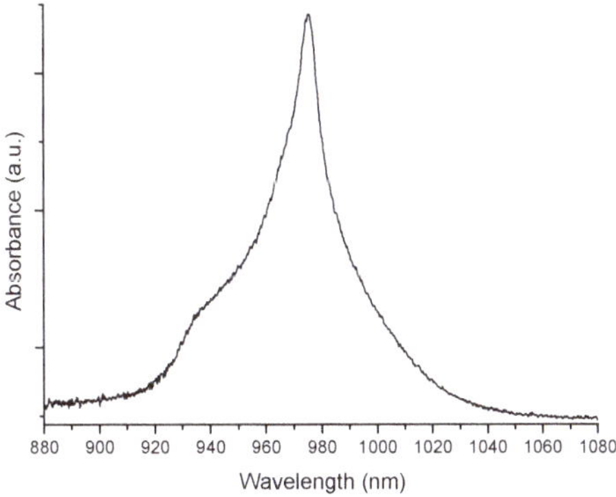

Figure 4. Near infrared absorption spectrum of 79 SiO$_2$-16 AlO$_{1.5}$-5 YbO$_{1.5}$ bulk gel.

2.4. Photoluminescence Spectroscopy

Figures 5 and 6 show the photoluminescence (PL) spectra of bulk Yb/Er co-doped phosphosilicate and aluminosilicate-matrix samples, respectively. The spectra have been normalized to the intensity of the Er^{3+} peak emission at 1532 nm. Erbium was used here as an internal reference [13,14] and its concentration, the same in all cases (0.002 mol% Er), was kept to a minimum in order to avoid any significant $Yb^{3+} \rightarrow Er^{3+}$ energy transfer through the so-called "antenna effect".

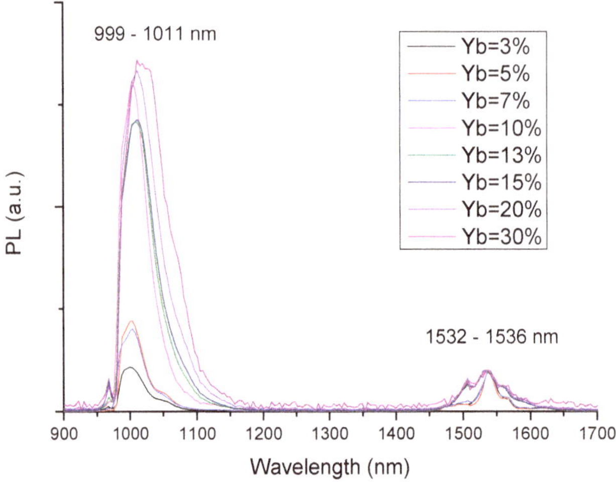

Figure 5. Normalized photoluminescence (PL) emission spectra of Yb/Er co-doped phosphosilicate bulk glass samples, excited at 970 nm.

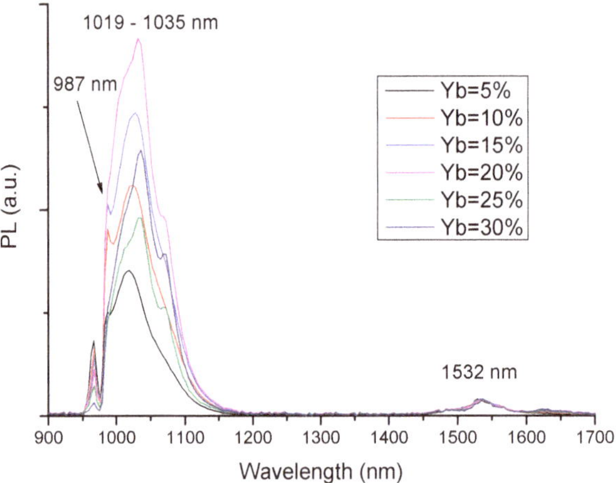

Figure 6. Normalized PL emission spectra of Yb/Er co-doped aluminosilicate bulk glass samples, excited at 970 nm.

In Figure 5, two main peaks can be observed for the phosphosilicate glasses at 999–1011 and 1532 nm, representing Yb^{3+} and Er^{3+} PL emissions due to $^2F_{5/2} \rightarrow {}^2F_{7/2}$ and $^4I_{13/2} \rightarrow {}^4I_{15/2}$, respectively. The Yb^{3+} PL peak position varied with the Yb content between 999 nm (for 3 mol% Yb) and 1011 nm (for 30 mol% $YbO_{1.5}$). The normalization at 1532 nm revealed a general increase in the PL peak intensity with increasing Yb content. A residual peak is also observed at 970 nm, due to the incompletely filtered excitation laser light.

In Figure 6, the main emission peak of Yb-doped aluminosilicate glasses is located between 1019 and 1035 nm and its intensity increased for Yb contents up to 20 mol%, whereas it decreased for concentrations above that. An additional Yb^{3+} emission peak can be observed at 987 nm, which was not resolved in the phosphosilicate glass spectra of Figure 5.

The $Yb^{3+}:^2F_{5/2}$ metastable level lifetimes were also measured for the phosphosilicate and aluminosilicate compositions. The PL decay curves had single exponential behavior as a function of time, t, as shown by the logarithmic plots of Figure 7 for aluminosilicate bulk glass samples doped with 5–30 mol% $YbO_{1.5}$. The corresponding 1/e lifetime, τ, was found to vary between 1043 and 542 µs for phosphosilicates and between 624 and 482 µs for aluminosilicates, as shown in Tables 1 and 2, respectively.

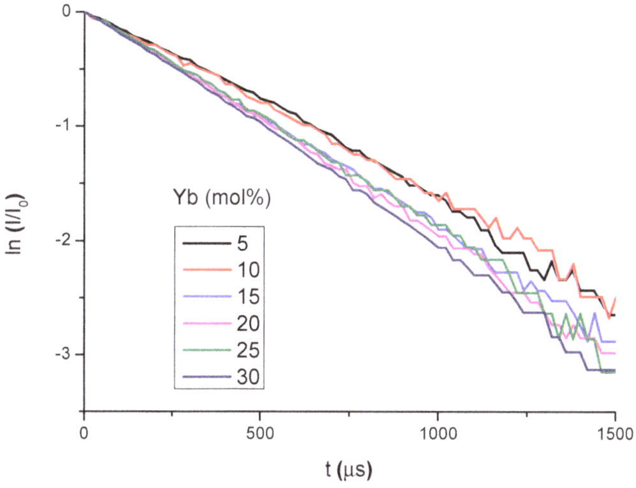

Figure 7. PL decay curves of aluminosilicate bulk glass samples doped with 5–30 mol% $YbO_{1.5}$.

Table 1. Lifetime, τ and $\sigma \times \tau$ values for Yb-doped phosphosilicate glasses.

$YbO_{1.5}$ (mol%)	τ (µs)	$\sigma \times \tau$ (µs)
3	1043	1126
5	782	1718
7	763	1528
10	735	5931
13	828	5845
15	586	4173
20	485	4039
30	542	4652

Table 2. Lifetime, τ and $\sigma \times \tau$ values for Yb-doped aluminosilicate glasses.

YbO$_{1.5}$ (mol%)	τ (μs)	$\sigma \times \tau$ (μs)
5	624	5509
10	614	8586
15	525	9648
20	511	11699
25	528	6349
30	482	7762

2.5. Active Disk

The active disk may consist of a thin disk (or thick film) of Yb-doped material, sandwiched between a highly reflective coating or Bragg Mirror (BM) and an anti-reflective coating (ARC). The characteristics of the BM, the ARC and active disk were all simulated and designed using the Transfer Matrix Method (TMM) software, developed in our group [11].

The high-reflective coatings, or BMs, were prepared by depositing seven pairs of low/high index layers of aluminosilicate glass (AS) and titania (T), respectively, directly onto the Si substrate. The molar composition 90 SiO$_2$-10 AlO$_{1.5}$ was chosen for the low-index AS material, based on previously developed 91 SiO$_2$-9 AlO$_{1.5}$ similar compositions [11]. Next, a thick Yb-doped aluminosilicate glass film with the optimized composition (71 SiO$_2$-14 AlO$_{1.5}$-15 YbO$_{1.5}$) was deposited, followed by a photonic structure with anti-reflective properties at the top of the active disk. A small number of layers were used for this ARC, in order to avoid damaging the previously deposited Yb-doped multilayer disk through the additional heat treatment routines. The final ARC structure prepared consisted of three layers only (AS/T/AS) deposited on top of the Yb-doped active disk.

2.5.1. TMM Simulations

TMM simulations allowed the calculation of the transmission and reflection of a multilayer structure within a given wavelength range, based on the thickness and refractive index of each individual layer. Such simulations were performed in order to optimize both the BM and ARC, in terms of the number of low/high index pairs and individual layer thickness. The optimized BM consisted of seven pairs of AS (90 SiO$_2$-10 AlO$_{1.5}$)/T (TiO$_2$) layers. The basic criterion is the "quarter-wave" condition Equation (1) for the optical thickness of each layer (equal to the product of the physical thickness, x and the refractive index, n) for a peak wavelength λ. The thickness of the AS (low index, $n = 1.46$) and T (high index, $n = 2.22$) layers were 175 and 115 nm, respectively, for a BM stop band with a center wavelength $\lambda = 1020$ nm (near the Yb^{3+} PL emission maximum), according to:

$$x = \frac{\lambda}{4n} \tag{1}$$

The ARC, on the other hand, designed for optimum anti-reflectivity at a pumping wavelength of 940 nm, was formed by a 42 nm titania layer sandwiched between two 180 nm aluminosilicate glass layers, whose thickness values (not "quarter-wave") were obtained from TMM simulations by trial and error.

Figure 8 compares a BM reflectance measurement with its TMM simulation, while Figure 9 compares the transmission of the ARC with the corresponding TMM simulation. In the former case, high reflectivity is observed at both the 940 nm and the 1020 nm wavelengths; the observed values were slightly higher than 100%, since the BM reflected more than the Al mirror used as reference. In the latter case, the transmission is near 100%, evidencing the ARC properties and the interference fringes are due to the ARC thickness the agreement between theory and experiment appears reasonably good in both cases.

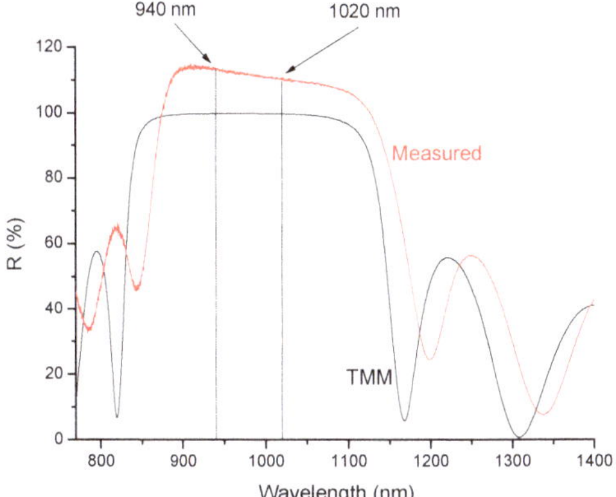

Figure 8. Measured reflectance spectrum of a Bragg Mirror (BM) with 7 pairs of AS/T layers deposited on a Si wafer, compared with Transfer Matrix Method (TMM) simulation.

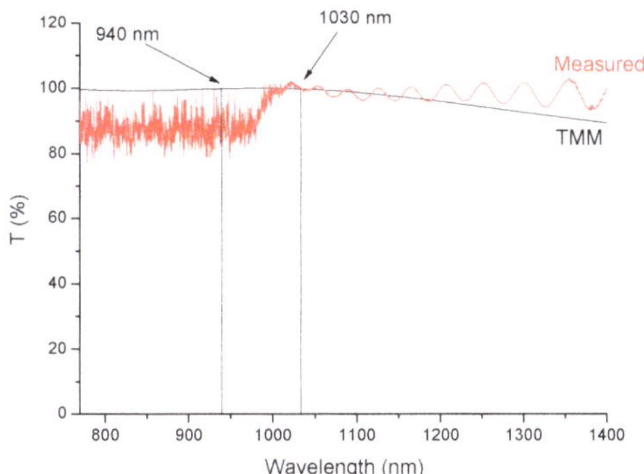

Figure 9. Transmission of an anti-reflective coating (ARC) with 3 layers deposited on 71 SiO_2-14 $AlO_{1.5}$-15 $YbO_{1.5}$ glass film deposited on silicon substrate, compared with TMM simulation.

2.5.2. Active Disk Structure

An active disk doped with 15 mol% $YbO_{1.5}$, whose structure consisted of a 7-pair BM, an active layer with molar composition 71 SiO_2-14 $AlO_{1.5}$-15 $YbO_{1.5}$ and a 3-layer ARC, deposited on single-crystal silicon, was observed by scanning electron microscopy (SEM) and Figure 10 shows the corresponding cross-sectional image. This active disk had an overall thickness of ~10 μm. In the highly reflective portion, it is possible to distinguish the individual BM layers, for a total of seven high/low index pairs. In addition, the fracture surface of the Yb-doped portion is typical of an amorphous material, without any visible grains at this magnification.

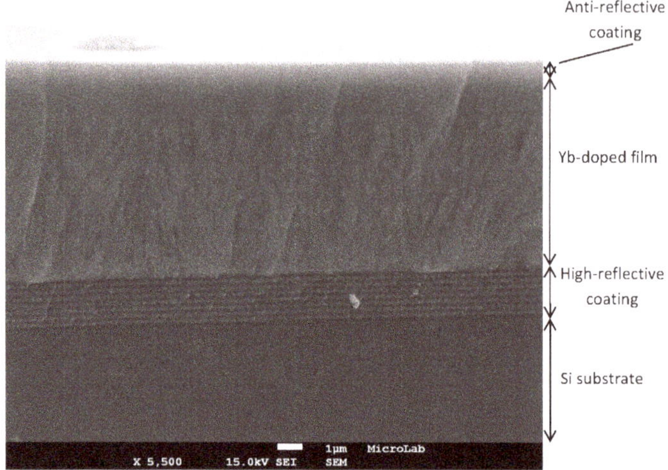

Figure 10. FEG-SEM cross section micrograph of a ~10 µm thick active disk doped with 15 mol% $YbO_{1.5}$, obtained in secondary electron mode. (Scale bar = 1 µm).

2.5.3. Active Disk Properties

Figure 11 shows the XRD patterns of a ~36 µm active disk, deposited on silicon, where there is no evidence of any residual crystalline material.

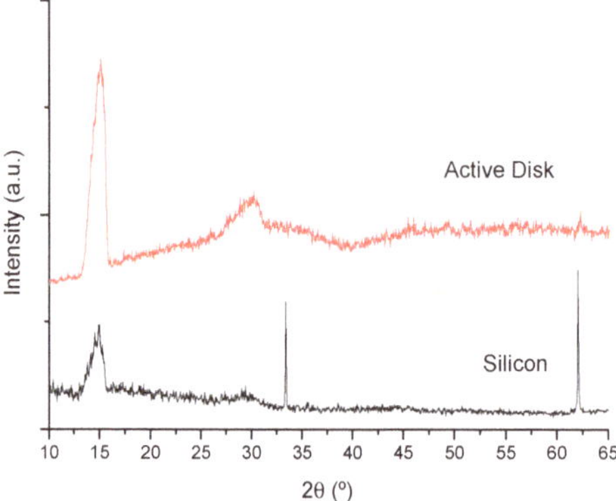

Figure 11. X-ray diffraction patterns of an active disk (BM/36.3 µm thick 71 SiO_2-14 $AlO_{1.5}$-15 $YbO_{1.5}$ film/ARC) on a silicon substrate.

The reflectance of the same active disk was measured in the NIR region (Figure 12) and a broad stop band of high reflectivity is visible over the range of ~750–1200 nm, due to the BM. A reflection-absorption effect is also observed at ~900–1000 nm, due to the Yb^{3+} absorption within the active layer.

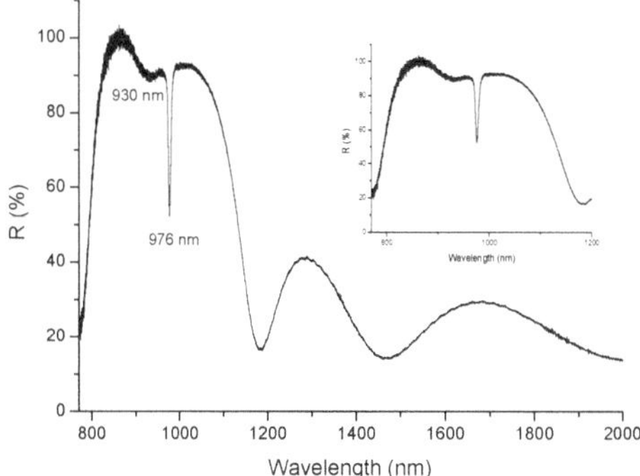

Figure 12. Reflectance of an active disk composed of a BM, an Yb-doped layer (36.3 micron thick) and an ARC coating, deposited on silicon.

The PL spectrum of the active disk is presented in Figure 13, which clearly shows the 987 nm and the major Yb^{3+} PL emission peaks, in addition to the residual excitation laser line.

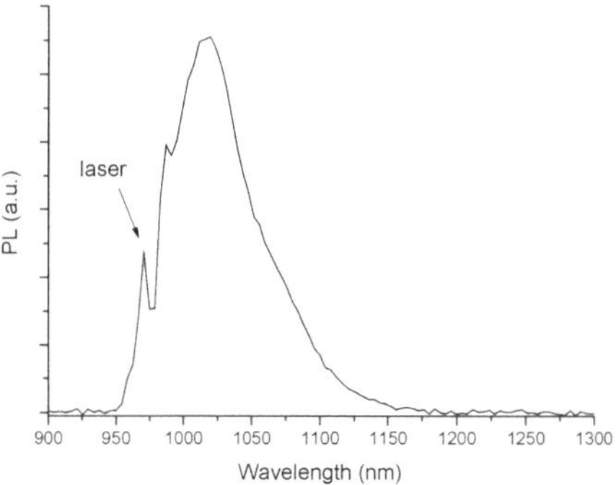

Figure 13. PL emission spectrum of an active disk (BM/71 SiO_2-14 $AlO_{1.5}$-15 $YbO_{1.5}$ film, 36.3 μm thick/ARC), excited at 970 nm.

3. Discussion

The SG method used in the present work enables the synthesis of the active material for an Yb-doped thin-disk laser through spin-coating deposition of disk-like films more than 30 μm thick. While the nature of the Yb-doped material was assessed by XRD plus FTIR and Raman spectroscopies, the morphology of the multilayer films and their optical quality were determined based on SEM analysis. The PL emission spectra and Yb^{3+} metastable level lifetimes of the doped material were also studied in detail.

XRD and Raman data indicate the amorphous nature of these aluminosilicate bulk samples and films with up to 15 mol% Yb, reveal a partially crystalline nature for 20 to 30 mol% Yb. However, an increase of the final heat treatment temperature of the bulk glasses from 1050 to 1100 °C is enough to promote some crystallization in the composition with 15 mol% Yb. Therefore, the Yb content, the glass matrix, and the heat treatment conditions are critical for the amorphous character of the samples. In fact, while aluminosilicate compositions remain amorphous for up to 15 mol% Yb, the phosphosilicates become partially crystallized with just 5 mol% Yb [14], revealing the different capabilities of Al and P to disperse the Yb^{3+} ions in the matrix, since the pentavalent P is less easily incorporated in the silicate glass structure compared to Al, which readily substitutes for Si.

The PL emission peaks at 987 and 1019–1035 nm correspond to transitions between the Stark levels of the $^2F_{5/2}$ excited state and the $^2F_{7/2}$ ground state of Yb^{3+}: the sharp peak at 987 nm corresponds to the transition $5 \rightarrow 1$, between the lowest Stark levels of each J-manifold, whereas the intense peak at 1019–1035 nm is attributed to longer wavelength transitions from level 5 to higher lying Stark levels of the ground state, followed by thermalization to the lower J-level [15], without well resolved Stark splitting due to the amorphous nature of the host matrix. The continuous decrease of the measured PL lifetime with increasing Yb content, especially in the case of the phosphosilicate glass matrix (Table 1), indicates the occurrence of concentration quenching phenomena. This effect was already observed in previous studies [14,16,17], involving non-radiative energy transfer phenomena between closely spaced Yb^{3+} ions, despite the common assumption that the simple electronic energy level structure of Yb^{3+} excludes Excited State Absorption and a variety of quenching processes [18]. While the PL spectra (Figures 5 and 6) refer to Yb/Er co-doped samples, the lifetime values were measured in Yb-doped samples without Er, to avoid any possible influence of $Yb^{3+} \rightarrow Er^{3+}$ energy transfer phenomena on the measured lifetimes. It is important to add that most samples were in the form of small plates, a few millimeters on the sides and a fraction of a mm thick. Since this area was still much larger than the laser spot size (typically of the order of ~0.5 mm), we did not observe any size effects. We also did not consider the possibility of self-absorption and photon trapping phenomena of the type reported by Mattarelli et al. [19] and Koughia and Kasap [20], given the fact that the samples were opaque and quite thin.

The potential performance of the present Yb-doped sol-gel thick films as thin-disk laser materials was evaluated based on the PL results, through a figure of merit (FOM) defined as the $\sigma\tau$ product between the normalized spontaneous emission cross section σ, taken as the ratio of Yb^{3+} (1000–1035 nm)/Er^{3+} (1532 nm) PL peak intensities and the corresponding $^2F_{5/2}$ metastable level lifetime, τ. Tables 1 and 2 indicate that the aluminosilicate composition 67 SiO_2-13 $AlO_{1.5}$-20 $YbO_{1.5}$ had the highest FOM. However, XRD data (Figures 1 and 2) show the occurrence of some incipient crystallization in this material. Therefore, the aluminosilicate composition 71 SiO_2-14 $AlO_{1.5}$-15 $YbO_{1.5}$, which is completely amorphous and had the second highest FOM value, is chosen as the best composition for the doped film portion of the active disk. It is expected that an FOM based on stimulated emission cross-sections would lead to a similar conclusion.

4. Materials and Methods

4.1. Sample Preparation

Heavily Yb-doped aluminosilicate and phosphosilicate glass films and monoliths were prepared by the acid (HCl) catalyzed SG method, with the compositions shown in Table 3, where the aluminosilicate and phosphosilicate matrices had Si:Al and Si:P molar ratios approximately constant and equal to ~5:1 in both cases in order to keep a similar starting glass matrix structure, while the $YbO_{1.5}$ content varied between 3 and 30 mol% on a cation basis. A small quantity of Er (0.002 mol%) was added as an internal reference for the PL measurements [13,14]. The precursors used for silica, phosphorus oxide, alumina, ytterbia, and erbia were tetraethyl orthosilicate (Alfa Aesar, Karlsruhe, Germany, 98%), P_2O_5 (Sigma-Aldrich Gmbh, Munich, Germany, $\geq$98%), aluminium nitrate nonahydrate (Alfa Aesar, Karlsruhe, Germany, 98%–102%), ytterbium nitrate pentahydrate (Strem Chemicals, Bischheim, France, 99.9%) and erbium nitrate pentahydrate (Aldrich Chemical Co., Inc., Milwaukee, WI, USA, 99.9%), respectively. Processing details for these materials are given elsewhere [14]. Titania films, used as high index material in the photonic bandgap (PBG) structures, were also synthetized by the SG method, using titanium (IV) isopropoxide (TiPOT) (Alfa Aesar, Karlsruhe, Germany, $\geq$97%) as precursor. Acid catalysis was also used here, through the mixing of glacial acetic acid (GAA), (Merck, Darmstadt, Germany, 100%) with TiPOT and stirring at room temperature over 1 h. After mixing, ethanol (Merck, $\geq$99.5%) was added slowly and the titania solution was stirred at room temperature for an additional hour.

Table 3. Rare-earth doped phosphosilicate and aluminosilicate glass compositions (mol%).

SiO_2	$PO_{2.5}$	$AlO_{1.5}$	$YbO_{1.5}$
81	16	–	3
79	16	–	5
77	16	–	7
75	15	–	10
73	14	–	13
71	14	–	15
67	13	–	20
58	12	–	30
79	–	16	5
75	–	15	10
71	–	14	15
67	–	13	20
63	–	12	25
58	–	12	30

4.1.1. Bulk Sample Preparation

Transparent silicate glass monoliths were obtained by aging the final SG solution for several days in an open container, followed by a heat-treatment of the obtained gel up to 1050 °C at a rate of 0.5 °C/min and slow cooling inside the furnace. During the heat treatment, most samples crumbled into a powder.

4.1.2. Film Deposition

Silicate or titania precursor solutions were aged for at least 24 h and 3 h, respectively, in a closed container and they were then spin-coated onto single crystal Si wafer substrates, at 2000 rpm for 30 s, using a Chemat KW-4A spin-coater. For thick multilayer deposition, each individual layer was heat treated at 1100 °C for 30 s in a muffle furnace.

4.1.3. Active Disk Preparation

In a Yb thin-disk laser, the active element may consist of a thin disk (or thick film) of Yb-doped material, sandwiched between a highly reflective coating or Bragg Mirror with peak reflectance at the signal wavelength (~1020 nm) and an anti-reflective coating with lowest reflectance at the pump wavelength (usually ~940 nm). When prepared by SG, the active disk will be a multilayered structure formed by several individual layers deposited in sequence according to the scheme shown in Figure 14.

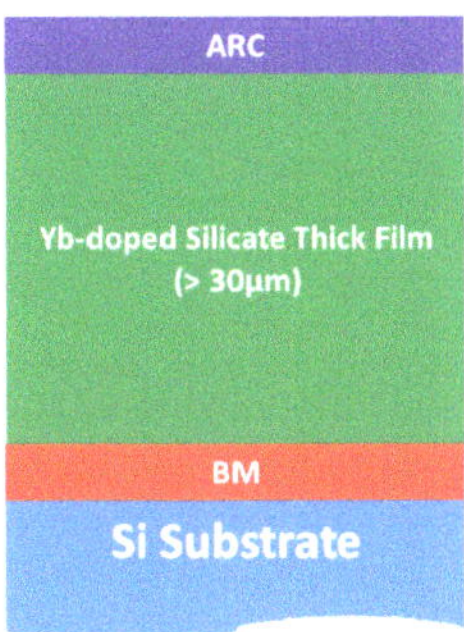

Figure 14. Schematic structure of the active disks prepared.

4.2. Sample Characterization

Bulk powdered samples and thick films were analyzed by X-ray diffraction (XRD) with a Philips PW3020 powder diffractometer, at room temperature, using Cu Kα radiation ($\lambda = 1.541874$ Å) generated at 40 kV and 30 mA, in the 2θ range of 10°–90°, with a step of 3°/min. The average crystal size was estimated by X-ray line broadening, using the Scherrer formula [21] and taking into account the instrumental broadening.

Surface and cross-section images of active disk samples were obtained by Scanning Electron Microscopy (SEM) using a 7001F FEG-SEM (JEOL, Zaventem, Belgium) in secondary and/or back-scattered electron modes, at 15 kV. A ~15 nm layer of chromium was deposited onto the film samples, in order to promote the electrical conductivity at the surface of the insulating materials under observation.

A Nicolet 5700 FT-IR spectrometer (Thermo Electron Corporation, Madison, WI, USA) was used to record near infrared spectra, at a resolution of 4 cm^{-1}. The 975 nm Yb^{3+} absorption peak was recorded in the NIR, in transmission, for transparent glass monolithic samples and in reflection mode, for highly reflective and ARC layers deposited on silicon wafers.

Raman spectra were collected with a LabRam HR 800 Evolution confocal micro-Raman spectrometer (Horiba, Villeneuve d'Ascq, France). The measurements were carried out with 532 nm laser excitation, using a 600 gr/mm grating and 100× objective lens. The incident laser power on the samples was ~10 mW and the spot diameter was ~1 µm.

The PL of Yb-doped silicate materials was excited with a 970 nm laser (Lumics High Power Module), using 3 W power and a filter to block the laser radiation. The emitted light was analyzed by an Avaspec-NIR256-1.7 fiber-optic spectrometer (Avantes, Apeldoorn, The Netherlands), at a resolution of 4 nm. The Yb^{3+} PL lifetimes were measured for chopped light with a PM3370B 60 MHz digital oscilloscope (Fluke, Eindhoven, The Netherlands).

5. Conclusions

A Ytterbium-doped active material for a thin-disk laser was developed using sol-gel processing. Ytterbium doped aluminosilicate and phosphosilicate glass matrices were investigated with a doping level between 3 and 30 mol% YbO$_{1.5}$. Films with a thickness up to ~36 µm were prepared by spin-coating and bulk samples were obtained by sol-gel processing as well. The amorphous nature of the materials prepared was assessed by XRD and Raman spectroscopy, with some incipient crystallization observed in aluminosilicate samples with $\geq$20 mol% Yb and in phosphosilicate samples with $\geq$5 mol% Yb. The PL properties of Yb^{3+} were measured, with lifetime values between 1.0 and 0.5 ms when the Yb concentration increased from 3 to 30 mol%, with a concentration quenching effect for >3 mol% Yb. The Yb^{3+} PL intensities at λ ~1 µm, normalized to that of Er^{3+} internal standard at ~1.5 µm, together with the lifetime values, were combined in an FOM which led to the selection of the 71 SiO$_2$-14 AlO$_{1.5}$-15 YbO$_{1.5}$ composition (mol%) as the best material for the active layer. An active disk ~36 µm thick, with a structure consisting of a Bragg mirror, an aluminosilicate layer doped with 15 mol% Yb and an anti-reflective coating was fabricated.

Acknowledgments: We gratefully acknowledge the "Fundação para a Ciência e a Tecnologia (FCT)", for financial support under contract UID/QUI00100/2013. We would also like to thank ANI for support under project Multilaser (QREN project (SI I & DT) NO. 30179 (2013) and Paulo Morais for useful discussions.

Author Contributions: The samples were prepared by Tiago Ribeiro. Characterization of the samples was performed by Tiago Ribeiro, Rui M. Almeida and Luís F. Santos. The manuscript was written by Tiago Ribeiro, Rui M. Almeida and Luís F. Santos, and approved by all authors. All authors contributed to discussions and reviewed the manuscript.

References

1. Heckl, O.H.; Kleinbauer, J.; Bauer, D.; Weiler, S.; Metzger, T.; Sutter, D.H. Ultrafast Thin-Disk Lasers. In *Ultrashort Pulse Laser Technology—Laser Sources and Applications*; Nolte, S., Schrempel, F., Dausinger, F., Eds.; Springer International Publishing AG: Cham, Switzerland, 2016; Volume 195, pp. 93–115, ISBN 978-3-319-17659-8.
2. Giesen, A.; Speiser, J. Fifteen Years of Work on Thin-Disk Lasers: Results and Scaling Laws. *IEEE J. Sel. Top. Quantum Electron.* **2007**, *13*, 598–609. [CrossRef]
3. TRUMPF. Disk Lasers. Available online: http://www.trumpf-laser.com/en/products/solid-state-lasers/disk-lasers.html (accessed on 5 July 2017).
4. BOEING. Boeing Thin Disk Laser Exceeds Performance Requirements During Testing. Available online: http://boeing.mediaroom.com/Boeing-Thin-Disk-Laser-Exceeds-Performance-Requirements-During-Testing (accessed on 5 July 2017).
5. Giesen, A.; Hugel, H.; Voss, A.; Wittig, K.; Brauch, U.; Opower, H. Scalable concept for diode-pumped high-power solid-state lasers. *Appl. Phys. B* **1994**, *58*, 365–372. [CrossRef]
6. Huang, Y.; Zhu, X.; Zhu, G.; Shang, J.; Wang, H.; Qi, L.; Zhu, C.; Guo, F. A multi-pass pumping scheme for thin disk lasers with good anti-disturbance ability. *Opt. Express* **2015**, *23*, 4605–4613. [CrossRef] [PubMed]
7. Marchese, S.V.; Baer, C.R.E.; Peters, R.; Kränkel, C.; Engqvist, A.G.; Golling, M.; Maas, D.J.H.C.; Petermann, K.; Südmeyer, T.; Huber, G.; et al. Efficient femtosecond high power Yb:Lu$_2$O$_3$ thin disk laser. *Opt. Express* **2007**, *15*, 16966–16971. [CrossRef] [PubMed]
8. Peters, R.; Kränkel, C.; Petermann, K.; Huber, G. Broadly tunable high-power Yb:Lu$_2$O$_3$ thin disk laser with 80% slope efficiency. *Opt. Express* **2007**, *15*, 7075–7082. [CrossRef]
9. Francini, R.; Giovenale, F.; Grassano, U.M.; Laporta, P.; Taccheo, S. Spectroscopy of Er and Er-Yb-doped phosphate glasses. *Opt. Mater.* **2000**, *13*, 417–425. [CrossRef]
10. Marques, A.C.; Almeida, R.M. Rare earth-doped photonic crystals via sol-gel. *J. Mater. Sci. Mater. Electron.* **2009**, *20*, 307–311. [CrossRef]
11. Li, Y.G.; Almeida, R.M. Simultaneous broadening and enhancement of the 1.5 micron photoluminescence peak of Er^{3+} ions embedded in a 1-D photonic crystal microcavity. *Appl. Phys. B* **2010**, *98*, 809–814. [CrossRef]
12. Li, Y.G.; Almeida, R.M. Elimination of porosity in heavily rare-earth doped sol–gel derived silicate glass films. *J. Sol-Gel Sci. Technol.* **2012**, *61*, 332–339. [CrossRef]
13. Ferreira, J.; Santos, L.F.; Almeida, R.M. Sol-gel derived Yb:YAG polycrystalline ceramics for laser applications. *J. Sol-Gel Sci. Technol.* **2017**, *83*, 436–446. [CrossRef]
14. Ribeiro, T.V.; Santos, L.F.; Gonçalves, M.C.; Almeida, R.M. Heavily Yb-doped silicate glass thick films. *J. Sol-Gel Sci. Technol.* **2017**, *81*, 105–113. [CrossRef]
15. Perumal, R.N.; Subalakshmi, G. Near-infrared down-conversion in Yb^{3+}:TiO$_2$ for solar cell applications. *J. Mater. Sci. Mater. Electron.* **2017**. [CrossRef]
16. Paschotta, R.; Nilsson, J.; Barber, P.R.; Caplen, J.E.; Tropper, A.C.; Hanna, D.C. Lifetime quenching in Yb-doped fibers. *Opt. Commun.* **1997**, *136*, 375–378. [CrossRef]
17. Malinoski, M.; Kaczkan, M.; Piramidowicz, R.; Frukacz, Z.; Sarnecki, J. Cooperative emission in Yb^{3+}: YAG planar epitaxial waveguides. *J. Lumin.* **2011**, *94–95*, 29–33. [CrossRef]
18. RP Photonics Encyclopedia. Available online: http://www.rp-photonics.com/ytterbium_doped_gain_media.html (accessed on 5 July 2017).
19. Matarrelli, M.; Montagna, M.; Zampedri, L.; Chiasera, A.; Ferrari, M.; Righini, G.C.; Fortes, L.M.; Gonçalves, M.C.; Santos, L.F.; Almeida, R.M. Self-absorption and Radiation trapping in Er^{3+}-doped TeO$_2$-based glasses. *Europhys. Lett.* **2005**, *71*, 394–399. [CrossRef]
20. Kouhia, C.; Kasap, S.O. Excitation diffusion in GeGaSe and GeGaS glasses heavily doped with Er^{3+}. *Opt. Express* **2008**, *16*, 7709–7714. [CrossRef]
21. Langford, J.; Wilson, A. Scherrer after Sixty Years: A Survey and Some New Results in the Determination of Crystallite Size. *J. Appl. Crystallogr.* **1978**, *11*, 102–103. [CrossRef]

 *materials* 

Article

1T1R Nonvolatile Memory with Al/TiO₂/Au and Sol-Gel-Processed Insulator for Barium Zirconate Nickelate Gate in Pentacene Thin Film Transistor

Ke-Jing Lee, Yu-Chi Chang, Cheng-Jung Lee, Li-Wen Wang and Yeong-Her Wang *

Institute of Microelectronics, Department of Electrical Engineering, National Cheng-Kung University, Tainan 701, Taiwan; hugh_2224@hotmail.com (K.-J.L.); s0964348@hotmail.com (Y.-C.C.); s.w.l.f.dd@gmail.com (C.-J.L.); jk220052@gmail.com (L.-W.W.)
* Correspondence: yhw@ee.ncku.edu.tw; Tel.: +886-6-275-7575 (ext. 62352); Fax: +886-6-208-0598

Received: 30 October 2017; Accepted: 6 December 2017; Published: 9 December 2017

Abstract: A one-transistor and one-resistor (1T1R) architecture with a resistive random access memory (RRAM) cell connected to an organic thin-film transistor (OTFT) device is successfully demonstrated to avoid the cross-talk issues of only one RRAM cell. The OTFT device, which uses barium zirconate nickelate (BZN) as a dielectric layer, exhibits favorable electrical properties, such as a high field-effect mobility of 2.5 cm²/Vs, low threshold voltage of −2.8 V, and low leakage current of 10^{-12} A, for a driver in the 1T1R operation scheme. The 1T1R architecture with a TiO₂-based RRAM cell connected with a BZN OTFT device indicates a low operation current (10 μA) and reliable data retention (over ten years). This favorable performance of the 1T1R device can be attributed to the additional barrier heights introduced by using Ni (II) acetylacetone as a substitute for acetylacetone, and the relatively low leakage current of a BZN dielectric layer. The proposed 1T1R device with low leakage current OTFT and excellent uniform resistance distribution of RRAM exhibits a good potential for use in practical low-power electronic applications.

Keywords: one transistor and one resistor (1T1R); organic thin-film transistor (OTFT); resistive random access memory (RRAM); sol-gel

---

1. Introduction

Resistive random access memory (RRAM) has attracted considerable attention due to its advantages of high-density integration, high switching speed, low power consumption, extended retention time, high operational speed, and non-destructive readout. However, a cross-point array built with only a RRAM cell suffers from unavoidable cross-talk interference because of its leakage current paths, which pass through neighboring unselected cells with low resistance. To solve this issue, another functional device should be combined with the memory device. Recently, Wan et al. [1] reported that the cross-talk issue could be reduced by an internal transistor. Organic thin-film transistors (OTFTs) have been studied intensively for their potential application in integrated circuits, their lightweight roll-to-roll printing compatibility, their flexibility, and the ease of tailoring their optoelectronic properties. Pentacene has been extensively studied as a p-type semiconductor in organic transistors, due to its favorable semiconducting behavior and high hole carrier mobility [2]. In addition, a gate oxide material with a high-κ is urgently required to achieve high mobility and low operation voltage OTFTs. However, the disadvantage of most high-κ materials is their relatively narrow bandgap, which results in aggravated gate leakage current. Owing to the leakage current of the transistor caused by a high resistive state (HRS) current of the 1T1R RRAM device, a suitable high-κ material that satisfies the sufficient wide bandgap and relatively low leakage current of the transistor is necessary to reduce the power consumption of the 1T1R RRAM device. The present study proposes a high-κ barium zirconate nickelate (BZN) dielectric layer for OTFT applications [3]. The gate leakage of the BZN-based OTFTs is reduced compared with barium zirconate oxide (BZO). This result may be attributed to adding Ni (II) acetylacetone in BZO, which increases the barrier height and reduces the leakage current. Thus, the 1T1R architecture with low leakage current BZN-based OTFT device connected to a RRAM cell is designed to reduce the power consumption and avoid cross-talk issues present when using only one RRAM cell. In addition, several 1T1R devices still require a forming process, although the cross-talk issue can be reduced by the internal transistor [4,5]. A high-voltage forming process is avoided in the 1T1R device, as a result of the sufficient oxygen vacancies possessed by the TiO_2 thin film. The 1T1R RRAM devices exhibit low HRS currents, low operation currents, low power consumption, and reliable data retention.

2. Results

Figure 1a illustrates the switching *I–V* curves of the $Al/TiO_2/Au$ structure under the direct current (DC) voltage sweep. The voltage was applied to the top Al electrode in a sequence of +4 V → 0 V → −4 V → 0 V → +4 V. Figure 1b depicts the logarithmic plots of the *I–V* curves for the positive and negative voltage sweep regions, which were created to investigate the switching mechanism of the TiO_2-based RRAM. The slope is unequal to 1, although the $ln(I)–V^{1/2}$ and $ln(I/V)–V^{1/2}$ plots of the *I–V* curves at the HRS were also investigated.

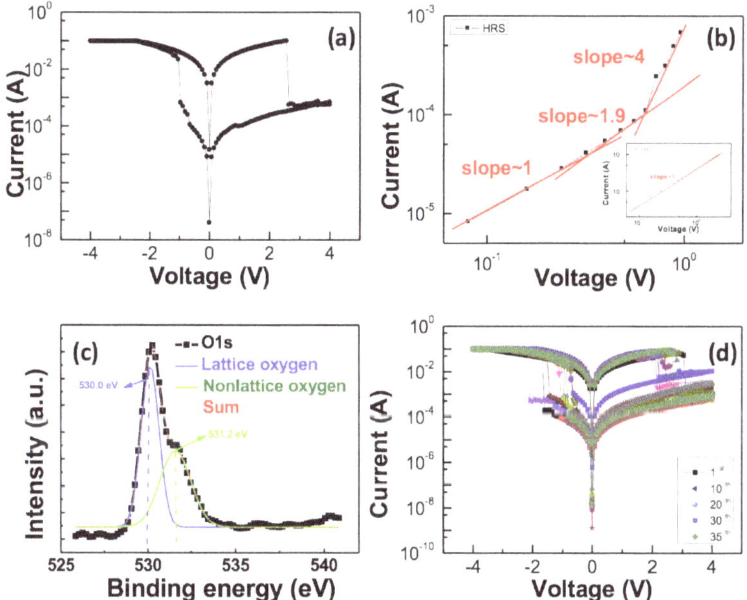

Figure 1. (**a**) *I–V* curves of the TiO_2 RRAM (resistive random access memory) device; (**b**) *logI–logV* characteristics for the switching mechanisms; (**c**) Displays the normalized XPS (X-ray photoelectron spectroscopy) spectra for O1s of the TiO_2 thin film; (**d**) Switching behavior of the TiO_2 RRAM device after 35 DC (direct current) sweep cycles.

In the low-voltage region of the HRS, the current was linearly proportional to the applied bias and showed a typical Ohmic behavior. In the high-voltage region of the HRS, the *I–V* slopes of the fitting line were approximately 2. The current density abruptly increases with *I–V* slopes greater than 2 when all the available traps are filled [6,7]. X-ray photoelectron spectroscopy (XPS) was used to analyze the TiO_2 thin films as presented in Figure 1c. The peaks located at 529.5 and 531.1 eV were associated with lattice and non-lattice oxygen ions, respectively. In the XPS results, the TiO_2 thin film contains defects, such as non-lattice oxygen ions. Defects in the TiO_2 thin film can form trap sites below the conduction band, where the injected charge carriers can be entrapped. Therefore, the fitting results at the HRS indicate that the space charge limited current (SCLC) model is the dominant mechanism. In the positive-voltage region, the current state maintains a low resistive state (LRS) and demonstrates an Ohmic conduction behavior with a slope of approximately 1, thereby confirming the formation of a conductive filament [8]. Figure 1d illustrates the TiO_2-based RRAM that can maintain an on/off ratio of over 10^2 after continuous DC voltage switching cycles of 35.

The output and transfer characteristics of the BZN-based OTFT in the 1T1R device measured at V_{dg} = −10 V are depicted in Figure 2a,b. In terms of the transfer characteristics of the devices, subthreshold slope swing (S.S.), the on/off current ratio threshold, voltage (V_{th}), and high field-effect mobility were approximately 1 V/decade, 10^3, −2.8 V, and 2.5 cm^2 V^{-1} S^{-1}, respectively. The inset of Figure 2b shows the current-voltage (*C–V*) characteristic of the Au/BZN/Al capacitor. The frequency in the *C–V* measurement is set to 1 MHz. The capacitance of the BZN layers was 1.5 nF for a device area of 3 mm^2. The TiO_2-based RRAM cell connected to the solution BZN OTFT device was demonstrated to successfully fabricate for the 1T1R architecture. The merging of the Au/pentacene/BZN/Al transistor device with Al/TiO_2/Au memory device as shown in the inset of Figure 3a resulted in a superimposed *I–V* characteristics, as illustrated in Figure 3a. As the results show, the 1T1R structure somehow presents analog behavior, while RRAM (Figure 1a) shows digital behavior, with an obvious On/Off ratio. This might be due to the effect of the pentacene transistor, and further study is needed. Although the shapes of Figures 1a and 3a look somehow different, the current levels are all effectively suppressed by two orders of magnitude under the 1T1R structure. In the set process, a positive voltage is applied on the top electrode sweeping from 0 to a set voltage of 3.8 V, and a set current of 0.23 μA is obtained with the source grounded. This process causes the current to suddenly increase to reach LRS, where the value of the set current is controlled by modulating the gate voltage of the transistor. The inset of Figure 3a shows the resistance resistance-voltage (*R–V*) curve corresponding to Figure 3a. A read voltage of −0.8 V indicates a resistance of 60 MΩ for the HRS and 1 MΩ for the LRS, corresponding to a ratio of 60.

Figure 3b demonstrates the 1T1R memory device without obvious deterioration after 60 successive DC resistive switching cycles. The shapes of the *I–V* curves and the first sweep were similar, and only minimal voltage and resistance variations were observed, thereby indicating an excellent stability in its bipolar resistive switching behavior. The memory cell can be switched to the on-state for V_{dg} at 10 V by applying a positive voltage V_{dg} at the gate of the control transistor. The cell can be switched back to the off-state for V_{dg} = 0 V.

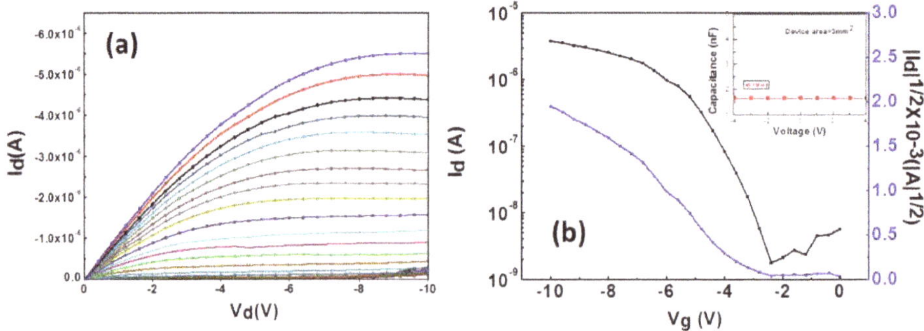

Figure 2. (**a**) *C–V* (current–voltage) output and (**b**) transfer characteristics of the BZN (barium zirconate nickelate)-based OTFT (organic thin film transistor) in the 1T1R device. The inset displays the *C–V* characteristic of the Au/BZN/Al capacitor. The frequency in the *C–V* measurement is set to 1 MHz.

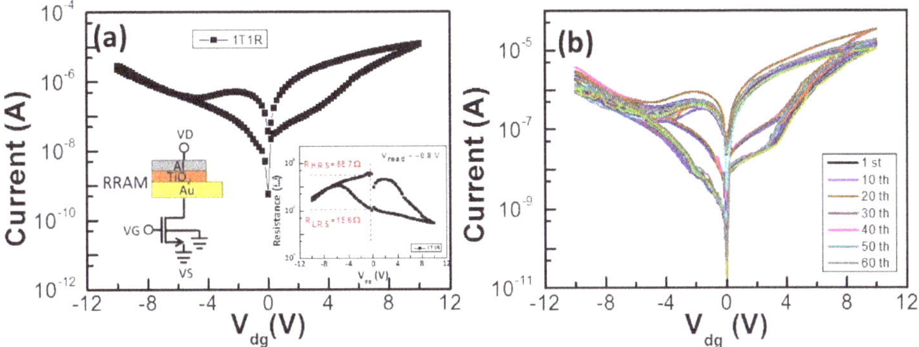

Figure 3. (**a**) *I–V* characteristics and schematics of the 1T1R (one-transistor and one-resistor) device. The inset shows the *R–V* (resistance-voltage) curve corresponding to Figure 3a; (**b**) Switching behavior of the 1T1R device cycles after 60 DC sweep cycles.

Figure 4a presents the retention capabilities of the $Al/TiO_2/Au/pentacene/BZN/Al$ 1T1R devices measured at room temperature (RT) obtained at a voltage of 0.1 V. For the $Al/TiO_2/Au/$ pentacene/BZN/Al 1T1R device, a 10-year on/off ratio of over 10^2 can be extrapolated at RT. The inset in Figure 4a illustrates the surface topography of the TiO_2 surface. The root-mean-square roughness (R_{rms}) value of the TiO_2 thin film is approximately 1.3 nm. The smooth roughness facilitates the stable resistive switching of the device [9]. Thus, the favorable retention capability can be partly attributed to the smooth TiO_2 thin film.

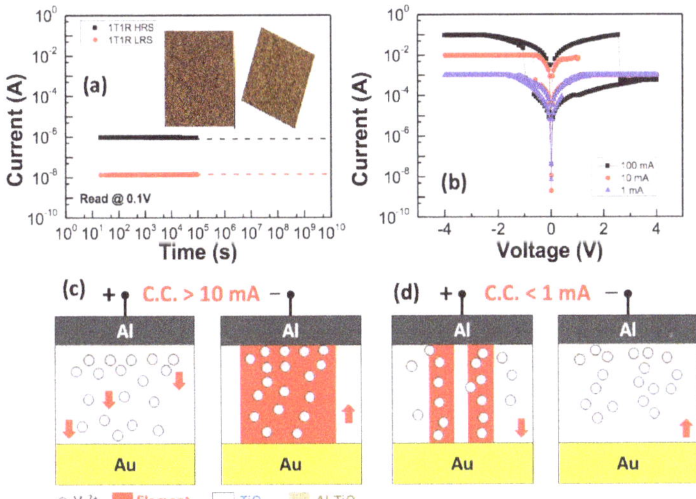

Figure 4. (**a**) Data retention results of HRS (high resistive state) and LRS (low resistive state) for $Al/TiO_2/Au/pentacene/BZN/Al$ 1T1R devices at RT (room temperature) obtained at 0.1 V. The inset image displays the AFM (atomic force microscopy) topography image of the TiO_2 thin film. (**b**) Dependence of the *I–V* curves on the TiO_2 RRAM device with the different compliance currents. The schematic of the proposed switching mechanism indicates that the compliance current of (**c**) >10 mA and (**d**) <1 mA was applied to the $Al/TiO_2/Au$ RRAM device.

The RRAM device exhibits bipolar behavior (the set process was triggered by a negative voltage, while the reset process occurred under a positive bias) when the compliance current is more than 10 mA. The device performs a reversed switching polarity when using a low compliance current of 1 mA, as displayed in Figure 4b. Similarly, the 1T1R device can be turned on/off by applying a positive/negative voltage, respectively. The switching current of the 1T1R device is lower than 10 μA. The distinct switching is attributed to different compliance currents. This phenomenon is due to the large oxygen vacancy in the Al-TiO$_x$ layer, and the resulting asymmetric oxygen vacancy profile.

The redox reactions of active electrodes at the electrode/insulator interface are considered to have an important role in the resistive switching behavior [10]. The spontaneous oxidation of Al at the Al-TiO$_2$ interface causes the formation of AlO$_x$, because Al is chemically active. The interface AlO$_x$ layer acts as a diffusion barrier that prevents oxygen from infusing into the atmosphere, and stores and releases oxygen during switching, similar to an oxygen reservoir [11,12].

Figure 4c,d depicts the schematics of the proposed switching mechanism. An oxygen-deficient layer was preferentially produced naturally in the TiO$_x$ layer near the Al electrode when the compliance current was more than 10 mA. Moreover, a certain fraction of the Al in the top electrode was diffused into the TiO$_x$ layer. The oxygen-deficient layer can be considered an Al-TiO$_x$ layer. Al^{3+} substitutes for Ti^{4+} within the TiO$_x$ for the Al-TiO$_x$ layer, and oxygen vacancies can be produced by diffused Al^{3+}. Thus, more oxygen vacancies were observed in the Al-TiO$_x$ layer than in the TiO$_x$ layer. Oxygen vacancies migrate toward the Al top electrode when a negative voltage is applied to the Al top electrode and form the filament in layers after a sufficient electric field is reached, thereby bringing the device into the LRS. By contrast, oxygen vacancies from the Al-TiO$_x$ layer move toward the Au bottom electrode when a positive voltage is applied to the Al top electrode, thus causing the filament to rupture, placing the device into the HRS. The HRS is difficult to reach during normal device operations when the compliance current is less than 1 mA. The thin film behaves similarly to a conductor, because the compliance current or positive bias is not high enough to sufficiently deplete the oxygen vacancy in the Al-TiO$_x$ layer. Thus, the polarity of bipolar switching is found to be determined by the compliance current with different distributions of the internal oxygen vacancies used in the RRAM and 1T1R devices.

3. Materials and Methods

The OTFT was fabricated onto a Corning glass Eagle 2000 substrate. First, the glass substrate was cleaned with acetone, ethanol, and de-ionized water. An approximately 38 nm-thick Aluminum (Al) was deposited as the gate electrode via a radio-frequency (RF) magnetron sputtering system.

The barium acetate, zirconium *n*-propoxide, and nickel II acetylacetone were synthesized by using the various sol-gel techniques. The required amount of the barium acetate (536.8 mg) was dissolved in glacial acetic acid (5 mL) by stirring, and then was heated on a 120 °C hot plate until complete dissolution was achieved (A1 solution). The nickel II acetylacetone (543.7 mg) and zirconium *n*-propoxide (65 μL) were mixed with acetylacetone (470 μL) and dissolved in 2-methoxyethanol (3 mL) by stirring (A2+B solution). Subsequently, A1 was slowly dropped into A2+B until a completely transparent solution was obtained. 2-methoxyethanol (1.5 mL) was added to acquire a 0.5 M BZN sol precursor. The BZN solution was spin-coated onto the Al/glass substrate with a thickness of 310 nm as the gate dielectric after the solution was completely prepared using the sol-gel process. The BZN layer was baked at 100 °C for 10 min under ambient air conditions. The pentacene thin film (76 nm) was thermally evaporated on the BZN layer at a deposition rate of 0.08 to 0.15 nm/s. Finally, a gold (Au) electrode was deposited using the RF magnetron sputtering system and defined through shadow masks as the source and drain. The channel length and width of the transistors were 150 and 1500 μm, respectively. For manufacturing the 1T1R configuration, the RRAM device was fabricated directly on the drain electrodes of the TFT device. A simple metal/insulator/metal structure, which consists of Al/TiO$_2$/gold (Au) structures, was fabricated. First, Au was deposited to be a bottom electrode of the RRAM device through the RF magnetron sputtering system. Then, the 17 nm TiO$_2$ layer was deposited onto the Au electrode by sputtering. Finally, a 68 nm-thick Al was deposited as the top electrode. The devices were measured at RT (300 K) with the humidity conditions of 40–50%. All measured data were extracted from the devices with an area of 3 mm^2, as demonstrated by the diagram of the Al/TiO$_2$/Au/pentacene/BZN/Al/glass in Figure 5. The electrical properties of the transistors were analyzed using an Agilent B1500 semiconductor parameter analyzer.

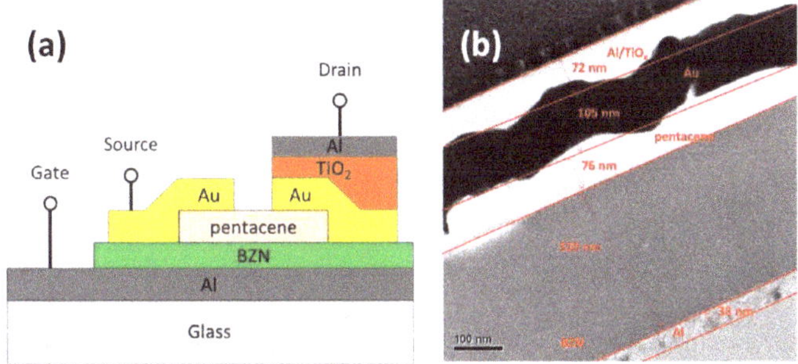

Figure 5. (**a**) Cross-sectional view of the 1T1R memory device. (**b**) Corresponding TEM (transmission electron microscopy) image of the Al/TiO$_2$/Au/pentacene/BZN/Al/glass stacked structure.

4. Conclusions

The 1T1R architecture with a RRAM cell connected to the solution BZN OTFT device was successfully demonstrated. The TiO$_2$-based RRAM device with bipolar mode shows an on/off ratio of 10^2 and a uniform resistance distribution. The favorable performance of the 1T1R device can be attributed to the low leakage current of the BZN-based OTFT. The 1T1R devices exhibit a low operation current of 10 μA and a reliable data retention of over ten years, because the TiO$_2$-based RRAM and BZN-based OTFT demonstrate a stable device yield and favorable device performance. The low power consumption and excellent uniform resistance distribution of the proposed 1T1R RRAM device indicate a considerable potential for practical applications.

Acknowledgments: This work was supported in part by the National Science Council of Taiwan under Contracts NSC102-2221-E-006-182-MY3 and MOST 105-2221-E-006-193-MY3.

Author Contributions: Ke-Jing Lee handled the device preparation and characterization, data analysis, and paper writing. Cheng-Jung Lee and Li-Wen Wang prepared the material/device and discussion. Yu-Chi Chang organized the modeling discussion and paper editing. Yeong-Her Wang was the advisor who monitored the progress and editing of the paper.

Conflicts of Interest: The authors declare no conflict of interest.

References

1. Yun, M.J.; Kim, S.; Kim, H.D. Resistive switching characteristics of Al/Si3N4/p-Si MIS-based resistive switching memory devices. *J. Korean Phys. Soc.* **2016**, *69*, 435–438.
2. Chang, Y.C.; Wei, C.Y.; Chang, Y.Y.; Yang, T.Y.; Wang, Y.H. High mobility pentacene based thin film transistors with synthesized strontium zirconate nickelate gate insulators. *IEEE Trans. Electron Devices* **2013**, *60*, 4234–4239. [CrossRef]
3. Lee, C.-J.; Lee, K.-J.; Chang, Y.-C.; Wang, L.-W.; Chou, D.-W.; Wang, Y.-H. Barium zirconate nickelate as the gate dielectric for low-leakage current organic transistors. *IEEE Trans. Electron Devices* **2017**. revised.
4. Walczyk, D.; Walczyk, C.; Schroeder, T.; Bertaud, T.; Sowińska, M.; Lukosius, M.; Fraschke, M.; Tillack, B.; Wenger, C. Resistive switching characteristics of CMOS embedded HfO2-based 1T1R cells. *Microelectron. Eng.* **2011**, *88*, 1133–1135. [CrossRef]
5. Wu, M.-C.; Lin, Y.-W.; Jang, W.-Y.; Lin, C.-H.; Tseng, T.-Y. Low-power and highly reliable multilevel operation in ZrO$_2$ 1T1R RRAM. *IEEE Trans. Electron Devices Lett.* **2011**, *32*, 1026–1028. [CrossRef]

6. Ji, Y.; Choe, M.; Cho, B.; Song, S.; Yoon, J.; Ko, H.C.; Lee, T. Organic nonvolatile memory devices with charge trapping multilayer graphene film. *Nanotechnology* **2012**, *23*, 105202. [CrossRef] [PubMed]
7. Chang, Y.C.; Lee, K.J.; Lee, C.J.; Wang, L.W.; Wang, Y.H. Bipolar resistive switching behavior in sol-gel MgTiNiOx memory device. *IEEE J. Electron Devices Soc.* **2016**, *4*, 321–327. [CrossRef]
8. Chang, Y.C.; Xue, R.Y.; Wang, Y.H. Multilayered barium titanate thin films by sol-gel method for nonvolatile memory application. *IEEE Trans. Electron Devices* **2014**, *61*, 4090–4097. [CrossRef]
9. Hwang, Y.H.; An, H.M.; Cho, W.J. Performance improvement of the resistive memory properties of InGaZnO thin films by using microwave irradiation. *Jpn. J. Appl. Phys.* **2014**, *53*. [CrossRef]
10. Younis, A.; Chu, D.; Shah, A.H.; Du, H.; Li, S. Interfacial redox reactions associated ionic transport in oxide-based memories. *ACS Appl. Mater. Interfaces* **2017**, *9*, 1585–1592. [CrossRef] [PubMed]
11. Chiang, K.K.; Chen, J.S.; Wu, J.J. Aluminum electrode modulated bipolar resistive switching of Al/fuel-assisted NiO$_x$/ITO memory devices modeled with a dual-oxygen-reservoir structure. *ACS Appl. Mater. Interfaces* **2012**, *4*, 4237–4245. [CrossRef]
12. Jung, S.; Kong, J.; Song, S.; Lee, K.; Lee, T.; Hwang, H.; Jeon, S. Resistive switching characteristics of solution-processed transparent TiO$_x$ for nonvolatile memory application. *J. Electrochem. Soc.* **2010**, *157*, H1042–H1045. [CrossRef]

Article

Wet-Chemical Synthesis of 3D Stacked Thin Film Metal-Oxides for All-Solid-State Li-Ion Batteries

Evert Jonathan van den Ham [1], Giulia Maino [1], Gilles Bonneux [1], Wouter Marchal [1], Ken Elen [1,2], Sven Gielis [1], Felix Mattelaer [3], Christophe Detavernier [3], Peter H. L. Notten [4,5], Marlies K. Van Bael [1] and An Hardy [1,*]

[1]   Inorganic and Physical Chemistry and Imec, Division Imomec, Institute for Materials Research, Hasselt University, Martelarenlaan 42, 3500 Hasselt, Belgium; jonathan.vandenham@uhasselt.be (E.J.v.d.H.); giulia.maino@uhasselt.be (G.M.); gilles.bonneux@uhasselt.be (G.B.); wouter.marchal@uhasselt.be (W.M.); Ken.elen@uhasselt.be (K.E.); gielissven@hotmail.com (S.G.); marlies.vanbael@uhasselt.be (M.K.V.B.)

[2]   Imec vzw, Division Imomec, Wetenschapspark 1, B-3590 Diepenbeek, Belgium

[3]   Department of Solid State Sciences, Ghent University, Krijgslaan 281 S1, 9000 Gent, Belgium; felix.mattelaer@ugent.be (F.M.); christophe.detavernier@ugent.be (C.D.)

[4]   Energy Materials & Devices, Eindhoven University of Technology, 5600 MB Eindhoven, The Netherlands; p.h.l.notten@tue.nl

[5]   Forschungszentrum Jülich, Fundamental Electrochemistry (IEK-9), D-52425 Jülich, Germany

*   Correspondence: An.hardy@uhasselt.be

Received: 28 June 2017; Accepted: 5 September 2017; Published: 12 September 2017

Abstract: By ultrasonic spray deposition of precursors, conformal deposition on 3D surfaces of tungsten oxide (WO_3) negative electrode and amorphous lithium lanthanum titanium oxide (LLT) solid-electrolyte has been achieved as well as an all-solid-state half-cell. Electrochemical activity was achieved of the WO_3 layers, annealed at temperatures of 500 °C. Galvanostatic measurements show a volumetric capacity (415 mAh·cm^{-3}) of the deposited electrode material. In addition, electrochemical activity was shown for half-cells, created by coating WO_3 with LLT as the solid-state electrolyte. The electron blocking properties of the LLT solid-electrolyte was shown by ferrocene reduction. 3D depositions were done on various micro-sized Si template structures, showing fully covering coatings of both WO_3 and LLT. Finally, the thermal budget required for WO_3 layer deposition was minimized, which enabled attaining active WO_3 on 3D TiN/Si micro-cylinders. A 2.6-fold capacity increase for the 3D-structured WO_3 was shown, with the same current density per coated area.

Keywords: ultrasonic spray deposition; tungsten oxide; lithium lanthanum titanium oxide; conformal coating; Li ion batteries

---

1. Introduction

Finding smart solutions for sustainable energy harvesting and storage is often opted as the main challenge of the near future. Lithium ion (Li-ion) batteries are major candidates for energy storage due to their superior energy and power density compared to other battery technologies. However, contemporary Li-ion batteries suffer from a number of intrinsic issues, mostly related to the use of liquid electrolyte: (i) safety risks; (ii) limited lifetime; and (iii) operating temperature limitations. Efforts are made to tackle these issues by stabilizing polymer and gel-based electrolytes, in combination with contemporary battery design, increasing the thermal stability and lifetime of the battery [1,2]. However, these issues can be dealt with to a greater extent by adopting a solid-electrolyte, yielding an all-solid-state battery [3,4]. However, all-solid-state batteries suffer from intrinsic issues as well, due to the lower conductivity of the solid-electrolyte. To avoid internal resistance, resulting in an ohmic drop, a thin film approach is often adopted [5]. Although this yields rate capability enhancements, it comes at a cost: thin film electrodes are intrinsically limited in capacity due to their limited volume. Simply thickening the electrodes would lead to high resistance as well, as Li-ions and electrons have to diffuse through a thicker layer. Therefore, the concept of 3D all-solid-state batteries was proposed [6]. More specifically, the combination between the thin-film approach and 3D batteries was proposed by Notten et al. leading to the integrated 3D all-solid-state Li-ion battery [7].

To achieve this challenging goal, several positive and negative electrode materials (often referred to as cathodes and anodes, respectively) have been studied for 3D deposition using high aspect ratio scaffolds [7]. For electrode materials, conformal coatings of $Li_4Ti_5O_{12}$ [8], $LiCoO_2$ [9,10], and $(Li)FePO_4$ were shown before [11], leading to an increased capacity due to the use of 3D geometries [6]. For solid-electrolytes, reports were found on the deposition of LiPO(N), $LiTaO_3$ and lithium silicates [12–17]. Recently, stacked electrodes and solid-electrolytes were shown, creating an all-solid-state 3D Li-ion battery [18,19]. In all these cases, vacuum based methods were required. Despite good performance of these materials, from an upscaling perspective, this is a costly route. In addition, the choice of materials is dominated by specific diffusion behavior of gas precursors used. In this study, an alternative is presented: conformal coatings of oxide materials for Li-ion batteries using ultrasonic spray deposition. Based on a breakthrough reported previously regarding non-planar deposition using wet-chemical methods [20,21], 3D all-solid-state Li-ion batteries could be within reach at much lower costs and a wider choice of oxide materials. Based on previous results using this approach [21], this study aims to investigate the stacking of tungsten oxide (WO_3) as a negative electrode in combination with amorphous lithium lanthanum titanium oxide ($Li_{3x}La_{(2/3)−x}TiO_3$, referred to as LLT) solid-electrolyte to compile a 3D all-solid-state half-cell.

Although WO_3 is mostly known within the field of electrochromic devices [22–28], and photovoltaics [29], it also serves as a negative electrode material for Li-ion batteries [22,30–33]. Though the gravimetric capacity of this electrode material is mediocre, the volumetric energy density is higher than most other oxide electrode materials. For instance, the volumetric density of WO_3 is 604 mAh·cm^{-3} [33,34], whereas the $Li_4Ti_5O_{12}$ (LTO) has a capacity of only 228 mAh·cm^{-3} [5]. For all-solid-state batteries the latter is arguably the most important device parameter. Furthermore, as is the case for LTO, WO_3 only shows a small volume change during lithiation [35]. The volume change proved to yield no practical issues for thin films up to 100 cycles [33]. In addition, the choice for WO_3 is related to the fact that the structure can crystallize at temperatures below 300 °C [24,36], which makes it interesting from a device integration point view. Finally, WO_3 exhibits an intrinsic advantage related to the compatibility with the solid-electrolyte material. With a relative high (de)intercalation voltage for a negative electrode material—ranging from 1.5 to 3 V vs. Li/Li$^+$ [22,29,30,33]—WO_3 is compatible with the electrochemically relatively unstable LLT electrolyte [33]. The latter material suffers from a Ti^{4+} reduction while cycling below 1.5 V vs. Li$^+$/Li, making it unsuitable for many negative electrode materials such as lithium and graphite [3]. In fact, all Ti-based electrolytes are susceptible to this problem, meaning that various materials NASICON electrolyte class suffers from this reduction as well [37,38]. Indeed, more stable solid-electrolytes—compatible with metallic lithium—are available such as "Li-stuffed" garnets [39–41]. However, besides vacuum based deposition of LiPO(N) and $LiTaO_3$ [12–15,17], no 3D compatible deposition methods have been established for these stable electrolyte materials.

Notably, the choice was made to prepare LLT in its amorphous state. Previous studies indicated that crystallization of the highly Li$^+$ conductive perovskite lithium lanthanum titanium solid-electrolyte leads to serious issues regarding cracks and pinholes [42,43]. Therefore, at the cost of a lower Li-ion conductivity of 10^{-8} S·cm^{-1}, the amorphous form of this material is chosen. This leads to enhanced morphology, which is of crucial importance to prevent short circuits over the electrolyte layer [43]. In addition, the more mild annealing conditions required for the amorphous phase eases integration with other materials, which is crucial for 3D all-solid-state Li-ion batteries.

2. Materials and Methods

2.1. Synthesis of the Tungsten Precursor (W-Precursor)

A W-precursor solution was prepared by adding tungstic acid (H_2WO_4, $\geq$99%, Sigma Aldrich, Overijse, Belgium) and citric acid hydrate (CA, $\geq$99%, Sigma Aldrich, Overijse, Belgium) to a round bottom flask, dispersed with a small amount of water. The H_2WO_4 to CA ratio was 1:4. This yellow suspension was stirred and heated at 120 °C for 2 h under reflux conditions. Subsequently, the pH was raised to pH > 12 with ammonia (NH_3, 32%, Merck, Overijse, Belgium) and left to stir for 24 h. After cooling, a transparent, grey colored solution was obtained. The final pH was 8, with a concentration of 0.35 mol·L^{-1}, as was determined by inductively coupled plasma-atomic emission spectroscopy (ICP-AES, Optima 3300, PerkinElmer, Zaventem, Belgium). Before ultrasonic spray deposition, the precursor was diluted and mixed with ethanol (10:9 water/ethanol volume ratio) to yield a 25 mM concentration [20,21].

2.2. Synthesis of the Lithium Lanthum Titanium Precursor (Li-La-Ti Precursor)

The Li-La-Ti-precursor used was reported earlier based on work in our labs [43]. The citrate-nitrate precursor was combined with ethanol (10:9 water/ethanol volume ratio) with a final (total) metal-ion concentration of 10 mM.

2.3. Substrates

Three different types of substrates were used for this study: (i) Pt (sputtered) and TiN (CVD)-coated silicon wafer; (ii) trench-structured Si wafer, prepared by reactive ion etching (Philips, Amsterdam, The Netherlands); and (iii) TiN coated silicon micro-cylinders, prepared by reactive ion etching (IMEC, Leuven, Belgium). All these substrates were cleaned by a UV/O_3 treatment at 60 °C (30 min, PSD Pro Series, Novascan, Ames, IA, USA) prior to deposition.

2.4. Thin Film Synthesis

The precursors were deposited via ultrasonic spray deposition (Exacta Coat, Sono-Tek Cooperation, Milton, NY, USA) with a deposition temperature set at 180 °C for the tungsten precursor and 200 °C for the Li-La-Ti-precursor. The liquid was dispensed at 0.2 mL·min^{-1} and the carrier gas (N_2) pressure was set at 1.5 psi. The spray nozzle to the substrate distance was 2.7 cm. The nozzle moved with a speed of 100 mm·s^{-1}. The number of deposition cycles was varied (2 to 20) with 5 s dry time between the deposition cycles, including intermediate heat treatments every 5 cycles. The deposited W-precursor was annealed at 600 °C for 1 h on the Pt planar substrate, 10 min at 500 °C for 10 min on the other substrates. Deposited Li-La-Ti-precursor was annealed at 500 °C for 1 h on a hotplate for Pt planar substrates, 10 min at 500 °C for on the other substrates.

2.5. Characterization

The thermal decomposition profile of the dried W- and Li-La-Ti-precursor gel, obtained by drying the precursor solution at 60 °C, was investigated by thermogravimetric analysis with coupled differential scanning calorimetry (TGA-DSC, SDT Q600, TA instruments, Asse, Belgium). Six milligrams of the gel was heated at 10 °C·min^{-1} using dry air (0.1 L·min^{-1}) in an alumina crucible. The thermal decomposition profile of the W-precursor films was investigated by thermogravimetric analysis coupled with mass spectrometry (TG-MS, Q5000 with Pfeiffer quadrupole MS, Asse, Belgium). The film was obtained by W-precursor deposition on thin borosilicate glass (Micro Cover Glasses, thickness 0.08 mm, VWR, Oud-Heverlee, Belgium). The coated glass was heated in a Pt sample crucible, ramped at 10 °C·s^{-1} from room temperature to 500 °C in static air.

Crystallization of the WO_3 films deposited on Pt was investigated by XRD via a Bruker D8 equipped with a linear strip detector, using a step size of 0.040° and a counting time of 0.3 s per step at room temperature. Crystallization of non-annealed W-precursor films on TiN coated Si micro-cylinders, was studied by in-situ heating XRD (D8 Discover, Bruker, Champs-sur-Marne, France) with experimental heating chamber [44]. The morphology and film thicknesses were investigated via scanning electron microscopy (SEM, Quanta 200F, FEI, Zavetem, Belgium), cross-sections were made by cutting the substrates with a diamond pen. In addition, the average film thickness of WO_3 films was determined by ICP-AES, after dissolving the deposited films in NH_3 (overnight). Cyclic voltammetry and galvanostatic measurements were done with an Autolab PGSTAT128N, using a three-electrode setup with a custom made Teflon cell similar to the setup used before [10]. The counter and reference electrodes consisted of metallic lithium (99.9% Sigma Aldrich, Overijsse, Belgium). The measurements were done in 1.0 M $LiClO_4$ in propylene carbonate (Soulbrain, Northville, MI, USA). After 3 cycles of CV at 1 $mV{\cdot}s^{-1}$ between 2 and 3.5 V, galvanostatic measurements were done with the same cut-off voltages with various (dis)charge currents. The test cell was operated at 20.0 °C in a control chamber, inside an Ar-filled glovebox with H_2O and O_2 concentrations < 1 ppm. In addition, cyclic voltammetry was done at 10 $mV{\cdot}s^{-1}$ for 5 cycles in 0.1 M of ferrocene (Sigma Aldrich, $\geq$98%, Overijsse, Belgium) in anhydrous acetonitrile (ATN; $\geq$99.9%, VWR, Oud-Heverlee, Belgium), using a Ag/$AgNO_3$ (0.01 M) in 0.1 M Bu_4NPF_6 as reference electrode, and acetonitrile with 0.1 M Bu_4NPF_6 as counter electrode. The ferrocene solution was bubbled in N_2 for 2 to 3 min before starting of the experiment.

3. Results and Discussion

3.1. Planar Tungsten Oxide and Lithium Lanthanum Titanium Oxide Stacks

3.1.1. Thermal Decomposition of the Precursor

The decomposition of the W-precursor is shown in Figure 1. After evaporation of residual water from the precursor gel due to incomplete drying, a first weight loss is observed at 140 and 180 °C with endothermal features. This is followed by two minor exothermal decomposition steps at 365 and 485 °C. Finally, a last major weight loss is observed at 560 °C during a strongly exothermic reaction. No significant weight loss is observed at higher temperatures, implying that the precursor is fully decomposed at 595 °C. Since the tungsten precursor was prepared with pH 8, the solution consists of tungsten complexes with three oxo-groups, one aqua group and a single citrate ligand coordinated to the tungsten, where the citrate co-exists in protonated and deprotonated form [45]. Since an excess of citric acid was initially required to completely dissolve the tungstic acid (4:1 citric acid to tungsten ratio), the excess of citric acid is converted to ammonium citrate due to the addition of ammonia. The presence of ammonium citrate can most clearly be observed during the endothermic peaks at 140 and 185 °C, ascribed to melting and decarboxylation reactions of the ammonium citrate, respectively [46]. Other features of the decomposition profile resemble decompositions of metal-citrate complexes studied before [47,48], although the temperature at which these reactions occur are shifted for this tungsten based complex, due to specific tungsten-oxygen interactions.

Since the Li-La-Ti-precursor is based on a different composition (citrates and nitrates) [43], this precursor decomposition shows different features than the W-precursor (solely citrates). First, the first major weight loss occurs at slightly higher temperatures, during a endothermic reaction at 220 °C, immediately followed by an exothermic reaction at 270 °C. Next, weight loss occurring after a small exothermal event at 420 °C continues towards the final major weight loss accompanied by strong heat generation at 535 °C. While comparing the separate decomposition profiles of the nitrate based-precursors (Figure S1) and titanium-citrato-peroxo precursor with small amounts of ammonium citrate [49], it becomes clear that the final decomposition temperature of the Li-La-Ti-precursor is lower than the separate starting products. Citrates and nitrates readily interact, forming a fuel-oxidizer mixture known from propellant chemistry [50]. Although the exact mechanism goes beyond the scope of this work, the interactions between citrates and nitrates at temperatures up to 250 °C may prevent formation of high temperature-stable lithium and lanthanum carbonates [51], yielding a lower final decomposition temperature.

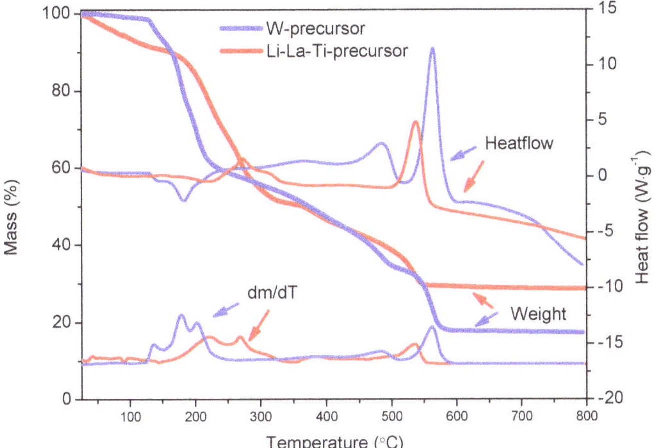

Figure 1. TGA-DSC analysis of the W- and Li-La-Ti-precursors dried at 60 °C, recorded at 10 °C·min^{-1} in dry air.

3.1.2. Morphology and Thickness of Deposited Films

Regarding the morphology, ultrasonic spray deposition of W-precursor yields 250 to 300 nm thick films, which look rather smooth, although porous features are visible (Figure 2a). The backscattering image shows a clear difference between the Pt and WO_3. ICP-AES analysis of dissolved films show that an average thickness of 280 nm is obtained for 10 deposition cycles with the 25 mM W-precursor (27–28 nm/cycle). Upon deposition of the LLT on top, a slightly darker layer of 120 nm can be found on top of the WO_3 (Figure 2b). To ensure a dense morphology, a lower precursor concentration was chosen to lower the growth rate (6 nm/cycle). The resulting LLT has a very smooth and dense morphology, due to the fact that the material remains amorphous at 500 °C [43]. Further heating towards the crystalline perovskite phase was not attempted, since previous attempts regarding a spin-coated film led to crack formation [43].

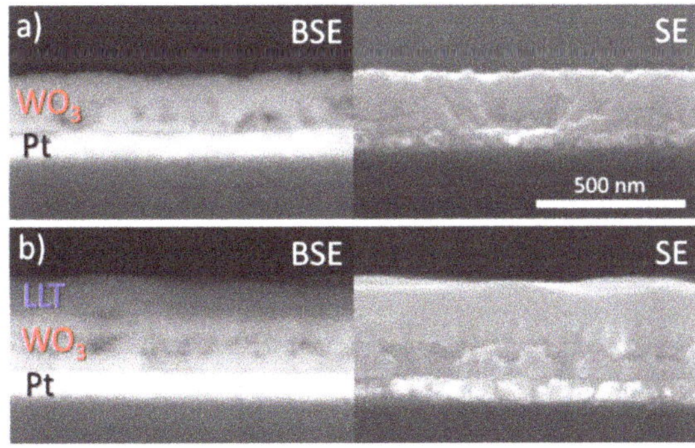

Figure 2. Cross-section SEM micrograph showing backscattered (BSE) and secondary electron (SE) images: (**a**) 10 cycles of W-precursor deposition on a planar Pt substrate, annealed at 600 °C; and (**b**) 20 cycles of Li-La-Ti-precursor deposition on WO_3/Pt substrate, annealed at 500 °C.

3.1.3. Functional Properties of Planar Films—Electrochemical Activity

Electrochemical activity of the WO_3 films as function of annealing temperature was probed by means of cyclic voltammetry, indicating that at least 400–600 °C annealing is required to obtain active films (not shown). With a 600 °C anneal, the films appeared to be most active (Figure 3a). Two distinct reduction peaks can be observed while cycling to lower potentials, associated with W^{6+} to W^{5+} (2.71 V) and W^{5+} to W^{4+} (2.40 V) reductions [52]. Although less intense, the oxidation reactions occur at 2.67 V (W^{4+} to W^{5+}) 2.95 V (W^{5+} to W^{6+}). The relative large currents at low voltages indicate kinetic hindrance effects due to mediocre Li^+ and/or e^- conductivity. In principle, degenerate energy levels of amorphous WO_3 phases could also be the cause of this [53], although the XRD analysis of the phase clearly indicates that significant amounts of crystalline WO_3 are present (Figure 3b).

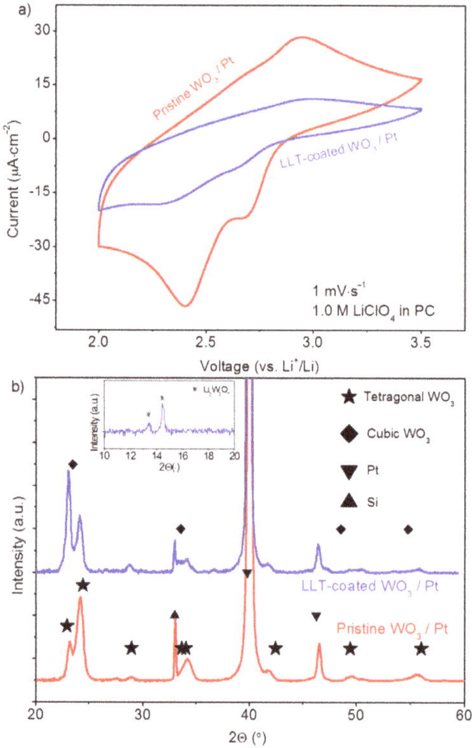

Figure 3. WO_3-coated planar Pt substrate, annealed at 600 °C for 1 h in static air (red) and LLT-coated WO_3 on a planar Pt substrate (prepared under the same conditions) annealed at 500 °C for 1 h in static air (blue), showing: (**a**) CV of the second cycles; and (**b**) XRD. Tetragonal WO_3 (JCPDS 1-85-807), cubic WO_3 (JCPDS 20-1323), Pt (JCPDS 1-87-647), Si (JCPDS 1-72-1088) and $Li_2W_2O_7$ (JCPDS 28-598) phases are all indicated.

Upon coating the WO_3 with LLT, the XRD analysis (Figure 3b) indicates that predominantly tetragonal WO_3 remains present after the LLT coating. As the LLT is amorphous after the 500 °C anneal [43], no additional peaks were expected for this material. However, a close analysis reveals that the intensities of the peaks located at 23 and 24° (2θ) are reversed. Since the cubic phase of WO_3 can be formed upon lithiation of WO_3 [33], this change in intensity suggests a mixture of cubic and tetragonal WO_3 is present in the LLT coated sample. The occurrence of this phase change is assigned to a reaction between Li^+ and WO_3. This is confirmed by additional analysis of the lower diffraction angles, revealing that minor amounts of $Li_2W_2O_7$ are present due to a solid-state reaction as a result of lithium diffusion from the LLT layer to the WO_3 underneath. Although $Li_2W_2O_7$ is known as an electrode material with a larger storage capacity than WO_3 [54], the intercalation voltage (1.55 V vs. Li^+/Li) is too low to be compatible with the LLT electrolyte, which suffers from a Ti^{4+} reduction at lower voltages [3]. Hence, for this study, the $Li_2W_2O_7$ is regarded as an undesirable but unavoidable secondary phase for the LLT-WO_3 half-cell. With respect to the electrochemical activity, the current measured in the CV (Figure 3a) for the LLT-WO_3 stack drops as compared to pristine WO_3. In addition, a larger over-potential (approximately 50 mV) is noted in case of the LLT-coated WO_3 sample. Although passivating effects of the electrolyte deposition and anneal are not excluded, possibly related to the formation of the $Li_2W_2O_7$ phase, both the drop in current and the increase in over-potential can be related to the resistance induced by the amorphous solid-electrolyte layer on top of the WO_3. Since amorphous LLT was reported to have a relatively low Li-ion conductivity (10^{-8} S·cm^{-1}) using comparable synthesis conditions [43], these features are not surprising, but merely serve as a proof of the limited Li-ion conductivity of the electrolyte layer.

Additional experiments were carried out to check the electron-blocking properties of the deposited electrolyte [16]. Figure 4 shows a CV of pristine and LLT-coated samples in the presence of ferrocene in a non-aqueous liquid electrolyte. Upon lowering the potential, ferrocene is reduced to ferrocenium because of electron-transfer at the surface of the sample. For pristine WO_3, this clearly occurs, expressed in the large reduction oxidation peaks. The WO_3 film clearly leads to electron-transfer, facilitating a ferrocene reduction. While comparing this with the LLT-coated WO_3, no reduction of the ferrocene can be observed. This implies that no surface is available for electron transfer although the ferrocene is fully exposed to the LLT surface. Hence, the lack of electron transfer indicates that the LLT electrolyte film is strongly electron insulating, which is crucial for application in an all-solid-state Li-ion battery.

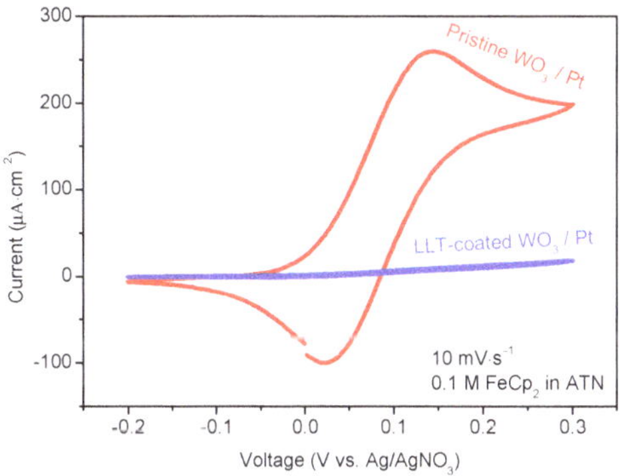

Figure 4. CV showing ferrocene reduction for 10 deposition cycles W-precursor deposition on a planar Pt substrate, annealed at 600 °C for 1 h in static air (red) and LLT-coated WO_3 (prepared under the same conditions) annealed at 500 °C for 1 h in static air (blue). The samples were measured at 10 mV·s^{-1}. The third cycles are shown.

Galvanostatic lithiation/delithiation curves of the deposited WO_3, as shown in Figure 5, exhibit typical behavior observed for WO_3 [33–35]; a short first plateau (2.7 V) followed by a longer plateau (2.6 to 2.4 V). Both plateaus are associated with the intercalation of Li^+ in the tetragonal and cubic phase, respectively [35,53]. As was expected based on the CV results (Figure 3a), kinetic factors and/or amorphous phases appear to be present and responsible for intercalation at lower voltages, expressed in long slopes instead of charge plateaus [25]. Since the capacity drops to lower values at high current densities (Figure 5b), kinetic factors seem the most plausible reason for the occurrence of these slopes.

At a first glance, the lithiation capacity of 1405 mAh·cm^{-3} (197 mAh·g^{-1}) for the films measured at low current densities of 1.2 µA·cm^{-2} (0.07C) (Figure 5a) appeared to be very high (232% of WO$_3$ theoretical capacity), especially compared to films measured with the 10-fold current density of 12 µA·cm^{-2} (0.7C) (Figure 5b) yielding 492 mAh·cm^{-3} (68 mAh·g^{-1}, 81% of WO$_3$ theoretical capacity). However, a closer inspection of samples cycled at high and low current densities showed a striking macroscopic difference. As an electrochromic material, the lithiation of WO$_3$ is macroscopically visible on the film due to dark blue coloration of reduced tungsten [52]. After a single lithiation down to 2.0 V, with a high current density of 12 µA·cm^{-2} (0.7C), a dark blue coloration of the sample measured was confined within the O-ring of the electrochemical cell (Figure S2a), as was expected. However, at a low current density of 1.2 µA·cm^{-2} (0.07C), the dark blue coloration was visible over all the substrate, including the area outside of the O-ring. XRD of selected areas (Figure S2b) indicated the same, as the cubic WO$_3$ (implying lithiation occurred) was observed outside the O-ring area for low current densities. Since electron transfer of the Pt to the WO$_3$ is possible all over the substrate, it is speculated that the relative high Li-ion diffusion through the WO$_3$ film is responsible for this.

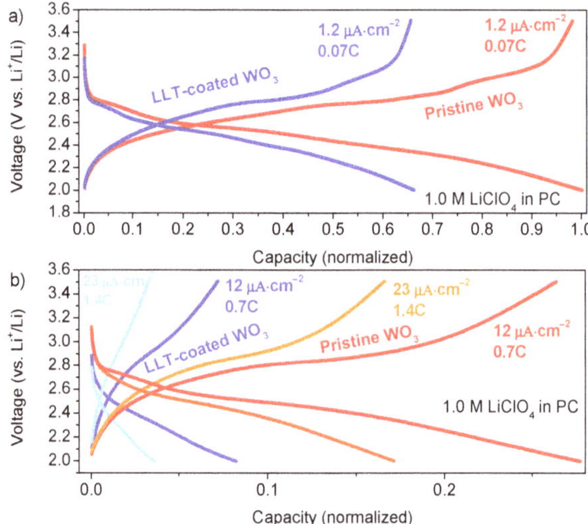

Figure 5. Lithiation/delithiation curves of 10 cycles W-precursor deposition on a planar Pt substrate, annealed at 600 °C for 1 h in static air and LLT-coated WO$_3$ (prepared under the same conditions) annealed at 500 °C for 1 h in static air. The figure shows samples measured at: (**a**) low; and (**b**) high current densities. The capacity is normalized according to pristine WO$_3$ cycled at 1.2 µA·cm^{-2} (0.07C). In all cases, the 10th cycle at the specified C-rate is shown.

With diffusion constants of lithium in (cubic) WO$_3$ in the order of 10^{-8} to 10^{-7} cm^2·s^{-1} [22,55], a simple calculation (stochastic diffusion without convection) indicates that an average diffusion length in the millimeter range is feasible within the timescale of the experiment. However, unknown effects caused by (long) exposure to the liquid electrolyte are not excluded, but more detailed analysis of this topic goes beyond the scope of this study. Hence, the total capacity of the planar films reported in this study are normalized based capacity measured at the lowest current density, since exact determination of the capacity at low current densities is impossible without the confined volume approximation. The lithiation/delithiation curves of the LLT-coated WO$_3$ (Figure 5) exhibit similar features as the pristine WO$_3$—including the charge plateaus of the WO$_3$ phases, although total capacity is significantly smaller. The capacity reduction can partly be ascribed to the presence of the inactive Li$_2$W$_2$O$_7$ phase, as was shown by XRD (Figure 3b). However, the capacity reduction can also be related to a decreased rate performance induced by the LLT layer, as was already shown by the fact that a larger over-potential and smaller peak current were measured during CV (Figure 3a). The lack of confined volume explained also applies here at low current densities, but the effect is probably less severe because of the resistive character of the LLT electrolyte on top. Although the confined volume approximation applies at higher current density for the LLT coated samples, a larger capacity reduction is observed compared to pristine WO_3 (Figure 5b). Thus, the smooth morphology and electronic blocking ensure crucial functions of the LLT electrolyte, but the amorphous character comes at a cost: reduced rate performance of the half-cell.

3.2. Tungsten Oxide and Lithium Lanthanum Titanium Oxide Stacks 3D Deposition

Now that activity of both materials has been shown, the next step is taken towards forming an integrated 3D all-solid-state Li-ion battery: the non-planar deposition of WO_3-LLT stacks. To study this extraordinary deposition requirement, Si substrates with trenches were used which were approximately 30 μm deep and 30 μm wide, i.e., aspect ratio 1 [8,11]. Upon deposition of the W-precursor at the optimized deposition temperature of 180 °C [21], the heavy tungsten is clearly visible all over the trench using backscattered imaging (Figure 6a), i.e., a fully covering WO_3 film is formed. The coating is thicker in the corners, which can be explained by a capillary pressure gradient, leading to enhanced flow towards the corners of the trench geometry [56].

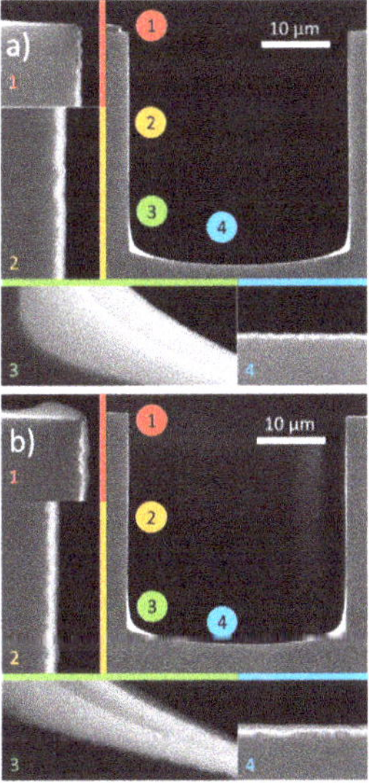

Figure 6. (**a**) SEM micrograph of 10 cycles of W-precursor deposition on a trench of 30 by 30 μm, annealed at 500 °C for 10 min in static air; and (**b**) additional deposition of 10 layers of Li-La-Ti-precursor deposition at 200 °C with an anneal at 500 °C for 10 min. Numbers indicate enlarged image locations.

However, the exact layer thickness quantification for the bottom corner is relatively difficult because of a possible reaction between the Si substrate and the tungsten oxide layer (forming, e.g., tungsten silicates) [57] at the thickest part of the deposited layer. This is observed in a local difference in tungsten concentration as expressed in intensity difference of the backscattered image. Similar results were found for the deposition of a W-precursor on a planar Si substrate (Figure S3a), while this phenomenon did not occur in presence of a buffer layer such as Pt (Figure 2). We therefore quantify, within the uncertainty explained above, that the conformality is only 2%, i.e., the minimum layer thickness (top edge of the trench) divided by the maximum layer thickness (bottom corner). Further increasing the aspect ratio of the trenches, i.e., decreasing the width of the trench, no longer allows a fully covering deposition (cf. Figure S3a,b). It is speculated that three main phenomena are causing the difference in conformality with respect to the high aspect ratio trenches: (i) available surface; (ii) thermophoretic forces; and (iii) capillary forces and surface tension.

(i)   The probability of a droplet reaching the bottom of the trenches scales with the area of the trench opening (top). As the width of the trench reduces, less material is able to reach and spread at the bottom of the trench.

(ii)   Thermophoretic forces, i.e., the repelling force caused by a huge pressure gradient near to the hot surface [58], are larger in the case of small trenches. Carrier gas diffusion into a small trench is more difficult than in to a broad trench, leading to an increased thermophoretic force for smaller trench. Therefore, layer deposition into deeper parts of the trenches is becoming more difficult.

(iii)   A necking mechanism is observed, especially for the smallest type of trench measured. The surface tension of the gel leads to necking (contacting similar material) instead of wetting the alien (Si) material down the trench.

In view of materials stacking, only the trench with aspect ratio 1, coated with a fully covering WO_3 coating, was used for the LLT-precursor deposition. The result in Figure 5b shows that a conformal LLT solid-electrolyte layer is deposited on top of the WO_3 layer. In addition, the quantified conformality (33%) of the LLT layer is much better than for the WO_3 deposition (2%), as the difference in thickness at the edge and bottom are considerably smaller. Besides possible differences in the surface chemistry, the large difference is probably generated by the fact that the sharp edge present in the uncoated trench is no longer present after coating with WO_3. Hence, the LLT layer covers a curved surface instead of a sharp corner, which is beneficial for the deposition homogeneity. Although there is room for optimization of layer thicknesses, these results indicate that stacked fully covering coatings are possible using a wet-chemical approach.

3.3. Towards Functional WO_3 3D Coatings

3.3.1. WO_3 Deposition on High Aspect Ratio Substrates

The activity of the materials was demonstrated for planar substrates, in addition to full coverage of the non-planar trench structures with WO_3. However, the capacity gain due to the use of 3D architectures is yet to be shown using this approach [3]. In order to do so, a substrate with a high aspect ratio and a suitable current collector was required. For this study, TiN coated micro-cylinders of 50 μm height, approximately 1 μm radius and inter-cylinder distance of 5 μm were available, enabling a potential 8-fold capacity increase. Figure S4 shows that good coatings could be obtained while submitting these micro-cylinders to the W-precursor spray, with an aspect ratio of 10. Although only limited aspect ratios were achieved for the trenches—with a "closed" structure (Figure 6a)—apparently higher aspect ratio "open" micro-cylinder can indeed be coated with this approach (Figure S4). It is speculated that the large openings (at the top) and beneficial capillary forces of these micro-cylinder structures attribute to this difference. In addition, no material accumulates at the top of the substrate, preventing the "necking" occurring in case of the trenches (Figure S3c). Besides SEM micrographs (Figure S4), the galvanostatic measurements also indicate the micro-cylinders are coated (cf. Section 3.3.3). Unlike the trench structures (Figure 5), cross sections by SEM were not possible on these micro-cylinders. ICP-AES based thickness estimations for the coated TiN/Si micro-cylinders indicate that the average WO_3 film thickness for 16 deposition cycles is 32 nm.

3.3.2. Thermal Decomposition and Crystallization

Although TiN (present on the micro-cylinders) is a good electronic conductor and is very interesting for this specific application because of its excellent (Li) barrier properties [59], the TiN current collector is highly susceptible to oxidation at elevated temperatures in air, which leads to the formation of TiO_2 with a low electronic conductivity [60]. This therefore initially resulted in unmeasurable samples (not shown). Optimized annealing conditions were therefore sought, in order to find an optimum between precursor decomposition, crystallization and electrochemical activity of the material (WO_3) on the one hand, and preservation of the current collector (TiN) on the other. Isothermal heating at 500 °C was applied to decompose the W-citrate, preserving the TiN current collector for a limited time. The non-isothermal TGA profile of the W-precursor suggests that part of the citrate complex is not fully decomposed at 500 °C, even while applying an isotherm at this temperature (not shown). To make sure that heat transfer, volume and geometry differences concerning the decomposition analysis with in-situ XRD (isXRD) and TGA were minimized, deposited films instead of powders were measured [61]. The result is a weight loss profile (Figure 7a) showing continuous weight loss, but to a smaller extent than in the case of gels (not shown), which is mostly attributed to higher heat transfer rates and a larger surface for gas evolution for film-based TGA. Using coupled MS methods, the weight loss could be assigned to CO_2 and H_2O, meaning that a continuous combustion of organic groups occurs during the isothermal period at 500 °C. The slow combustion occurs together with a gradual increase of crystallinity, which was monitored in real-time using isXRD (Figure 7b,c). Since W–O–W bonds are required for crystallization of WO_3, W–O–C bonds of the citrate complex should be broken. The gradual but immediate crystallization was to be expected, as amorphous WO_3 is reported to crystalize at temperatures as low as 300 °C [24,36]. However, this does not fully exclude occurrence of decomposing residual organic groups after crystallization, since ammonium citrate (partly decomposed) residues or detached citrate ligand groups can decompose after crystallization. The crystallization increases over time to form the tetragonal phase of WO_3.

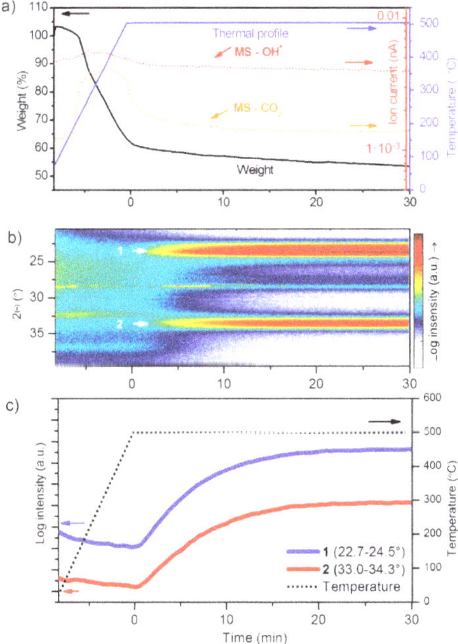

Figure 7. Analysis of as-deposited W-precursor films on: (**a**) glass showing TGA coupled with MS (10 °C·min^{-1} in static air, m/z 17/OH and 44/CO_2) and isXRD on coated TiN/Si micro-cylinders showing: (**b**) peak intensity as function of peak position and time; and (**c**) integrated peak intensities of specified diffraction peaks as function of temperature, where blue is related to peak 1 and red to peak 2 to in (**b**).

A relatively large gain in intensity is found during the first 10 min of the isothermal anneal as approximately 75% of the maximum intensity is reached within this time for the 23.5 and 33.5 peaks (Figure 7c). This matches with the H_2O and CO_2 evolution, which show a decreasing intensity over this period of time. Therefore, 10 min of isothermal heating was chosen as an optimum between film crystallization and preservation of the current collector (TiN) present on the micro-cylinders.

3.3.3. Functional Measurements of 3D WO_3 Coatings

Despite the presence of residual organic fragments and incomplete crystallization after only 10 min of isothermal heating at 500 °C, active WO_3 present on the micro-cylinders could be cycled as shown in Figure 8. To study the effect of 3D deposition more thoroughly than earlier reports by our labs [21], the micro-cylinder sample was compared with a planar substrate exhibiting the same amount of material per footprint area (i.e., 2 deposition cycles for 2D, compared to 16 deposition cycles for 3D with an area enhancement factor of 8). In addition, the current applied to the 3D samples is eight fold the current passed through the 2D samples (Figure 8). The lithiation/delithiation curves look similar for 2D and 3D, although the capacity is significantly larger for the 3D sample. All curves in Figure 8a lack plateaus as observed on planar samples which were annealed longer at higher temperature. Instead, slopes are observed, especially during delithiation. This is an indicator that primarily amorphous WO_3 is present in the samples [25], despite indications of isXRD that part of the material is crystalline (Figure 7b). Since these diffraction peaks are broad compared to a non-isothermal anneal at higher temperature (Figure S5), it can be concluded that the crystallites are rather small, i.e., occurrence of nano-crystallinity. It is speculated that these nanocrystals might lead to comparable, (quasi) amorphous lithiation/delithiation behavior. At 0.5C, the two 2D sample is able to store a charge of 3.35 μAh·cm^{-2}, whereas the 3D sample stores 8.90 μAh·cm^{-2} with a comparable current per equivalent coating area. This is a 2.6-fold capacity improvement which can be attributed to the benefit that 3D geometries exhibit over 2D electrodes. This result equals the capacity enhancement obtained for LTO deposited using CVD, on similar high aspect ratio substrates (TiN coated micro-cylinders) which achieved a 2.5 capacity enhancement as well as previous reports from our labs reaching a near three fold capacity increase [8,21]. However, these results were achieved by applying the equal current density per footprint area, whereas the current study shows that a more critical equal current per coating area. In addition, it enables to study the kinetic performance of the 3D structured films in more detail (Figure 8b). Nevertheless, the eight-fold capacity enhancement, which theoretically should be achievable, is not met. There are several reasons for not achieving the theoretical capacity increase:

(i)   The deposition of W-citrate on the trench structures (Figure 6a) illustrates that layer thickness cannot be controlled perfectly, which leads to local differences in capacity of the layer, therefore lowering the capacity enhancement.

(ii)   Although the coating is present on the micro-cylinders, certain defects can be observed (Figure S4).

(iii)   The current collector may remain an issue; upon cycling, the voltage drop at the start of the measurement is significantly larger for the 3D samples than for the 2D samples (Figure 8a). This illustrates that higher resistance is presence in these 3D samples, which is primarily attributed to the partial oxidation of the TiN current collector (20 nm). As the TiN current collector is much thicker for the 2D samples (80 nm), and the applied currents are much lower, this leads to lower IR-drop for planar films, compared to the 3D counterparts.

Nevertheless, the de-lithiation kinetics of the 2D and 3D film are comparable between 0.5C and 10C (Figure 8b). The same slope of the lines in the double log plot indicate the same kinetic regime acts on both films [11], where the 2.6 fold capacity enhancement is maintained up to high current densities.

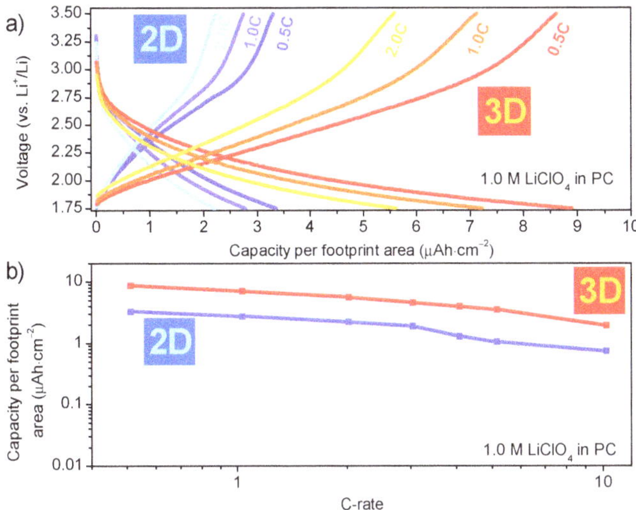

Figure 8. (**a**) Lithiation/delithiation curves; and (**b**) delithiation capacity-rate log-log plot of 2D (Pt) and 3D (TiN) WO$_3$ coated substrates, annealed at 500 °C for 10 min in static air. 2D samples were subjected to two deposition cycles, and 3D sampled (with a factor 8 area enhancement) were subjected to 16 deposition cycles. In all cases, data of the 10th cycles are shown and volumetric capacities are determined with ICP-based (2D and 3D) coating thicknesses.

4. Conclusions

All things considered, the stacking of lithium lanthanum titanate (LLT) with tungsten oxide (WO$_3$) yielded an active all-solid-state half-cell, showing both the Li-ion conducting and electron blocking function of the electrolyte layer. With smooth morphologies and relatively mild thermal budgets (500 °C), the combination of citrate-nitrate chemistry led to active 3D structured materials within the thermal budget limitations. This is another step forward within the field of wet-chemical synthesis, since 3D depositions were previously limited to all-citrate based precursors. This adds to the understanding of this complicated, fascinating process to establish high aspect ratio oxide coatings. Although additional progress is required to suppress secondary phase formation and enhancement of the Li-ion conductivity of the half-cell, the results of this study form the beginning of tackling problems related to the Achilles' heel of the all-solid-state 3D Li-ion battery research: 3D deposition of suitable solid-electrolyte materials. Besides the advancements made for the solid-electrolyte, the near threefold capacity enhancement of the 3D-structured WO$_3$ negative electrode is an important indicator of the versatility of wet-chemical method in combination with ultrasonic spray deposition. Future improvements should be booked to obtain a suitable current collector with a larger thermal budget, as this is the limiting factor while probing the performance of 3D-structured electrode materials. Nevertheless, stacking of oxide materials using ultrasonic spray deposition—in 3D—is a major step forward in terms of possible new material architectures, especially because both layers were deposited with an easy-to-upscale process that can possibly be applied to many other applications as well.

Supplementary Materials: The following are available online at www.mdpi.com/1996-1944/10/9/1072/s1, Figure S1: TGA analysis of (a) the lithium nitrate precursor and (b) lanthanum nitrate precursor dried at 60 °C, recorded at 10 °C·min^{-1} in dry air (0.1 ml·min^{-1}), Figure S2: Comparison between WO$_3$ coated (10 cycles, 25 mM) Pt samples after a single lithiation down to 2.0 V vs Li$^+$/Li, at high and low current density. (a) Photograph of the samples and O-ring used in the electrochemical cell, showing dark blue coloration of lithiated areas. The colored squares relate to (b), showing XRD on isolated parts after cutting the samples, Figure S3: SEM micrograph showing backscattered (BSE) image of 10 cycles of W-precursor deposition on a (a) planar Si substrate, (b) trench of 10 by 27 μm and (c) trench of 3.5 by 22.5 μm, all annealed at 500 °C for 10 min in static air, Figure S4: SEM micrograph of 10 cycles of W-precursor deposition at 180 °C on 50 μm high micro-cylinders, with an average diameter of 2.5 μm, with 5 μm inter-cylinder spacing. The sample was annealed at 500 °C for 10 min in static air, Figure S5: In-situ XRD results showing applied temperature profile (top), as well as (bottom) diffraction intensity as function of peak position and time. Both graphs are based on the same sample; W-citrate deposited using 10 cycles on TiN micro-cylinders at a deposition temperature of 180 °C.

Acknowledgments: Huguette Penxten is acknowledged for her assistance during the ferrocene reduction experiments. Imec (Leuven, Belgium) and Phillipe Vereecken are acknowledged for providing micro-cylinder substrates. This study was partly supported by the IWT SBO SOS Lion project.

Author Contributions: E.J.v.d.H., G.M., W.M., K.E., F.M., P.H.L.N., A.H. conceived and designed the experiments; E.J.v.d.H., G.M., G.B., W.M., K.E., S.G., F.M., performed the experiments; E.J.v.d.H., G.M., W.M., F.M., P.H.L.N., M.K.V.B., A.H. analyzed data; C.D., P.H.L.N., M.K.V.B, A.H. contributed materials and analysis tools; E.J.v.d.H. and A.H. wrote the paper.

Conflicts of Interest: The authors declare no conflict of interest.

References

1. Radzir, N.N.M.; Hanifah, S.A.; Ahmad, A.; Hassan, N.H.; Bella, F. Effect of lithium bis(trifluoromethylsulfonyl)imide salt-doped UV-cured glycidyl methacrylate. *J. Solid State Electrochem.* **2015**, *19*, 3079–3085. [CrossRef]

2. Nair, J.R.; Destro, M.; Bella, F.; Appetecchi, G.B.; Gerbaldi, C. Thermally cured semi-interpenetrating electrolyte networks (s-IPN) for safe and aging-resistant secondary lithium polymer batteries. *J. Power Sources* **2016**, *306*, 258–267. [CrossRef]

3. Knauth, P. Inorganic Solid Li Ion Conductors: An Overview. *Solid State Ion.* **2009**, *180*, 911–916. [CrossRef]

4. Mahmoud, M.M.; Cui, Y.; Rohde, M.; Ziebert, C.; Link, G.; Seifert, H.J. Microwave Crystallization of Lithium Aluminum Germanium Phosphate Solid-State Electrolyte. *Materials* **2016**, *9*, 506. [CrossRef] [PubMed]

5. Oudenhoven, J.F.M.; Baggetto, L.; Notten, P.H.L. All-Solid-State Lithium-Ion Microbatteries: A Review of Various Three-Dimensional Concepts. *Adv. Energy Mater.* **2011**, *1*, 10–33. [CrossRef]

6. Long, J.W.; Dunn, B.; Rolison, D.R.; White, H.S. Three-Dimensional Battery Architectures. *Chem. Rev.* **2004**, *104*, 4463–4492. [CrossRef] [PubMed]

7. Notten, P.H.L.; Roozeboom, F.; Niessen, R.A.H.; Baggetto, L. 3-D Integrated All-Solid-State Rechargeable Batteries. *Adv. Mater.* **2007**, *19*, 4564–4567. [CrossRef]

8. Xie, J.; Peter-Paul, R.M.L.H.; Li, D.; Raijmakers, L.H.J.; Notten, P.H.L. Planar and 3D Deposition of Li4Ti5O12 Thin Film Electrodes by MOCVD. *Solid State Ion.* **2016**, *287*, 83–88. [CrossRef]

9. Donders, M.E.; Oudenhoven, J.F.M.; Baggetto, L.; Knoops, H.C.M.; van de Sanden, M.C.M.; Kessels, W.M.M.; Notten, P.H.L. All-Solid-State Batteries: A Challenging Route Towards 3D Integration. *ECS Trans.* **2010**, *33*, 213–222. [CrossRef]

10. Donders, M.E.; Arnoldbik, W.M.; Knoops, H.C.M.; Kessels, W.M.M.; Notten, P.H.L. Atomic Layer Deposition of LiCoO$_2$ Thin-Film Electrodes for All-Solid-State Li-Ion Micro-Batteries. *J. Electrochem. Soc.* **2013**, *160*, A3066–A3071. [CrossRef]

11. Dobbelaere, T.; Mattelaer, F.; Dendooven, J.; Vereecken, P.; Detavernier, C. Plasma-Enhanced Atomic Layer Deposition of Iron Phosphate as a Positive Electrode for 3D Lithium-Ion Microbatteries. *Chem. Mater.* **2016**, *28*, 3435–3445. [CrossRef]

12. Nisula, M.; Shindo, Y.; Koga, H.; Karppinen, M. Atomic Layer Deposition of Lithium Phosphorus Oxynitride. *Chem. Mater.* **2015**, *27*, 6987–6993. [CrossRef]

13. Xie, J.; Oudenhoven, J.F.M.; Peter-Paul, R.M.L.H.; Li, D.; Notten, P.H.L. Chemical Vapor Deposition of Lithium Phosphate Thin-Films for 3D All-Solid-State Li-Ion Batteries. *J. Electrochem. Soc.* **2014**, *162*, A249–A254. [CrossRef]

14. Lin, C.-F.; Noked, M.; Kozen, A.C.; Liu, C.; Zhao, O.; Gregorczyk, K.; Hu, L.; Lee, S.B.; Rubloff, G.W. Solid Electrolyte Lithium Phosphous Oxynitride as a Protective Nanocladding Layer for 3D High-Capacity Conversion Electrodes. *ACS Nano* **2016**, *10*, 2693–2701. [CrossRef] [PubMed]

15. Kozen, A.C.; Pearse, A.J.; Lin, C.-F.; Noked, M.; Rubloff, G.W. Atomic Layer Deposition of the Solid Electrolyte LiPON. *Chem. Mater.* **2015**, *27*, 5324–5331. [CrossRef]

16. Perng, Y.-C.; Cho, J.; Sun, S.Y.; Membreno, D.; Cirigliano, N.; Dunn, B.; Chang, J.P. Synthesis of Ion Conducting LixAlySizO Thin Films by Atomic Layer Deposition. *J. Mater. Chem. A* **2014**, *2*, 9566–9573. [CrossRef]

17. Li, X.; Liu, J.; Banis, M.N.; Lushington, A.; Li, R.; Cai, M.; Sun, X. Atomic Layer Deposition of Solid-State Electrolyte Coated Cathode Materials with Superior High-Voltage Cycling Behavior for Lithium Ion Battery Application. *Energy Environ. Sci.* **2014**, *7*, 768–778. [CrossRef]

18. Pearse, A.J.; Schmitt, T.E.; Fuller, E.J.; El-Gabaly, F.; Lin, C.F.; Gerasopoulos, K.; Kozen, A.C.; Talin, A.A.; Rubloff, G.; Gregorczyk, K.E. Nanoscale Solid State Batteries Enabled by Thermal Atomic Layer Deposition of a Lithium Polyphospazene Solid State Electrolyte. *Chem. Mater.* **2017**, *29*, 3740–3753. [CrossRef]

19. Létiche, M.; Eustache, E.; Frexias, J.; Demortière, A.; De Andrade, V.; Morgenroth, L.; Tilmant, P.; Vaurette, F.; Troadec, D.; Roussel, P.; et al. Atomic Layer Deposition of Functional Layers for on Chip 3D Li-ion All Solid State Microbattery. *Adv. Energy Mater.* **2017**, *7*, 161402. [CrossRef]

20. Gielis, S.; Hardy, A.; Van Bael, M.K. Conformal Coating on Three-Dimenstional Substrates. Eur. Pat. Appl. 2947178, 2015.

21. Van den Ham, E.J.; Gielis, S.; Van Bael, M.K.; Hardy, A. Ultrasonic Spray Deposition of Metal Oxide Films on High Aspect Ratio Microstructures for 3D All-Solid-State Li-Ion Batteries. *ACS Energy Lett.* **2016**, *1*, 1184–1188. [CrossRef]

22. Yu, A.; Kumagai, N.; Liu, Z.; Lee, J.Y. Electrochemical Lithium Intercalation into WO_3 and Lithium Tungstates Li X WO_3 + X/2 of Various Structures. *J. Solid State Electrochem.* **1998**, *2*, 394–400. [CrossRef]

23. Niklasson, G.A.; Granqvist, C.G. Electrochromics for Smart Windows: Thin Films of Tungsten Oxide and Nickel Oxide, and Devices Based on These. *J. Mater. Chem.* **2007**, *17*, 127–156. [CrossRef]

24. Regragui, M.; Addou, M.; Outzourhit, A.; Bernede, J.C.; El Idrissi, E.; Benseddik, E.; Kachouane, A. Preparation and Characterization of Pyrolytic Spray Deposited Electrochromic Tungsten Trioxide Fims. *Thin Solid Films* **2000**, *358*, 40–45. [CrossRef]

25. Garciacanadas, J. Charging and Diffusional Aspects of Li+ Insertion in Electrochromic a-WO_3. *Solid State Ion.* **2004**, *175*, 521–525. [CrossRef]

26. Nishio, K.; Sei, T.; Tsuchiya, T. Preparation of Electrochromic Thin Film by Sol-Gel Process. *J. Ceram. Soc. Jpn.* **1999**, *107*, 199–203. [CrossRef]

27. Guo, C.; Yin, S.; Yan, M.; Kobayashi, M.; Kakihana, M.; Sato, T. Morphology-Controlled Synthesis of W18O49 Nanostructures and Their Near-Infrared Absorption Properties. *Inorg. Chem.* **2012**, *51*, 4763–4771. [CrossRef] [PubMed]

28. Denayer, J.; Aubry, P.; Bister, G.; Spronck, G.; Colson, P.; Vertruyen, B.; Lardot, V.; Cambier, F.; Henrist, C.; Cloots, R. Improved Coloration Contrast and Electrochromic Efficiency of Tungsten Oxide Films Thanks to a Surfactant-Assisted Ultrasonic Spray Pyrolysis Process. *Sol. Energy Mater. Sol. Cells* **2014**, *130*, 623–628. [CrossRef]

29. Ji, R.; Zheng, D.; Zhou, C.; Cheng, J.; Yu, J.; Li, L. Low-Temperature Preparation of Tungsten Oxide Anode Buffer Layer via Ultrasonic Spray Pyrolysis Method for Large-Area Organic Solar Cells. *Materials* **2017**, *10*, 820. [CrossRef] [PubMed]

30. Li, W.J.; Fu, Z.W. Nanostructured WO_3 Thin Film as a New Anode Material for Lithium-Ion Batteries. *Appl. Surf. Sci.* **2010**, *256*, 2447–2452. [CrossRef]

31. Perereira-Ramos, J.P.; Baddour-Hadjean, R.; Kumagai, N.; Tanno, K. Improvement of the Electrochemical Behaviour of WO_3 as Reversible Cathodic Material for Lithium Batteries. *Electrochim. Acta* **1993**, *38*, 431–436. [CrossRef]

32. Huang, K.; Zhang, Q. Rechargeable Lithium Battery Based on a Single Hexagonal Tungsten Trioxide Nanowire. *Nano Energy* **2012**, *1*, 172–175. [CrossRef]

33. Van den Ham, E.J.; Elen, K.; Kokal, I.; Yagci, M.B.; Peys, N.; Bonneux, G.; Ulu, F.; Marchal, W.; Van Bael, M.; Hardy, A. From Liquid to Thin Film: Colloidal Suspensions for Tungsten Oxide as an Electrode Material for Li-Ion Batteries. *RSC Adv.* **2016**, *6*, 51747–51756. [CrossRef]

34. Li, W.; Sasaki, A.; Oozu, H.; Aoki, K.; Kakushima, K.; Kataoka, Y.; Nishiyama, A.; Sugii, N.; Wakabayashi, H.; Tsutsui, K.; et al. Improvement of Charge/discharge Performance for Lithium Ion Batteries with Tungsten Trioxide Electrodes. *Microelectron. Reliab.* **2015**, *55*, 402–406. [CrossRef]

35. Zhong, Q.; Dahn, J.R.; Colbow, K. Lithium Intercalation into WO_3 and the Phase Diagram of LixWO3. *Phys. Rev. B* **1992**, *46*, 2554–2560. [CrossRef]

36. Nishio, K.; Tsuchiya, T. Electrochromic Thin Films Prepared by Sol-Gel Process. *Sol. Energy Mater. Sol. Cells* **2001**, *68*, 279–293. [CrossRef]

37. Delmas, C.; Nadiri, A.; Soubeyroux, J.L. The Nasicon-Type Titanium Phosphates $ATi_2(PO_4)_3$ (A = Li, Na) as Electrode Materials. *Solid State Ion.* **1988**, *28–30*, 419–423. [CrossRef]

38. Takada, K.; Tansho, M.; Yanase, I.; Inada, T.; Kajiyama, A.; Kouguchi, M.; Kondo, S.; Watanabe, M. Lithium Ion Conduction in $LiTi(PO_4)_3$. *Solid State Ion.* **2001**, *139*, 241–247. [CrossRef]

39. Kim, S.; Hirayama, M.; Taminato, S.; Kanno, R. Epitaxial Growth and Lithium Ion Conductivity of Lithium-Oxide Garnet for an All Solid-State Battery Electrolyte. *Dalton Trans.* **2013**, *42*, 13112–13117. [CrossRef] [PubMed]

40. Reinacher, J.; Berendts, S.; Janek, J. Preparation and Electrical Properties of Garnet-Type $Li_6BaLa_2Ta_2O_{12}$ Lithium Solid Electrolyte Thin Films Prepared by Pulsed Laser Deposition. *Solid State Ion.* **2014**, *258*, 1–7. [CrossRef]

41. Tadanaga, K.; Egawa, H.; Hayashi, A.; Tatsumisago, M.; Mosa, J.; Aparicio, M.; Duran, A. Preparation of Lithium Ion Conductive Al-Doped $Li_7La_3Zr_2O_{12}$ Thin Films by a Sol–gel Process. *J. Power Sources* **2015**, *273*, 844–847. [CrossRef]

42. Aaltonen, T.; Alnes, M.; Nilsen, O.; Costelle, L.; Fjellvåg, H. Lanthanum Titanate and Lithium Lanthanum Titanate Thin Films Grown by Atomic Layer Deposition. *J. Mater. Chem.* **2010**, *20*, 2877–2881. [CrossRef]

43. Van den Ham, E.J.; Peys, N.; De Dobbelaere, C.; D'Haen, J.; Mattelaer, F.; Detavernier, C.; Notten, P.H.L.; Hardy, A.; Van Bael, M.K. Amorphous and Perovskite $Li_{3x}La_{(2/3)−x}TiO_3$ (thin) Films via Chemical Solution Deposition: Solid Electrolytes for All-Solid-State Li-Ion Batteries. *J. Sol-Gel Sci. Technol.* **2015**, *73*, 536–543. [CrossRef]

44. Knaepen, W.; Detavernier, C.; Van Meirhaeghe, R.L.; Jordan Sweet, J.; Lavoie, C. In-Situ X-Ray Diffraction Study of Metal Induced Crystallization of Amorphous Silicon. *Thin Solid Films* **2008**, *516*, 4946–4952. [CrossRef]

45. Cruywagen, J.J.; Krüger, L.; Rohwer, E.A. Complexation of tungsten(VI) with citrate. *J. Chem. Soc. Dalton Trans.* **1991**. [CrossRef]

46. Rajendran, M.; Subba Rao, M. Formation of BaTiO3 from Citrate Precursor. *J. Solid State Chem.* **1994**, *113*, 239–247. [CrossRef]

47. Peys, N.; Ling, Y.; Dewulf, D.; Gielis, S.; De Dobbelaere, C.; Cuypers, D.; Adriaensens, P.; Van Doorslaer, S.; De Gendt, S.; Hardy, A.; et al. V_6O_{13} Films by Control of the Oxidation State from Aqueous Precursor to Crystalline Phase. *Dalton Trans.* **2013**, *42*, 959–968. [CrossRef] [PubMed]

48. Hardy, A.; Mondelaers, D.; Van Bael, M.K.; Mullens, J.; Van Poucke, L.C.; Vanhoyland, G.; D'Haen, J. Synthesis of $(Bi,La)_4Ti_3O_{12}$ by a New Aqueous Solution-Gel Route. *J. Eur. Ceram. Soc.* **2004**, *24*, 905–909. [CrossRef]

49. Truijen, I.; Hardy, A.; Van Bael, M.K.; Van den Rul, H.; Mullens, J. Study of the Decomposition of Aqueous Citratoperoxo-Ti(IV)-Gel Precursors for Titania by Means of TGA-MS and FTIR. *Thermochim. Acta* **2007**, *456*, 38–47. [CrossRef]

50. Jain, S.R.; Adiga, K.C.; Verneker, V.R.P. A New Approach to Thermochemical Calculations of Condensed Fuel-Oxidizer Mixtures. *Combust. Flame* **1981**, *40*, 71–79. [CrossRef]

51. Fortal'nova, E.A.; Mosunov, A.V.; Safronenko, M.G.; Venskovskii, N.U.; Politova, E.D. Electrical Properties of $(La_{1/2}Li_{1/3+x})TiO_3$ Solid Solutions. *Inorg. Mater.* **2006**, *42*, 393–398. [CrossRef]

52. Darmawi, S.; Burkhardt, S.; Leichtweiss, T.; Weber, D.A.; Wenzel, S.; Janek, J.; Elm, M.T.; Klar, P.J. Correlation of Electrochromic Properties and Oxidation States in Nanocrystalline Tungsten Trioxide. *Phys. Chem. Chem. Phys.* **2015**, *17*, 15903–15911. [CrossRef] [PubMed]

53. Cheng, K.H.; Whittingham, M.S. Lithium Incorporation in Tungsten Oxides. *Solid State Ion. Commun.* **1980**, *1*, 151–161. [CrossRef]

54. Pralong, V.; Venkatesh, G.; Malo, S.; Caignaert, V.; Baies, R.; Raveau, B. Electrochemical Synthesis of a Lithium-Rich Rock-Salt-Type Oxide. *Inorg. Chem.* **2014**, *53*, 522–527. [CrossRef] [PubMed]

55. Raistrick, I.D.; Mark, A.J.; Huggins, R.A. Thermodynamics of Electrochemical Insertion of Lithium into Tungsten Bronzes. *Solid State Ion.* **1981**, *5*, 351–354. [CrossRef]

56. Dong, M.; Chatzus, I. The Imbibition and Flow of a Wetting Liquid along the Corners of a Square Capillary Tube. *J. Colloid Interface Sci.* **1995**, *172*, 278–288. [CrossRef]

57. Ten Dam, J.; Badloe, D.; Ramanathan, A.; Djanashvili, K.; Kapteijn, F.; Hanefeld, U. Synthesis, Characterisation and Catalytic Performance of a Mesoporous Tungsten Silicate: W-TUD-1. *Appl. Catal. A Gen.* **2013**, *468*, 150–159. [CrossRef]
58. Perednis, D.; Gauckler, L.J. Thin Film Deposition Using Spray Pyrolysis. *J. Electroceram.* **2005**, *14*, 103–111. [CrossRef]
59. Baggetto, L.; Oudenhoven, J.F.M.; Van Dongen, T.; Klootwijk, J.H.; Mulder, M.; Niessen, R.A.H.; De Croon, M.; Notten, P.H.L. On the Electrochemistry of an Anode Stack for All-Solid-State 3D-Integrated Batteries. *J. Power Sources* **2009**, *189*, 402–410. [CrossRef]
60. Saha, N.C.; Tompkins, H.G. Titanium Nitride Oxidation Chemistry: An X-Ray Photoelectron Spectroscopy Study. *J. Appl. Phys.* **1992**, *72*, 3072–3079. [CrossRef]
61. Marchal, W.; De Dobbelaere, C.; Kesters, J.; Bonneux, G.; Vandenbergh, J.; Damm, H.; Junkers, T.; Maes, W.; D'Haen, J.; Van Bael, M.K.; et al. Combustion Deposition of MoO3 Films: From Fundamentals to OPV Applications. *RSC Adv.* **2015**, *5*, 91349–91362. [CrossRef]

 *materials*

Article

Transparent Glass-Ceramics Produced by Sol-Gel: A Suitable Alternative for Photonic Materials

Giulio Gorni [1], Jose J. Velázquez [1], Jadra Mosa [1], Rolindes Balda [2,3], Joaquin Fernández [2,3], Alicia Durán [1,*] and Yolanda Castro [1,*]

[1] Instituto de Cerámica y Vidrio (CSIC), 28049 Madrid, Spain; ggorni@icv.csic.es (G.G.); josvel@icv.csic.es (J.J.V.); jmosa@icv.csic.es (J.M.)

[2] Departamento de Física Aplicada I, Escuela Superior de Ingeniería, Universidad del País Vasco UPV/EHU, 48940 Bilbao, Spain; rolindes.balda@ehu.eus (R.B.); joaquin.fernandez@ehu.es (J.F.)

[3] Centro de Física de Materiales UPV/EHU-CSIC, 20018 San Sebastian, Spain

* Correspondence: aduran@icv.csic.es (A.D.); castro@icv.csic.es (Y.C.)

Received: 11 January 2018; Accepted: 27 January 2018; Published: 30 January 2018

Abstract: Transparent glass-ceramics have shown interesting optical properties for several photonic applications. In particular, compositions based on oxide glass matrices with fluoride crystals embedded inside, known as oxyfluoride glass-ceramics, have gained increasing interest in the last few decades. Melt-quenching is still the most used method to prepare these materials but sol-gel has been indicated as a suitable alternative. Many papers have been published since the end of the 1990s, when these materials were prepared by sol-gel for the first time, thus a review of the achievements obtained so far is necessary. In the first part of this paper, a review of transparent sol-gel glass-ceramics is made focusing mainly on oxyfluoride compositions. Many interesting optical results have been obtained but very little innovation of synthesis and processing is found with respect to pioneering papers published 20 years ago. In the second part we describe the improvements in synthesis and processing obtained by the authors during the last five years. The main achievements are the preparation of oxyfluoride glass-ceramics with a much higher fluoride crystal fraction, at least double that reported up to now, and the first synthesis of $NaGdF_4$ glass-ceramics. Moreover, a new SiO_2 precursor was introduced in the synthesis, allowing for a reduction in the treatment temperature and favoring hydroxyl group removal. Interesting optical properties demonstrated the incorporation of dopant ions in the fluoride crystals, thus obtaining crystal-like spectra along with higher efficiencies with respect to xerogels, and hence demonstrating that these materials are a suitable alternative for photonic applications.

Keywords: sol-gel; oxyfluoride glass-ceramics; nanocrystal; optical properties

---

1. Introduction

Phosphor materials emit light under exposure to external stimulation, such as an electron beam, light at different wavelengths, voltage or electric field, etc. These materials are widely applied in light-emitting diodes, solar cells, sensing, catalysis, integration in photovoltaic devices, and more recently in biosensing, bioimaging, or medical diagnosis [1–6].

The optical properties of these materials can change drastically depending on the processing parameters. Thus, in the past two decades, interest in studying the synthesis and processing of nanophosphors to develop luminescent materials with high efficiency has increased. The properties of existing devices can be improved by including luminescent phosphors and/or nanocrystalline oxyfluorides doped with lanthanide ions (Ln^{3+}). Ln^{3+} ions are commonly used as active ions because they show emissions in a wide spectral range, from UV to NIR, and their use in telecom, lasers, lightening, etc. makes them indispensable nowadays. Moreover, their most stable oxidation state is

3+, consisting of partially filled inner 4f levels screened by outer 5s and 5p orbitals. This allows for maintaining emission energies centered near the same values even in different hosts, making them suitable for applications where certain wavelengths are required. Several studies and applications deal with nanoparticles doped with Ln^{3+} ions suspended in liquid phase or with phosphor powders. However, for many applications as laser materials, waveguides and optical fibers, lightening devices, etc., solid and easy-to-handle samples are required.

In particular, oxyfluoride nano-glass-ceramics (OxGCs) are attractive materials for photonic applications, because they combine the very low phonon energy of fluoride nanocrystals (NCs) (300–450 cm^{-1}) with the high chemical and mechanical stability of oxide glass matrices, especially alumino-silicate ones [7–9]. Moreover, such materials can be cast in several forms to obtain the desired device. However, to maintain the transparency and avoid Rayleigh scattering produced by the quite big difference of refractive index between the oxide glass matrix and fluoride crystals, precise control of the crystal size is mandatory. Hence, to obtain materials with high transparency in the UV-Vis range, NCs with a size lower than 40–50 nm are required.

The pioneering work about OxGCs, published in 1993 by Wang and Ohwaki [10], in which $Pb_xCd_{1-x}F_2$ fluoride NCs doped with Er^{3+} and Yb^{3+} ions were precipitated in a silica-aluminate glass matrix after a proper heat treatment, showed an increase in the Up-Conversion (UC) emission of GCs with respect to precursor glass. Since that paper, the number of publications about transparent GCs has grown exponentially. In the last few decades, many compositions have been studied with the aim of obtaining different crystal phases and using different Ln^{3+} ions as dopants to obtain enhanced linear and nonlinear optical processes.

The usual method to prepare OxGCs is traditional glass melt-quenching (MQ). This method allows for obtaining several compositions; however, the most studied ones are those based on alumino-silicate matrices, even though there are also some studies of phosphate and fluoride GCs [11,12]. According to the definition of GCs [13], *"Glass-ceramics are ceramic materials formed through the controlled nucleation and crystallization of glass"*. In general, for MQ materials, the crystallization process is performed at temperatures slightly higher than T_g (T_g + 20–100 °C) to ensure the growth of crystals with a nanometric size that permits maintaining transparency. Moreover, phase separation commonly acts as a precursor for crystallization and the typical crystallization mechanism is a diffusion-controlled process.

The MQ method presents some drawbacks, most of them related to the high temperatures involved during the glass melting. In fact, high melting temperatures (1400–1700 °C) cause relevant fluorine loss, up to 30–40%, thus limiting the final content of crystal phase that can be obtained and resulting in uncontrollable compositions with respect to fluorine and Ln^{3+} ions. In addition, due to quite common phase separation of the precursor glass due to fluorine immiscibility in oxide glass matrices, it is a challenge to prepare high-quality samples for applications as laser devices or any other that requires high optical quality. Furthermore, heat treatments, in general, are quite long (from 3 h up to 120 h) and thin films cannot be directly obtained, thus limiting the possibility of application as integrated devices. Despite these disadvantages, the MQ method demonstrated the possibility of obtaining OxGCs as bulk materials with enhanced optical properties with respect to precursor glass, and in recent years, the preparation of novel OxGC optical fibers has also been achieved [14–17]. Therefore, MQ is still the most used processing method to obtain OxGCs and is also scalable to an industrial level. Indeed, transparent GC containing up to 75% of crystal fraction, with a crystal size below 40 nm, are extensively used in telescope mirrors [18].

To overcome previous limitations without giving up the field of OxGCs, many researchers tried different processing methods. In particular, the Sol-Gel (SG) route has been recognized and indicated as a promising alternative process to obtain transparent OxGCs. In fact, SG is a handy, flexible, and quite cheap method to fabricate novel and innovative materials at temperatures much lower than those used for MQ materials. Such low temperatures allow for introducing a higher amount of fluoride NCs with much better dispersion than in MQ compositions. Moreover, because SG is a chemical process using a bottom-up approach, high homogeneity can be obtained with no phase separation detected in SG glasses or GCs.

Another important feature of SG is the versatility of the processing method that permits obtaining thin films, powders, and bulk materials. However, many fewer papers about SG OxGCs are reported in the literature compared to those describing materials prepared by MQ [8]. This is mostly associated with the requirement of the optimization of the synthesis method, and in many cases the starting precursor can also affect the final crystal phase. Most of the materials studied have a simple composition that can be easily summarized with the formula:

$$(100 - x)SiO_2\text{-}x\ M_1F_2/M_1F_3/M_1M_2F_4/M_1M_2F_5,$$

where M_1 and M_2 are alkaline, alkaline-earth metals, or lanthanide elements, respectively. More complex glass matrices, where SiO_2 is partly substituted with Al_2O_3, are also reported in the literature but are quite rare, and even more complex structures have also been prepared but for different applications as biomaterials [19].

SG synthesis typically involves the hydrolysis and polycondensation of metal salts or metal-organic precursors, such as tetraethyl orthosilicate (TEOS), in an alcoholic medium. The active phase and Ln^{3+} ions are prepared separately using acetates, nitrates, or chlorates as precursors and dissolved using a fluorine acid. The mixing of both solutions, followed by a controlled crystallization, allows for obtaining transparent OxGCs.

Even though the SG method offers potential advantages, they have not been fully exploited, and all the papers published to date describe the preparation of similar compositions containing nominal small crystal fractions (5–10 mol %), with the real crystal content not being estimated. Furthermore, in many cases only slight changes are added to the already published compositions, the main modifications consisting of changing the dopants and/or their concentration.

In the following sections the most relevant results for several Ln^{3+} doped OxGCs for optical applications studied to date and prepared by SG are summarized. The results are separated into two sections depending on the crystal phase. A third section regarding oxides and oxychlorides GCs has also been added, while in the last sections we will present the results and improvements introduced by our group in the last five years.

1.1. Alkaline-Earth Oxyfluoride Glass-Ceramics, M_1F_2 (M_1 = Mg, Pb, Sr, Ca, Ba), $M_1M_2F_4$ ($BaMgF_4$)

Alkaline-earth fluoride GCs, with the formula M_1F_2 (M = Mg, Pb, Sr, Ca, Ba), are useful for photonic applications due to their high optical transparency in a wide spectral range, from (UV) ultraviolet to infrared (IR). MgF_2-based materials have been extensively studied for different applications, for example as protective coatings on glass optics due to their low chemical reactivity, low refractive index (1.38), and scratch- and weather-resistance. Even though this review is centered on materials with luminescence properties, MgF_2 materials deserve special attention because they have been among the first fluorides produced by SG. Even more interesting is that the synthesis of MgF_2 introduced important novelties in the synthesis of OxGCs.

One of the first works reported about MgF_2 NCs and SiO_2-MgF_2 materials prepared by SG as bulk and thin films dates back to 1996 [20]. $Mg(OCH_3)_2$ and HF were used as Mg and F sources, respectively, to obtain MgF_2 sol. SiO_2-MgF_2 sols were also prepared with ratios of (mol %) 20:80 and 40:60, using tetramethyl orthosilicate (TMOS) as the SiO_2 precursor. X-ray diffraction (XRD) showed that MgF_2 crystals precipitated in both sol and glass samples, but the crystal growth in SiO_2-MgF_2 samples was dramatically slowed (~15 nm) in comparison with that of pure MgF_2 samples (~115 nm). For $20SiO_2$:$80MgF_2$ samples, the crystal size maintained the same value. MgF_2 NCs have been obtained after treatment at different temperatures following the procedure described in [21]. In 1997 Fujihara et al. [22] reported on MgF_2 thin films using trifluoracetic acid (TFA) and $Mg(OC_2H_5)_2$ as a fluorine or magnesium precursor, respectively, with a molar ratio 1:2.6. The most important innovation is the use of TFA as a fluorine precursor, together with an explanation of the fluoride NCs' formation. It was assumed that $Mg(OC_2H_5)_2$ complexes were formed in the xerogel and upon heat treatment, Mg^{2+} ions react with thermally activated fluorine in CF_3COO^- ions to form MgF_2 crystals. This decomposition reaction was accompanied by a huge weight loss, around 75%, as confirmed by Differential Thermal Analysis (DTA) (Figure 1). Some years later it was demonstrated that, thanks to the screening effect of CF_3COO ions of TFA, metal cations with oxidation state 2+ or 3+ are prevented from being incorporated in the SiO_2 matrix, making possible subsequent fluoride NCs' precipitation upon heat treatment. Therefore, this work can be considered pioneering and a boost towards the preparation of OxGCs by SG.

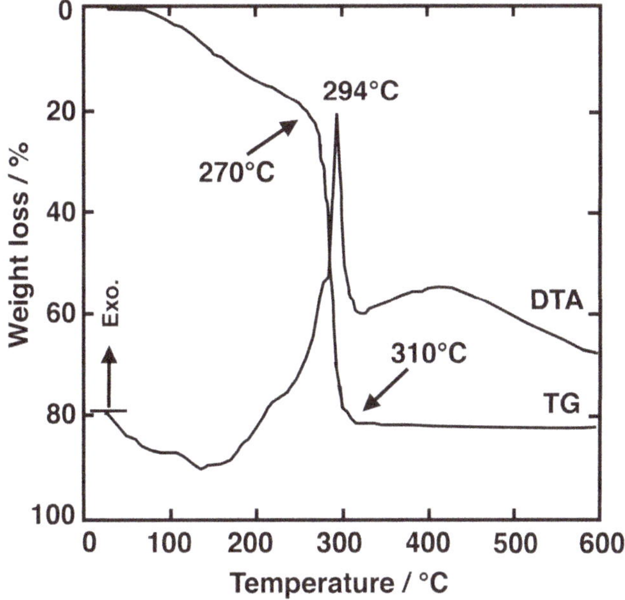

Figure 1. TGA/DTA curve of the MgF_2 gel obtained by drying the sol at 80 °C. Figure modified from Figure 4 of [22].

Another phase extensively studied is PbF_2. One of the first works associated with SiO_2-PbF_2 samples was reported in 2004 by Luo et al. [23]. Bulk materials of composition $90SiO_2$-$10PbF_2$ doped with Er^{3+} were prepared using TEOS and TFA as the SiO_2 or fluorine precursor, respectively. The molar ratio of precursors $TEOS:CH_3CH_2OH:DMF:Pb(CH_3COO)_2:TFA:H_2O:HNO_3$ was 1:2:2:0.1:0.6:4:0.4. Upon heat treatment at 300 °C and 480 °C β-PbF_2 NCs precipitated in the glass matrix. Er^{3+} ions seemed to be segregated at the surface of the crystallites and hindered the growth of PbF_2 NCs, thus delaying crystallization and reducing the crystal size from 20 to 9 nm. Other authors reported on UC measurements in bulk materials of the same composition doped with $0.1Er^{3+}$-$0.3Yb^{3+}$ [24]. A $TEOS:EtOH:H_2O:CH_3COOH$ ratio of 1:4:10:0.5 was used for the SiO_2 sol, while the ratio $Ln(CH_3COO)_3:TFA$ was 1:4. UC bands showed well-resolved Stark components indicating the incorporation of Ln^{3+} ions into PbF_2 crystals. Results for the same composition doped with Tm^{3+}-Yb^{3+}-Er^{3+} and Tm^{3+}-Yb^{3+}-Ho^{3+} were also reported [25]. In both cases, the color tunability of UC luminescence was obtained, allowing white UC generation. Following the same synthesis and composition of Luo et al. described before, Szpikowska-Sroka et al. published some papers about OxGCs with different dopants [26–28]. After heat treatment, Eu^{3+} and Tb^{3+} ions showed an increase in the emission spectra in GC samples with respect to the xerogels due to the incorporation of these ions into the NCs. In fact, the asymmetry ratio, *R*, described as the ratio between the two visible emissions $I(^5D_0-^7F_2)/I(^5D_0-^7F_1)$ of Eu^{3+}, gives information about the nature of the environment surrounding the ion. Moreover, the effect of the TFA on the optical properties was also analyzed, showing the best result for a Pb/TFA ratio of 5 and corresponding to a crystal fraction of 3 wt %.

$90SiO_2$-$10SrF_2$ transparent GCs doped with ErF_3 (1 mol %) were prepared by Yu et al. in 2006 [29]. Samples treated at 300 °C already showed the precipitation of SrF_2 crystals with a size of 10 nm. The same crystal size is maintained for heat treatment up to 800 °C. However, upon excitation at 378 nm no Er^{3+} emission was detected unless for samples treated at 800 °C. The authors attributed this phenomenon to the presence of –OH groups that quench the luminescence. Two years later, the same authors prepared transparent bulk samples of composition $90SiO_2$-$10SrF_2$-$0.5ErF_3$ and $97.5(90SiO_2$-$10SrF_2)$-$2.5Al_2O_3$-$0.5ErF_3$ (mol %) [30]. Al_2O_3 was added as the nitrate, with Al^{3+} acting as a glass network former, replacing Si^{4+}. The introduction of Al^{3+} to the SiO_2 network causes a decrease in non-bridging oxygens, hindering the crystallization of fluoride NCs. The SiO_2-Al_2O_3 glass matrix showed higher transparency in the UV region due to lower pore content and the final optical properties were better for this composition treated at 1000 °C. Very intense and visible UC luminescence was observed for GC based on SiO_2-Al_2O_3 matrix, as compared to the one with only SiO_2.

The first SiO_2-CaF_2 transparent GC prepared by SG was reported in 2007 by Zhou et al. [31]. $95SiO_2$-$5CaF_2$ bulk samples doped with 1 mol % ErF_3 were treated at 900 °C, allowing the precipitation of CaF_2 crystals of 20 nm in size, homogeneously distributed in the amorphous SiO_2 matrix. Er^{3+} incorporation in the crystal phase was tested by Energy-Dispersive X-ray Spectroscopy (EDXS). Only Red UC emission was detected upon 980 nm excitation. Georgescu et al. prepared $89SiO_2$-$5CaF_2$-$5YbF_3$-$1ErF_3$ (mol %) GCs upon heat treatment at 800 °C [32]. A solid solution of $(Ca_{1-x}Ln_x)F_{2+x}$ crystals was observed by XRD, and both red–green and Violet-UV UC were observed when exciting the samples at 973 nm. The process was described as a two-photon and three-photon absorption process for visible and UV emissions, respectively, upon IR excitation. On the other hand, a different approach was used to prepare $95SiO_2$-$5CaF_2$ GCs [33]. Er^{3+}-doped CaF_2 crystals were previously synthesized by co-precipitation using $Er(NO_3)_3$, $Ca(NO_3)_2$, and ammonium bi-fluoride NH_4HF_2 as precursors. Then, NCs were mixed to SiO_2 sol, dried at 50 °C during one month and heat-treated to obtain bulk materials. The authors considered that conventional SG route does not allow control of the size and quantity of crystals as well as the Ln^{3+} concentration into the NCs. The optical results showed that the UC emissions increased when passing from xerogels to GCs, the corresponding lifetimes of the $^4F_{9/2}$-$^4I_{15/2}$ emission being 1.3 µs and 1.7 ms, respectively. This effect was ascribed to the incorporation of Er^{3+} into CaF_2 NCs with phonon energy around 460 cm^{-1}.

Another interesting host for Ln^{3+} ions is BaF_2 due to its very low phonon energy, around 320 cm^{-1}. The first SiO_2-BaF_2 SG material was reported in 2006 by Chen et al. [34]. $95SiO_2$-$5BaF_2$ bulk materials were prepared and doped with 0.5 and 1 Er^{3+} (mol %). BaF_2 crystals, 2–4 nm in size, precipitated in the glass matrix after heat treatment at 300 °C. Up to 700 °C, no relevant effects in the crystallization were observed. Over this temperature, the crystal grew up to 10 nm. However, treatment at higher temperature also caused the crystallization of the glass matrix in the form of cristobalite. The incorporation of Er^{3+} ions into the NCs was revealed due to the shift of the XRD peaks toward higher angles in Er^{3+}-doped samples and by EDXS and Judd–Ofelt calculations. Photoluminescence (PL) emission was detected only for samples treated at 800 °C and green and red UC emissions were observed upon 980 nm excitation. The same composition and synthesis were used by Secu et al. when doping materials with Pr^{3+}, Sm^{3+}, Eu^{3+}, Dy^{3+}, and Ho^{3+} [35,36]. Small BaF_2 crystals (~7 nm) were precipitated upon heat treatment at 800 °C for 1 h. No Ln^{3+} emission was observed for xerogels except for Eu^{3+}, while for all the GCs typical 4f-4f transitions were observed showing the Stark splitting of the bands (Figure 2). Moreover, the typical Eu^{3+} emissions in GCs indicated that these ions are effectively incorporated into NCs. The Eu^{3+} lifetime, measured at 620 nm, increased from 0.3 to 4.7 ms passing from xerogel to GC.

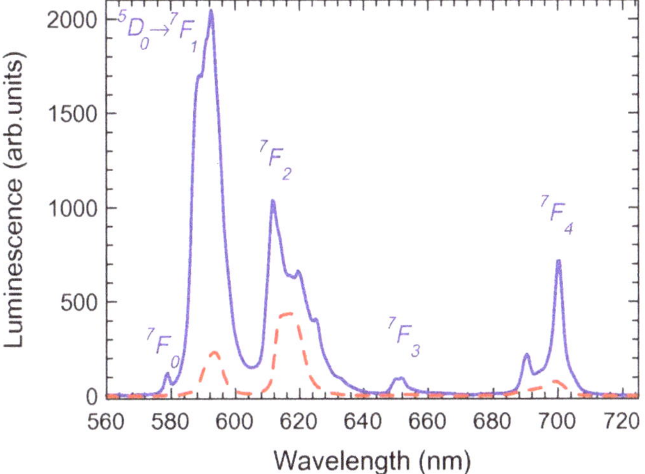

Figure 2. PL spectra recorded on Eu^{3+}-doped xerogel (dotted curve) and GC (solid curve) using $\lambda ex. = 394\,nm$ [35].

To conclude this first section, we mention the preparation of $90SiO_2$-$10BaMgF_4$ films doped with Eu^{3+} prepared by Fujihara et al. [37]. Upon heat treatment at 750 °C for 10 min small $BaMgF_4$ NCs 3 nm in size, were observed by High-Resolution Transmission Electron Microscopy (HRTEM). By N,N-dimethylformamide (DMF) addition a porous film was obtained, while, without DMF the film was dense. DMF addition also favors crystallization, reducing stresses in films [38]. The corresponding PL measurements only showed Eu^{2+} emission, meaning that Eu^{2+} ions are effectively incorporated in the crystal phase, substituting Ba^{2+} ions. Furthermore, Eu^{2+} emission was much stronger for porous films due to higher photon confinement, producing an increase in the absorption process.

1.2. Lanthanides Oxyfluoride Glass-Ceramics, M_1F_3 (M_1 = La, Y, Gd) and $M_1M_2F_4$, $M_1M_2F_5$ (M_1 = Na, K, Li; M_2 = Gd, Y)

1.2.1. M_1F_3

Oxyfluoride GCs in which one element of the NCs former is a Ln^{3+} ion are the most studied compositions. In fact, Ln^{3+} ions are easily replaced by other ions used as dopants, due to their similar ionic radius and equal charge. Moreover, some ions as Gd^{3+} are known to act as sensitizers for other ions as Eu^{3+}, Tb^{3+}, and Dy^{3+} [39–41], making possible energy transfer (ET) processes and thus making this NC phase attractive for several photonic applications.

Among all the lanthanide crystals, LaF_3 is the most studied. Several papers about LaF_3 OxGCs have been published since 1998, when Fuhijara et al. described the preparation of LaF_3 thin films on silica substrates [42]. In that work, the authors explained the importance of controlling the synthesis and heat treatment process to obtain high-purity LaF_3 NCs, avoiding the precipitation of other phases as LaOF. In the following years, the same authors worked extensively in the preparation of SiO_2–LaF_3 bulk samples, observing the crystallization of LaF_3 NCs with a size ranging from 10 to 30 nm [43]. They also studied the influence of TFA on the LaF_3 NCs' crystallization, proposing that such formation occurs by the following chemical reaction [44–46]:

$$La(CF_3COO)_3 \rightarrow LaF_3 + (CF_3CO)_2O + CO + CO_2.$$

However, the authors stated that it is not possible to raise the LaF_3 content beyond 10 mol % without losing transparency. Therefore, all the papers published after this work concentrated on

Fujihara's synthesis and composition, just changing the active phase. TMOS was partially or totally replaced by TEOS but the same stoichiometric ratios of H_2O and TFA were used, bringing little innovation to the synthesis and processing of OxGC materials.

The preparation of transparent Eu^{3+} doped SiO_2-LaF_3 and SiO_2-LaOF OxGC films, in which Eu^{3+} ions were incorporated into the LaF_3 crystals, resulted in interesting PL emissions [47]. Results showed important differences depending on the sintering temperature and, therefore, on the crystals phase. On the other hand, Ribeiro et al. [48] prepared films with the same composition, finding that for heat treatment between 600 and 900 °C both LaF_3 and LaOF phases appeared. Eu^{3+} ions emissions mainly from the $^5D_0$ excited state reflected the change in the environment of these ions. They used thin films as planar waveguides at Vis and IR wavelengths, reported losses ~1.8 dB/cm at 632.8 nm. However, the authors considered that the intensity of PL measurements depends on the nature of the crystal phase and the presence of –OH groups that are responsible for PL quenching. At such a high sintering temperature, –OH groups could be eliminated, but fluorine can be lost, producing a silica matrix rich of lanthanide ions and changing the emission spectra.

LaF_3 OxGC bulk materials were also prepared by other authors. Biswas et al. [49] described the preparation of transparent 95SiO_2-5LaF_3 GC, after densification at 1000 °C, using TEOS instead of TMOS as the SiO_2 precursor and a TEOS:H_2O:CH_3COOH molar ratio of 1:10:0.5. Optical results showed very good UC efficiency from the IR-to-UV region due to the Yb^{3+}–Er^{3+} ET process, which is enhanced in GCs, suggesting the incorporation of Er^{3+} ions in the low-phonon environment of LaF_3 NCs.

In order to optimize the optical properties, other authors have described the preparation of bulk GC materials with molar ratios $(100 - x)SiO_2$-$xLaF_3$ (x = 5 and 10 mol %), doped with (0.1–0.5) mol % of Ln^{3+} ions and heat treated at temperature up to 1000 °C [50–55] following a similar synthesis to Biswas et al. Optical results showed that Eu^{3+} ions could be selectively excited and, even more interesting, opened the way for the estimation of the percentage of ions that are incorporated in the NCs, concluding that at least 50% are effectively incorporated in LaF_3. Another paper [53] described the preparation of transparent 95SiO_2-5LaF_3 bulk GC, co-doped and tri-doped with Ln^{3+} ions. Good UC properties after excitation at ~980 nm were observed due to the interionic distance reduction between Ln^{3+} ions, as a consequence of their incorporation into the NCs. Moreover, by increasing the pump power the emitting color was tuned, obtaining white light generation with potential applications in multicolor solid-state displays and optical integrated devices (Figure 3). The incorporation of a high amount of dopant ions, around 75%, in the precipitated LaF_3 NCs was also reported by Velázquez et al. [54] for Tb^{3+}-Dy^{3+} co-doped GCs, showing enhanced UV absorption band and allowing a shift in such emission to the Vis range.

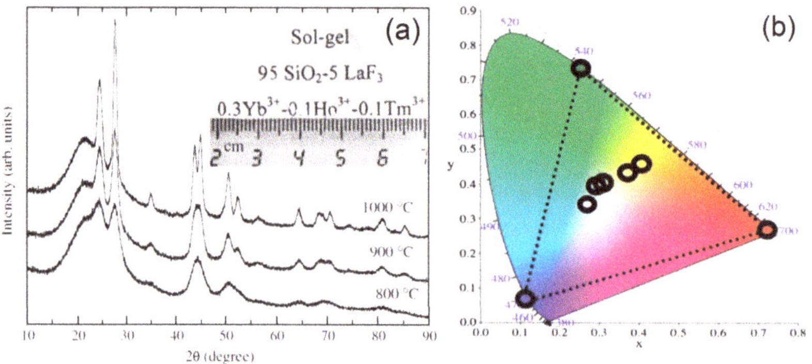

Figure 3. (a) XRD patterns of 95SiO_2-5LaF_3 GCs at different temperatures and (b) CIR standard chromaticity showing the up-conversion emission for GCs co-doped with 0.3Yb^{3+}, 0.1Ho^{3+} and 0.1 Tm^{3+} [53].

Other oxyfluoride GCs with special interest for photonics are those containing YF_3 NCs. This crystal phase can act as a high quantum efficient host lattice for Ln^{3+} ions. Since 1998, when Dejneka published one of the first works on Eu^{3+}-doped YF_3 OxGCs by the MQ method [56], many authors started preparing materials containing this crystal phase. However, it was necessary to wait until 2006 to find one of the first works related to SG bulk materials based on the composition $90SiO_2$-$10YF_3$ [57]. The authors precipitated YF_3 crystallites with sizes around 5 nm from the silica matrix when the xerogel was treated at 400 °C. According to the authors, NCs aggregated to form larger particles. By increasing the temperature to 600 °C, YF_3 NCs tended to separate without changing their size, finally resulting in a homogeneous distribution in the glass matrix. They proposed that the precipitation of these crystals in the glassy phase induces high stress in the local region, which was reduced when increasing the treatment temperature due to the separation of the NCs from the agglomerates. Méndez-Ramos and co-workers published results based on the same $90SiO_2$-$10YF_3$ composition [58–60]. In these works, using Ln^{3+} as probe ions of the local environment, the authors found that a large fraction of optically active ions is efficiently embedded into the YF_3 NCs, 11 nm in size (Figure 4a). When the samples were co-doped with Yb^{3+}-Tm^{3+}, bright and efficient UC was achieved (Figure 4b), along with intense high-energy emissions in the UV range, due to rare 4- and 5-infrared photon processes. Moreover, co-doping with Ho^{3+} or Er^{3+} ions showed white light generation.

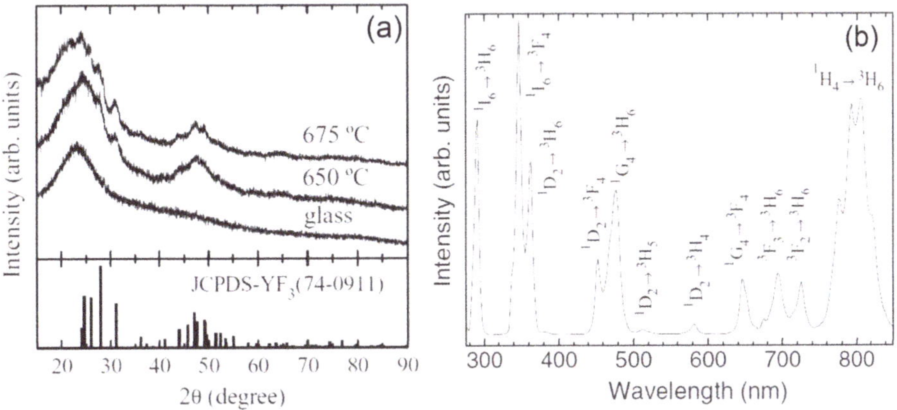

Figure 4. (a) XRD patterns of $90SiO_2$-$10YF_3$ Ln^{3+}-doped GCs heat treated at 650 and 675 °C together with the JCPDS-YF$_3$ and (b) Up-conversion emission spectrum of GCs co-doped with 1.5 Yb^{3+} and 0.1 Tm^{3+} heat-treated at 675 °C [60].

Differently to LaF_3 and YF_3, GdF_3 is considered a promising host because Gd^{3+} ions can act as s sensitizer for other Ln^{3+} ions and favor some ET processes [61], being suitable for application in Plasma Display Panels (PDPs) and mercury-free fluorescent lamps [62,63]. Only a few papers about SG GdF_3 OxGCs have been published, with the nominal crystal phase content being less than 10 mol %. Fuhijara et al. reported the synthesis of $90SiO_2$-$9GdF_3$ thin films doped with 1 mol % of EuF_3 [64]. Results confirmed that GdF_3 NCs, 5 nm in size, precipitated during heat treatment at 300 and 400 °C, Figure 5, but GdOF NCs appeared when the temperature was increased to 500 °C. More recently, Szpikowska-Sroka et al. studied the optical properties of Eu^{3+}-doped SiO_2-GdF_3 GCs, with a very low composition of active media, around 3–6 mol %, based on Fuhijara's synthesis [65–67]. The PL associated with Eu^{3+} ions in GdF_3 phase was more efficient due to the low phonon energy of the crystal phase.

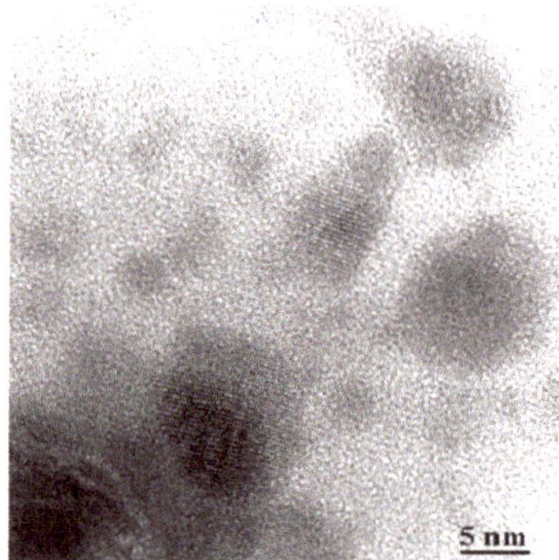

Figure 5. HRTEM image of Eu^{3+} doped SiO_2-GdF_3 GC film heat = treated at 400 °C [64].

1.2.2. $M_1M_2F_4/F_5$

Complex fluoride structures became increasingly interesting in the last decade due to their good optical properties and capability of producing intense and efficient nonlinear optical processes as UC emission or other ET processes. In most cases phosphors were synthesized using SG, solvo-thermal, or co-precipitation methods and the final materials were used as powders. Other works concern the preparation of bulk materials and only few are about thin films.

Phosphors based on $NaYF_4$ crystals have been known since the 1970s for their excellent UC properties when doped with Yb^{3+} and Er^{3+}. In particular, $NaYF_4$ phosphors' UC emission was found to be 4–5 times higher than that of LaF_3 crystals [68]. Kramer et al. proposed a route to synthesize only pure hexagonal $NaYF_4$ phosphors [69]. Moreover, transparent $NaYF_4$ OxGCs bulk materials have been obtained by the MQ method, showing interesting PL properties even more efficient than those of powder phosphors [70,71].

However, the first work on $NaYF_4$ OxGCs prepared by SG was published in 2009 by Yanes et al. [72]. $95SiO_2$–$5NaYF_4$ bulk materials doped with $0.1Er^{3+}$ and $0.3Yb^{3+}$ were obtained using acetates as precursor and the "Fujihara route". The ratio between red and green UC emission bands varied as a function of temperature of heat treatment and pump power resulting in color-tunable UC phosphors. The same authors also prepared $95SiO_2$–$5NaYF_4$ bulk materials doped with $0.1Eu^{3+}$ (mol %) [73]. $NaYF_4$ face-centered cubic NCs precipitated upon heat treatment between 550 and 650 °C, their size increasing from 5 to 10 nm with the treatment temperature. PL measurements demonstrated the incorporation of Eu^{3+} into the NCs, due to changes of emission bands using a selective excitation, together with an increase in the lifetime. The authors also reported bright white light generation achieved in similar materials tri-doped with Yb^{3+}-Ho^{3+}-Tm^{3+} [74] and Yb^{3+}-Er^{3+}-Tm^{3+} [75].

To the best of our knowledge, no KYF_4 GC or composite materials prepared by MQ have been reported in the literature. The first GC was prepared by Mendez-Ramos et al. using the SG method and co-doping with Yb^{3+}-Er^{3+}-Tm^{3+} [76]. The processing method was the same used by the authors for other phases such as $NaYF_4$, LaF_3, etc. K, Y, and Ln^{3+} acetates were dissolved using ethanol, TFA, and water, while hydrolyzed TEOS were used as the SiO_2 precursor, employing the well-known TEOS:EtOH:H_2O:CH_3COOH ratio 1:4:10:0.5. KYF_4 crystals, 14–20 nm in size, were observed treating $95SiO_2$-$5KYF_4$ bulk materials between 650 and 700 °C. Very well-resolved PL Stark components were observed for Er^{3+} and Tm^{3+} ions, indicating their incorporation into low-phonon KYF_4 crystals (Figure 6). UC was obtained upon Yb^{3+} excitation at 980 nm and tunable emission was achieved depending on the dopant concentration and excitation power, allowing white light generation. The same composition was also studied by doping the system with other Ln^{3+} ions [77–79]. In all cases, the incorporation of Ln^{3+} ions into cubic KYF_4 crystals was proved by PL measurements that showed defined Stark components, a better DC and UC process, and an increased lifetime as compared to emissions of Ln^{3+} ions remaining in the glass matrix. The possibility of obtaining both UC and DC simultaneously had interesting applications for a photovoltaic silicon solar cell and white light-emitting diodes.

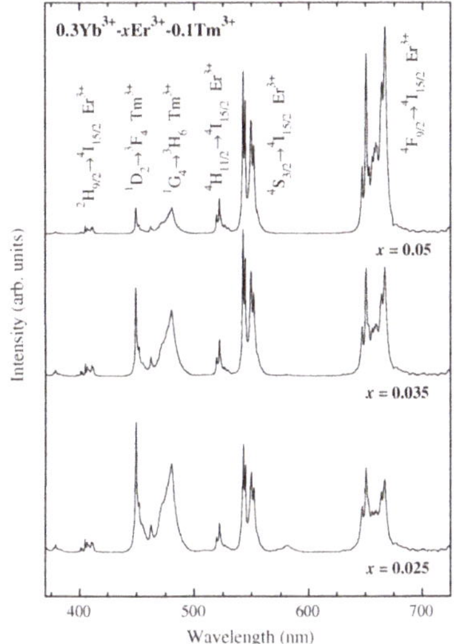

Figure 6. Up-conversion emission spectra of 95SiO$_2$-5KYF$_4$ GCs co-doped with Yb^{3+}-Er^{3+}-Tm^{3+} heat-treated at 700 °C [76].

Despite some works based on LiYF$_4$ OxGCs prepared by the MQ method appearing since 2009 [80], the first SG GCs containing LiYF$_4$ crystals dates back to 2013 with the works of Kawamura et al. [81] and Secu et al. [82]. In both cases the authors used Fujihara's synthesis, and Li and F were added in excess to avoid losses due to evaporation and guarantee LiYF$_4$ formation. In fact, a stoichiometric Y:Li ratio favored the crystallization of YF$_3$ instead of LiYF$_4$. Gel powders were then treated between 500 and 600 °C. Kawamura et al. obtained a mix of LiYF$_4$ and YF$_3$ phases when the Li:Y molar ratio was lower than 3, while only pure LiYF$_4$ phase was produced using this ratio. By EDXS analysis, Nd^{3+} incorporation into the fluoride NCs was proven even though observable amounts of F, Nd, and Y were also detected in the glass matrix. UC spectra were obtained when exciting the samples at 800 nm, showing crystal-like behavior. The Nd^{3+} lifetime in LiYF$_4$ was longer than in YF$_3$, suggesting a better Nd^{3+} distribution for the former. The same authors prepared the same materials with only LiYF$_4$ crystals and treated the samples with HF to remove the SiO$_2$ matrix [83]. However, relevant amounts of O and Si were detected after HF treatment and PL measurements did not show a significant difference after and before the HF treatment, thus ensuring the presence of Nd^{3+} ions in the fluoride NCs.

Secu et al. [82,84] prepared $(100 - x)$SiO$_2$-xLiYF$_4$ GC powders doped with Eu^{3+} and Er^{3+}-Yb^{3+} and compared the results for LiYF$_4$:Eu^{3+} and LiYF$_4$:Er^{3+}-Yb^{3+} crystal pellets. As compared to an untreated xerogel, Eu^{3+}-doped GC samples showed 7–8-fold stronger emission, a much longer lifetime of the ^{5}D$_0$-^{7}F$_2$ red emission (613 nm), and better resolved bands due to the Stark splitting and reduced inhomogeneous broadening typical of amorphous environments. Crystal pellets and GCs showed similar PL spectra and lifetimes. For Er^{3+}-Yb^{3+}-doped samples, UC emission was detected in both GCs and crystal pellets. A saturation effect of the red UC emission was observed for GC samples, associated with the high Ln^{3+} ions concentration, which caused a back-ET from Er^{3+} to Yb^{3+}.

BaYF$_5$ and BaGdF$_5$ are other interesting crystal phases suitable for photonic applications. The first 95SiO$_2$-5BaYF$_5$ GC obtained by SG was prepared recently using Eu^{3+} and Sm^{3+} as dopants [85]. After heat treatment at 750 °C cubic BaYF$_5$ NCs, 11 nm in size, were detected by HRTEM and XRD.

For comparison, $BaYF_5$ NCs were synthesized by the solvo-thermal method and dispersed in toluene. Similar PL features and lifetimes were obtained for the GC and the dispersed NCs. The same authors also prepared $95SiO_2$-$5BaGd_{(1-x)}RE_xF_5$ (x = 0 or 0.02 mol %, where RE = Eu^{3+}, Sm^{3+}, Dy^{3+} and Tb^{3+}) GC [86]. After heat treatment at 650 °C, $BaGdF_5$ NCs, with a size around 10 nm, precipitated in the SiO_2 matrix (Figure 7). Upon Gd^{3+} excitation at 272 nm or Eu^{3+} direct excitation at 393 nm, very similar Eu^{3+} emission spectra were obtained and the bands showed typical Stark components that were quite well resolved, thus confirming the incorporation of Eu^{3+} in the $BaGdF_5$ NCs and the ET from Gd^{3+} to Eu^{3+}.

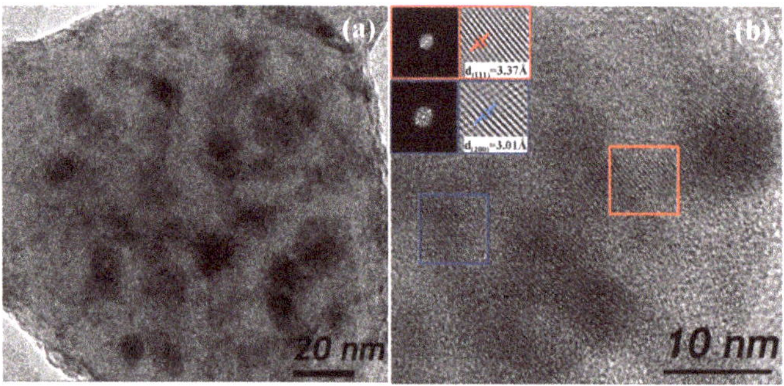

Figure 7. (**a**) TEM and (**b**) HRTEM images of $95SiO_2$-$5BaGd_{(1-x)}Eu_xF_5$. Inset show power spectrum (FFT pattern) and filtered higher-contrasted red and blue square nanoparticles [86].

1.3. Other Glass-Ceramics (Oxides and Oxyclorides)

In this section we resume work concerning GCs based on different compositions containing oxide and oxychloride NCs. Oxide GCs have received great attention because some oxide phases present interesting properties. For example, it is worth citing SnO_2, which is a wide gap n-type semiconductor with strong UV absorption (energy gap ~3.6 eV at 300 K) and tunable emission spectra depending on the crystal size. In fact, SnO_2 is known to act as a Quantum Dot (QD) when the crystal size is smaller than or comparable to the Bohr radius, thus showing properties between bulk semiconductor and discrete molecules that have many applications in several technological fields. Significant blue-energy shift of the intrinsic absorption edge can be obtained by strong quantum confinement of excitons inside the QDs [87,88]. The incorporation of SnO_2 NCs in an SiO_2 glass matrix using SG dates back to 2002 with the works of Chiodini et al. [87] and Nogami et al. [89] Chiodini et al. reported on $98SiO_2$-$2SnO_2$ materials obtained by mixing TEOS and dibutyltin diacetate ($Sn(CH_2CH_2CH_2CH_3)_2(OOCCH_3)_2$) in ethanol and using a $TEOS:H_2O$ molar ratio of 1:8. The samples were heated from 450 to 1050 °C in an O_2 atmosphere and tetragonal SnO_2 NCs with a size of 1.5–2 nm were observed by HRTEM between 950 and 1050 °C. However, the NCs' size and distribution are strongly dependent on the atmosphere. NCs above 10 nm (clusters) were obtained for heat treatment at 1050 °C in reducing atmosphere. The near-UV absorption edge shifts to high energies by decreasing the synthesis temperature due to a decrease in the SnO_2 NCs' size. The author concluded that a possible application of these materials could be as all-optical switching devices. The authors also reported that an increase in the SnO_2 content up to 15 mol % produces negative photorefractivity, activated by UV-Vis light [90]. A refractive index change of −4 and −6, measured at 980 nm, was obtained after sample irradiation at 266 and 532 nm, respectively.

Nogami et al. prepared materials with compositions $(100 - x)SiO_2$-$xSnO_2$ (x = 1, 3 and 5 mol %) doped with 1 mol % of Eu_2O_3 [89]. They used TEOS as the SiO_2 precursor and $SnCl_2·2H_2O$ together with $EuCl_3$ as SnO_2 and Eu^{3+} precursors, respectively. After a gelation period at room temperature for two weeks, bulk samples were treated from 500 to 1000 °C to ensure the densification of the glass matrix and allow SnO_2 NCs' precipitation. For compositions containing 3 and 5 mol % of SnO_2, NCs with a size of 6.9 and 8.5 nm were obtained, while no NCs were observed for $99SiO_2$-$1SnO_2$

composition. Absorption measurements showed a blue shift of the band-tail with respect to SnO_2 bulk sample. Higher blue-shift and energy gap were obtained when decreasing the SnO_2 content, related to a smaller NCs size. PL showed that Eu^{3+} emission intensity increases proportionally to the third power of SnO_2 concentration: Eu^{3+} emission in a $95SiO_2$-$5SnO_2$ sample was 150 times higher than in $99SiO_2$–$1SnO_2$. PL excitation spectra revealed an ET process between Eu^{3+} and SnO_2 NCs, thus indicating the incorporation of Eu^{3+} in the crystal phase and making possible its emission upon SnO_2 excitation. Other authors studied $(100 - x)SiO_2$-$xSnO_2$ (x = 1–10 mol %) GCs doped with Eu^{3+} and Eu^{3+}-Tb^{3+} following Nogami's synthesis [91–93]. By increasing the SnO_2 content up to 10 mol % (NCs size ~5 nm), a low quantum confinement effect was observed. Hence, most research was performed on GCs containing 5 mol % of SnO_2 and the optical characterization showed that in strong confinement conditions the energy gap has a high dependence on the NCs' size. Eu^{3+} ions incorporated in SnO_2 NCs were excited in the range 340–394 nm and, considering the variation of the band gap with the crystal size, the use of a certain excitation wavelength allowed for exciting only NCs with a defined crystal size, thus producing remarkable variation in the PL emission spectra. ET from SnO_2 NCs to Tb^{3+} ions was observed in Eu^{3+}-Tb^{3+}-doped GCs. It was observed that the ET depends on the crystal size, being favored for the smallest NCs. A relevant drawback related to SnO_2 NCs is the low solubility of Ln^{3+} limited to ~0.05%, the remaining ions being segregated at grain boundaries, probably in the form of $Ln_2Sn_2O_7$ crystals [94]. To overcome this limitation, Van Tran et al. [95,96] prepared SiO_2-SnO_2 GCs with a maximum nominal NCs concentration of 20 mol % to allow for higher Ln^{3+} amount incorporation in SnO_2 NCs. The authors prepared Er^{3+}-doped materials and PL measurements showed that an increase in SnO_2 concentration promotes Er^{3+} ions' incorporation in SnO_2 NCs.

Among SiO_2 glass matrices containing oxide NCs, those containing ZrO_2 crystals deserve special attention because ZrO_2 is a pretty cheap material, transparent over a wide range of wavelengths: from 300 nm to 8 µm, it has quite low phonon energy ~650 cm^{-1} and a high refractive index. Such properties have been exploited to develop Ln^{3+}-activated planar waveguides at telecom wavelength [97,98]. $(100 - x)SiO_2$-$xZrO_2$ (x = 10–30 mol %) homogenous and crack-free thin films doped with Er^{3+} were prepared following a synthesis route similar to SnO_2, using TEOS and $ZrOCl_2 \cdot 8H_2O$ as SiO_2 and ZrO_2 precursors, respectively. All waveguides showed the existence of one mode at 1550 nm with relatively low propagation losses; the refractive index increased from 1.492 to 1.609 for 10 and 30 mol % ZrO_2, respectively. Narrow Er^{3+} emissions were observed in thin films with respect to glass materials, thus indicating Er^{3+} incorporation in ZrO_2 NCs. Suhaimi et al. [99] prepared several $(100 - x)SiO_2$-$xZrO_2$ compositions (x = 30–70 mol %) doped with 0.58 mol % Er^{3+}. The refractive index at 1550 nm changed linearly from 1.6931 to 1.7334 by increasing ZrO_2 content. Much higher PL emission of Er^{3+} at 568 nm was observed for films containing 70 mol % of ZrO_2, thus indicating that ZrO_2 facilitates Er^{3+} to disperse homogeneously and the low phonon energy of the crystal phase reduces non-radiative losses. Other authors also studied SiO_2-ZrO_2 waveguides doped with Er^{3+}-Yb^{3+} with ZrO_2 contents up to a maximum of 25 mol % [100]. Low roughness, a crack-free surface, and a high confinement coefficient were observed for all the compositions. Er^{3+} NIR PL was enhanced when the waveguide was co-doped with Yb^{3+}, denoting an efficient ET between these ions. The authors considered the possibility of applying these materials as EDWA and WDM. Many GCs containing other oxide NCs have also been developed as waveguides [101–104].

To conclude this section, we mention the work of Secu et al. on oxychloride GCs [105]. $95SiO_2$-$5LaOCl$ GCs doped with 1 mol % Eu^{3+} were prepared following a similar synthesis to LaF_3 GCs but replacing TFA with trichloacetic acid (CCl_3COOH). $LaOCl$:Er^{3+} pellets have also been prepared using a conventional solid state reaction between lanthanum oxide and ammonium chloride. DTA analysis showed that several exothermic peaks appear for measurements performed in air, but these peaks disappear completely for measurements in Ar. However, XRD confirmed the formation of $LaOCl$ NCs, 20–60 nm in size, during heat treatment at 450–750 °C in air. For heat treatment in Ar, smaller crystal size and fraction were obtained. Judd–Ofelt analysis along with PL measurements showed that, as the annealing temperature increases, a higher amount of Eu^{3+} ions is incorporated in LaOCl NCs, thus producing a better resolved Stark component and a much longer lifetime with respect to xerogel. Similar results were obtained for $LaOCl:Eu^{3+}$ pellets, indicating the effective incorporation of Eu^{3+} in this crystal phase.

1.4. Prospects and Perspectives

In summary, most of the transparent SG OxGCs previously described are based on a unique synthesis developed by Fujihara et. al., using TMOS and/or TEOS as the SiO_2 precursor, TFA as the fluorine source, and with an upper limit of nominal crystal phase around 10 mol %. The real crystal fraction in GCs was never estimated by a reliable method as Rietveld refinement and the few works that reported such information only used optical results to estimate the crystal content. Furthermore, many papers describe materials with 5 mol % of nominal active phase, this being even less than that achievable by classical MQ and requiring an extremely long time for bulk sample preparation, from several weeks up to months. Moreover, quite high treatment temperatures, up to 1000 °C, are used to crystallize the fluoride NCs when the crystallization of fluorides detected by DTA occurs at around 300 °C. For all these reasons, research in the SG OxGCs field shows good and enhanced optical properties but no improvement of the synthesis, and no new processing methods have been reported in the literature since the first papers published 20 years ago.

In recent years, the GlaSS group of CSIC has been working on the optimization of the synthesis for significantly increasing the crystal content of SG OxGCs. In fact, it is worth noting that synthesis parameters such as molar ratios between precursors, temperature, and time of reaction are strictly dependent on the crystal phases and their content. Moreover, a unique synthesis, in general, is not suitable to obtain different compositions for the same crystal phase, and modifications of the synthesis are necessary to obtain novel materials. In addition, we partially replaced TEOS with methyltriethoxysilane (MTES) in the SiO_2 sol synthesis, reducing the sintering temperature to 550 °C and making it possible to obtain enhanced optical properties without a need for much higher treatment temperatures.

2. Experimental

2.1. Synthesis of $(100 - x)SiO_2$-$xLaF_3$

Our research started with LaF_3 GC films prepared by dip-coating. The precursors used were TEOS, H_2O (0.1 M HCl), EtOH, TFA, and $La(CH_3COO)_3$. Dopants were added as acetates. A first SiO_2 sol was prepared using a molar ratio $1TEOS:2H_2O$ (0.1 HCl):9.5EtOH stirred for 2 h at room temperature. Separately, $1La(CH_3COO)_3:5EtOH:5TFA:4H_2O$ were mixed and stirred for 2 h at 40 °C in a glycerine bath. By mixing different volumes of SiO_2 sol with the La solution, $90SiO_2$-$10LaF_3$ and $80SiO_2$-$20LaF_3$ compositions were obtained. To further increase the nominal LaF_3 concentration to 30 mol %, the La solution was modified, adding more TFA (7 mol %) to favor acetate dissolution. The last composition we obtained is $60SiO_2$-$40LaF_3$ and in this case a further increase of TFA up to 10 mol % was necessary to obtain transparent thin films. Thin films were deposited by dip-coating on silica or Si substrates using a withdrawal rate of 20–45 cm/min. The thickness and refractive index of thin films were measured by a 2000 U ellipsometer (J.A. Woollam Co., Inc., Lincoln, NE, USA), using a Cauchy dispersion relation as a model.

Self-supported layers (or bulk-like samples) were also prepared with compositions $90SiO_2$-$10LaF_3$ and $80SiO_2$-$20LaF_3$ using the modified ratio of SiO_2 sol $1TEOS:7.5H_2O$ (0.1 HCl):5EtOH while leaving unchanged the synthesis of La solution described before for thin films with the same composition. Samples were dried at 50 °C over two days and then the covering was removed, letting the solvent evaporate for seven days.

A further modification to obtain higher thickness of thin films and improve the mechanical resistance of self-supported layers was to partially replace TEOS with MTES in the ratio

40TEOS:60MTES [106]. A modification of the molar ratios was necessary for thin film preparation to obtain the highest film thickness along with good sol stability. In particular, for thin films the SiO_2 sol was prepared using 0.4TEOS:0.6MTES:1H_2O (0.1 HCl):2.5EtOH:0.2CH_3COOH, while the La solution was left unchanged. For self-supported layers, the previous TEOS:H_2O ratio was used, replacing TEOS with 0.4 TEOS + 0.6 MTES (mol %) to maintain the same ratio of SiO_2 precursor and H_2O, which is a crucial parameter for bulk-like samples.

2.2. Synthesis of $(100 - x)SiO_2-xGdF_3/NaGdF_4$

90SiO_2-10GdF_3 and 80SiO_2-20LaF_3 self-supported layers were also obtained following the two-step process and using the aforementioned TEOS:MTES ratio for SiO_2 sol. In this case Gd solution, using $Gd(CH_3COO)_3$ as precursor, was stirred for 24 h at 40 °C to obtain a homogeneous solution of the products, thus making possible the crystallization of GdF_3.

Finally, we prepared for the first time SiO_2-$NaGdF_4$ materials as self-supported layers and thin films. A SiO_2 sol was prepared using only TEOS as the SiO_2 precursor with the same ratios used for SiO_2-LaF_3 compositions. Then, $Na(CH_3COO)$, $Gd(CH_3COO)_3$, EtOH, TFA, and H_2O were mixed using similar ratios of TFA and EtOH and trying several Na:Gd ratios (1.125–0.80):1. The solution was stirred 24 h at 40 °C in a glycerin bath.

For all compositions, GCs were obtained after heat treatment at 550–750 °C from 1 min up to several hours using heating rates of 1–10 °C/min. In all cases, thermal quenching in air was performed to obtain good crystallization of the samples.

2.3. Thermal and Structural Characterization

DTA was performed in air and argon (inert) atmosphere on small pieces of 1–1.25 mm in size, with heating rates 10–40 °C using SDT Q600 (TA Instruments, New Castle, DE, USA) equipment. Measurements were performed in the range 25–800 °C using 15–30 mg of sample.

High-resolution XRD patterns were collected in the range 6–27° at the synchrotron SpLine BM25B of the ESRF (European Synchrotron Radiation Facility, Grenoble, France) using a wavelength of 0.619 Å and a step size of 0.02°.

HRTEM was performed using a JEOL 2100 microscope (Akishima, Tokyo, Japan). Samples were prepared using lacey formvar carbon film that had a small amount of scratched sample deposited on them.

Fourier Transform Infra-Red (FTIR) spectra in the range 2000–600 cm^{-1} were obtained, with a resolution of 4 cm^{-1}, using a Perkin Elmer Spectrum 100 instrument (Waltham, MA, USA).

^{19}F magic-angle spinning nuclear magnetic resonance (^{19}F MAS/NMR) spectra of xerogel and GCs were recorded using a NMR Spectrometer AVANCE II (BRUKER, Billerica, MA, USA) equipped with a 9.4 Tesla magnet (400 MHz) and a 2.5 mm rotor spinning at 20 kHz.

X-ray Absorption Spectroscopy (XAS) was measured at the SpLine BM25A of the ESRF collecting the spectra in fluorescence mode using a 13-element Si (Li) solid-state detector with the sample surface placed at an angle of 45° to the incident beam. Six scans were acquired to obtain an average spectrum. XAS data were processed using ATHENA software [107]. Eu_2O_3 and EuF_3 crystal powders were also measured as reference materials to compare their spectra with those of Eu^{3+}-doped xerogel and GC samples.

Photoluminescence measurements were recorded using a FS5 fluorescence spectrometer (Edinburgh Instruments Ltd., Livingston, UK) equipped with a 150 W Xenon lamp or Ti:sapphire ring laser (0.4 cm^{-1} linewidth) in the 770–920 nm spectral range. The emission was detected by Hamamatsu H10330A-75 or Hamamatsu R928P photomultiplier (Hamamatsu City, Shizuoka, Japan).

Even though several compositions were prepared, most of the results shown in the next sections deal with 80SiO_2-20LaF_3/GdF_3/$NaGdF_4$ compositions, which have been studied in detail in recent years, all containing 20% of fluoride phases, at least double that reported by other authors.

3. Results and Discussion

3.1. Materials

Transparent and crack-free GC thin films were obtained for all compositions, even for those containing 30 or 40 mol % of active phase. For single deposition at 30 cm/min and heat treatment at 550 °C for 1 h, thicknesses around 500 nm and 1 µm were obtained for TEOS and TEOS/MTES compositions, respectively. Higher thicknesses up to 900 nm and 1.7 µm were obtained for TEOS and TEOS/MTES, respectively, depositing two films or increasing the withdrawal rate. Hence, the addition of MTES in the SiO_2 sol allows a notable increase of the thickness, thus making this precursor attractive for application where a thin film of 1 µm or thicker is required. Transparent and high-quality films were obtained, as confirmed by a good agreement between ellipsometry measurements and fits.

Transparent and crack-free self-supported layers were also obtained after heat treatments. The addition of MTES allows increasing the heating rate up to 10 °C/min without the appearance of cracks.

3.2. Thermal and Structural Characterization

3.2.1. SiO_2–LaF_3

DTA measurements for $(100 - x)SiO_2$-$xLaF_3$ (x = 10–40 mol %) bulk samples obtained drying thin films sols are shown in Figure 8a. The curves show a first endothermic peak with a weight loss (not shown) associated with H_2O and other solvent removal. Such an endothermic process is much more intense for a composition with a higher amount of La. This could be associated with the increasing water content passing from x = 10 to x = 40.

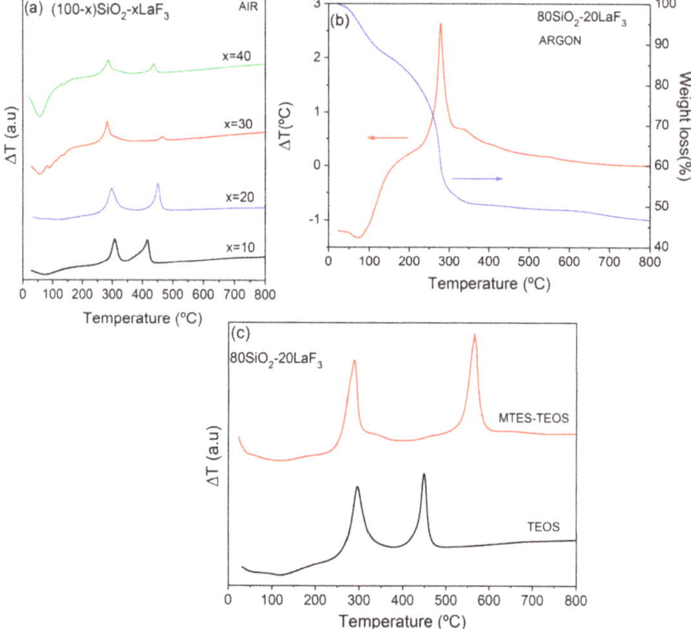

Figure 8. (**a**) DTA in air of $(100 - x)SiO_2$-$xLaF_3$ (x = 10–40 mol %) bulk samples prepared with TEOS; (**b**) DTA and TG in argon atmosphere of $80SiO_2$-$20LaF_3$ bulk samples prepared with TEOS; (**c**) DTA in air of $80SiO_2$-$20LaF_3$ bulk samples with TEOS and TEOS/MTES. All measurements were performed using a heating rate of 10 °C/min.

Then, a first exothermic peak appears around 300 °C, together with a big mass loss that increases passing from x = 10 to x = 40 mol %. The peak shifts towards lower temperatures when increasing the active phase content. As reported by other authors [43], such a peak is associated with the crystallization of LaF_3, favored by increasing the La content. Such crystallization takes place after a chemical decomposition with gas release, which is responsible for the notable weight loss. The intensity of the endothermic and exothermic peaks is affected by the sample amount used for the analysis, and therefore it cannot be compared quantitatively. The second exothermic peak in the range 400–500 °C is assigned to organic combustion and was not detected in an argon atmosphere (Figure 8b). The effect of the SiO_2 precursor on the DTA curve is shown in Figure 8c for $80SiO_2$-$20LaF_3$ samples prepared using TEOS and TEOS/MTES in the ratio 40:60 as SiO_2 precursor. The crystallization peak appears centered around the same value, near 300 °C, while the second exothermic peak is associated with organic combustion of different species. In fact, when MTES is used, CH_3 groups are introduced in the system and released around 550–600 °C. Such release is associated with a contraction of thin films and self-supported layers. Therefore, when MTES is used, treatment temperatures no higher than 550 °C should be used to avoid sample cracking. However, and concerning the crystallization mechanism, no relevant effects are introduced by the partial replacement of TEOS with MTES. The introduction of MTES improved the mechanical resistance during the treatment process and MTES bulk-like samples can be treated using a 10 °C/min heating rate instead of 1–2 °C/min, as is commonly used for samples prepared using only TEOS. Moreover, MTES addition facilitates –OH group removal [108], without requiring extremely high treatment temperatures (900–1000 °C).

The XRD of $80SiO_2$-$20LaF_3$ bulk sample prepared using TEOS/MTES and treated at 550 °C for 1 min is represented in Figure 9. Well-defined diffraction peaks are observed even for this fast treatment; the crystal size, estimated by Scherrer's equation, is around 8.5 nm. No relevant SiO_2 amorphous pattern is observed, different to the XRD results obtained by other authors [50–55,57], indicating a relevant increase of fluoride crystal fraction. Dopant incorporation, such as Er^{3+}, was confirmed by an appreciable shift of the diffraction peaks towards higher angles, associated with a shrinking of the unit cell due to the lower size of Er^{3+} with respect to La^{3+} [109]. By Rietveld refinement (not shown), we obtained a LaF_3 crystal fraction of 18 wt %, which is, to the best of our knowledge, the highest value of fluoride concentration ever reported in OxGCs. The crystallization mechanism of LaF_3 in SG OxGCs has been studied in a previous paper [91]. It was shown that the LaF_3 crystallization mechanism is very different to that for MQ GCs. In fact, for MQ GC a diffusion-controlled process is responsible for the crystallization of LaF_3 and many other fluoride phases; phase separation acts as a precursor for crystallization. On the contrary, LaF_3 crystallization in SG OxGCs is not a diffusion-controlled process but consists of a fast crystal precipitation taking place after a chemical decomposition. It was observed that LaF_3 NCs are not stable for heat treatment at crystallization or higher temperatures and amorphization was observed for a long treatment time (5–80 h) at 550 °C.

HRTEM micrographs of $80SiO_2$-$20LaF_3$ thin films and self-supported layers treated at 550 °C for 1 min are shown in Figure 10a,b, respectively. Very small LaF_3 NCs around 2–3 nm are observed in thin film samples, as confirmed by the crystal size distribution shown in the inset. For self-supported layers the crystal size is much bigger, around 8 nm, in agreement with the XRD results in Figure 3. In both cases homogeneously distributed NCs are observed, without the formation of clusters, even for such a fast heat treatment at 550 °C for 1 min.

NMR spectra of SG OxGCs are rarely encountered in the literature but the information that can be extracted is of relevant importance. In particular, ^{19}F NMR spectra allow for obtaining evidence about fluorine surrounding in the xerogel and how it changes after the crystallization process. Figure 11 shows ^{19}F MAS NMR of $80SiO_2$-$20LaF_3$ xerogel and GC treated at 550 °C for 1 min [110]. The spectrum of polycrystalline LaF_3 was also acquired for comparison. In the xerogel sample the fluorine surrounding is as in the precursor TFA acid and no bonding with Si is observed. After heat treatment at 550 °C for 1 min, the spectrum of the GC sample is practically the same as that of the LaF_3 polycrystalline powders. This further indicates that a chemical reaction accompanied by fast crystal precipitation is responsible for the LaF_3 NCs' formation.

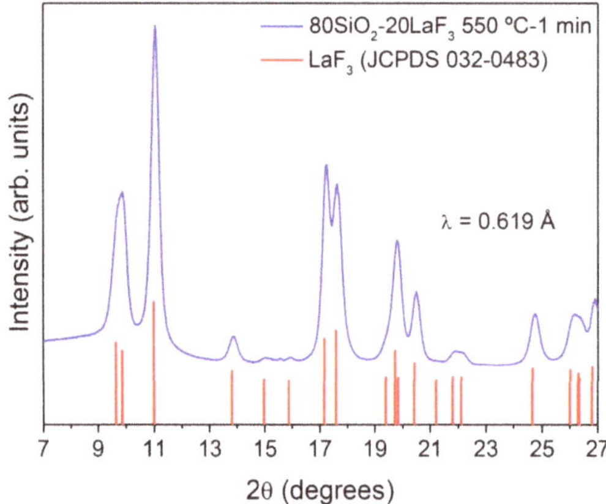

Figure 9. XRD of $80SiO_2$-$20LaF_3$ GC treated at 550 °C for 1 min performed at the synchrotron SpLine BM25B of the ESRF.

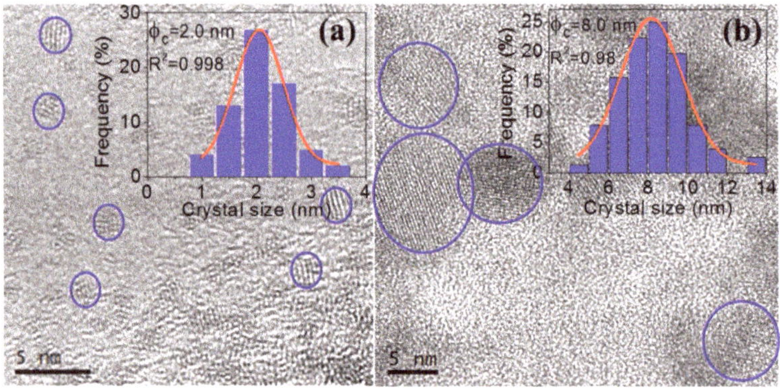

Figure 10. HRTEM of $80SiO_2$-$20LaF_3$ (**a**) thin film and (**b**) self-supported layer prepared using TEOS/MTES and treated at 550 °C for 1 min. The corresponding crystal size distributions are also shown.

These results are further confirmed by FTIR spectra of $80SiO_2$-$20LaF_3$ xerogel and GCs treated at 550 °C for 1 min and 1 h, as shown in Figure 12 [110]. In the xerogel, the bands centered at 1680 and 1650 cm^{-1} are associated with H-O-H bending and C=O stretching vibrations, respectively. In the range 1500–1400 cm^{-1} several absorption bands appear and are assigned to TFA, acetates, and/or derived ions [43]. A C-F stretching band is also identified between 1400 and 1000 cm^{-1}. All of these bands disappear in the GC sample, accompanied by LaF_3 NCs' precipitation. In the GC the bands in the range 1100–800 cm^{-1} are associated with Si-O-Si asymmetric and symmetric stretching vibrations of $[SiO_4]$ units. A small band at 1279 cm^{-1}, also present in the xerogel sample but much more intense, is assigned to C-O vibration and due to incompletely removed organic compounds. Such a band disappears over longer treatment times at 550 °C.

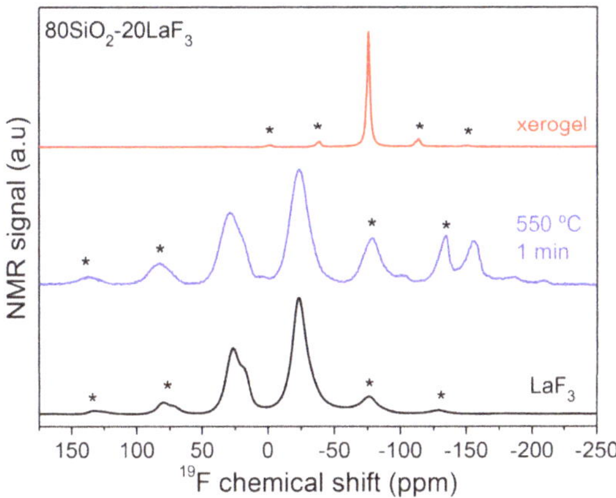

Figure 11. ^{19}F MAS-NMR spectra of 80SiO$_2$-20LaF$_3$ xerogel and GC treated at 550 °C for 1 min. The spectrum of pure LaF$_3$ crystal powder is also given for comparison. Stars indicate spinning sidebands.

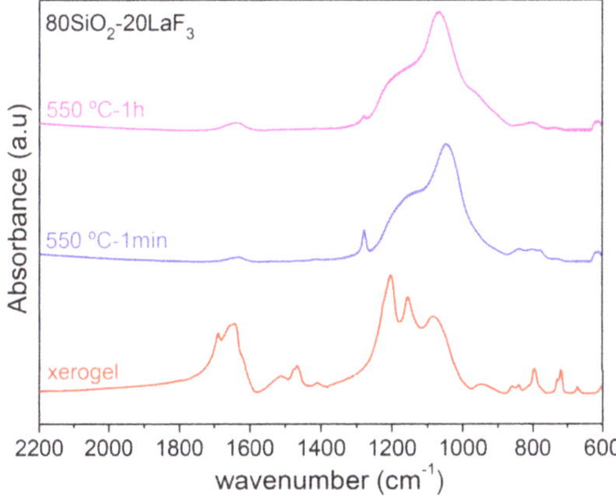

Figure 12. FTIR of 80SiO$_2$-20LaF$_3$ xerogel and GC self-supported layer treated at 550 °C for 1 min and 1 h.

Photoluminescence (PL) measurements are of great importance in most papers about SG OxGCs; in particular, enhanced properties are obtained when dopants are embedded into the fluoride NCs with low phonon energy. As an example, low-temperature (9 K) PL emission and excitation spectra of 80SiO$_2$-20LaF$_3$ xerogel and GC treated at 650 °C for 3 h and doped with 0.5 Nd^{3+} are shown in Figure 13 [111]. Both the emission and excitation spectra of the xerogel show broad and less structured bands, indicating a predominant amorphous environment for Nd^{3+} ions. Instead, for the GC sample sharp peaks and well-resolved Stark components are observed and associated with Nd^{3+} emission in LaF$_3$ NCs. Similar features are observed for the excitation spectrum, where well-resolved peaks are observed for the ^{4}I$_{9/2}$→^{4}F$_{5/2}$ band; moreover, the ^{4}I$_{9/2}$→^{4}F$_{3/2}$ doublet narrows and splits into two main single components, as expected for a well-defined crystal field site. Therefore, Nd^{3+} incorporation into LaF$_3$ NCs was unambiguously confirmed.

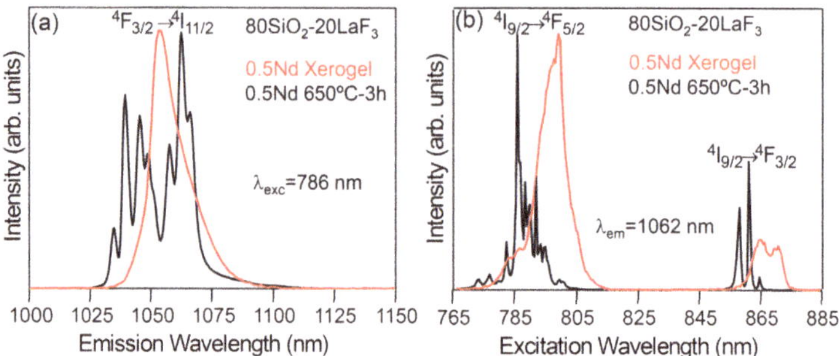

Figure 13. PL (**a**) emission and (**b**) excitation spectra of 80SiO$_2$-20LaF$_3$ bulk xerogel and GC treated at 650 °C for 3 h. All spectra were recorded at 9 K.

3.2.2. SiO$_2$–GdF$_3$/NaGdF$_4$

Figure 14 shows DTA curves of the 90SiO$_2$-10NaGdF$_4$ bulk-like sample. The weight loss between 70 and 200 °C, of less than 10%, is ascribed to solvent removal. Then, as usually occurs for SG oxyfluoride compositions, a strong and sharp exothermic peak appears around 300 °C along with a mass loss of 30%. Such an exothermic peak is associated with chemical decomposition with NaGdF$_4$ precipitation, similar to that described by other authors for other fluoride crystal phases [43,57,82]. Further weight loss for heat treatment in the range 400–600 °C can be associated with the combustion of organic compounds. The same sample measured in an argon atmosphere showed similar characteristics; therefore, the exothermic peak is associated with NaGdF$_4$ crystallization. Similar features are obtained for SiO$_2$-GdF$_3$ composition and described elsewhere [112].

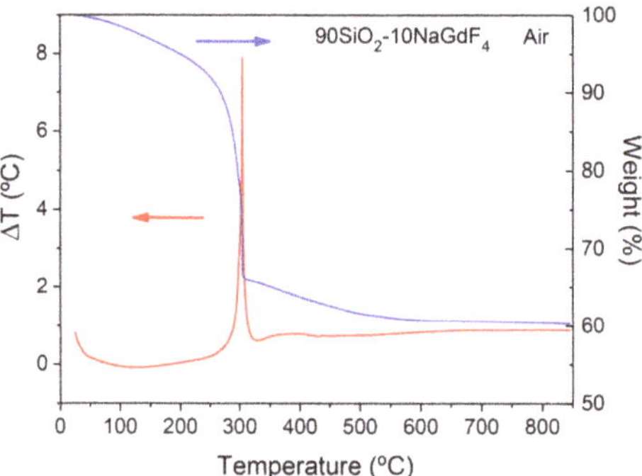

Figure 14. DTA (red) and TG (blue) curve of 90SiO$_2$-10NaGdF$_4$ bulk sample acquired in air using a heating rate of 10 °C/min.

Diffractograms of (100 − x)SiO$_2$-xGdF$_3$ (x = 10 and 20 mol %) treated at 550 °C for 1 min are shown in Figure 15. As observed, both phases of GdF$_3$, orthorhombic and hexagonal, appear even for such fast heat treatment. It was shown that from LaF$_3$ to EuF$_3$, the hexagonal phase is preferred, while elements heavier than Gd are organized in orthorhombic structures [113]. Considering that Gd is right in the middle of a lanthanide series, it seems reasonable that a mixture of both phases appear. The relative intensity of hexagonal and orthorhombic structures is quite similar for all heat treatments, thus suggesting that there is no one preferential crystal structure but that both configurations coexist. Crystal size is around 8 and 9 nm for hexagonal and orthorhombic phases, respectively. Small changes in relative peak intensities can be associated with the deformation or preferential incorporation of Eu^{3+} ions at certain crystallographic sites. Moreover, the fact that similar sizes are obtained for both compositions suggests that NC formation is not related to the amount of the initial precursors, the mechanism being explained as fast crystal precipitation after a chemical reaction similar to that described by Fujihara for LaF_3 [43]. However, the final crystal fraction should be affected by the initial content as observed for LaF_3 compositions. Nevertheless, for GdF_3 crystallization a quite different scenario is observed with respect to LaF_3 crystallization, where an amorphization was observed over a long heat treatment time at 550 °C. In fact, no changes in the crystal size are observed for treatment times up to 8 h; the diffractograms of GCs treated for 1 min up to 8 h being practically the same. It seems that GdF_3 NCs are more stable against decomposition as opposed to LaF_3 ones. Further work is still necessary to clarify this point.

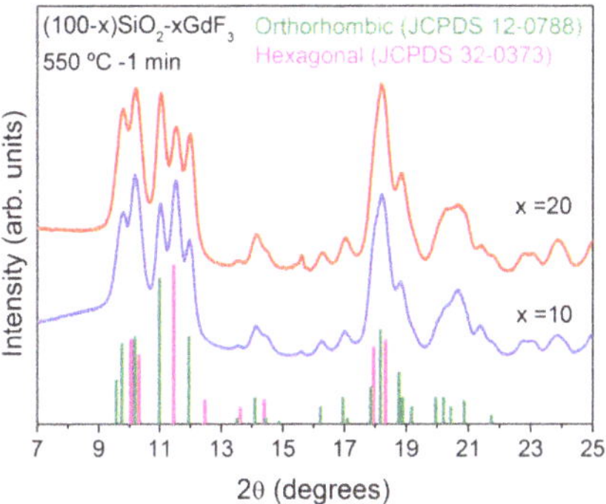

Figure 15. XRD of $(100 - x)SiO_2$-$xGdF_3$ (x = 10 and 20 mol %) GC treated at 550 °C for 1 min performed at the synchrotron SpLine BM25 of the ESRF.

It is also worth noting that SiO_2 precursors affect the final crystal phase and the crystal size. For example, for $80SiO_2$-$20GdF_3$ samples prepared using TEOS, only hexagonal crystals are detected but it was necessary to raise the temperature to 750 °C to obtain GdF_3 crystals. Instead, for TEOS/MTES samples, even for treatment temperature as low as 550 °C for 1 min, a good crystallization was obtained but in this case both a hexagonal and an orthorhombic phase appeared. Therefore, as was shown, the synthesis precursor and the different ratios between them can strongly affect the final crystal size, crystal fraction, and symmetry of the crystal phase. However, the influence of the SiO_2 precursors and the synthesis route on the crystallization tendency of the systems has not yet been totally elucidated.

XRD of $80SiO_2$-$20NaGdF_4$ are shown in Figure 16. This is the first time that $NaGdF_4$ crystals have been obtained in SG OxGCs. Diffractograms obtained for super-stoichiometric Na:Gd ratios caused the crystallization of silicates, along with $NaGdF_4$. Therefore, precise control of the Na:Gd ratio was necessary to avoid the formation of silicates during heat treatment and obtain only the precipitation of fluoride NCs. It is known that Na^+ acts as a network modifier, producing more open glass structures. In fact, Na^+ ions can break SiO_4 units, thus producing non-bridging oxygens with subsequent softening of the glass network. Hence, a sub-stoichiometric ratio Na:Gd 0.95:1 was necessary to ensure the complete reaction of Na with Gd, avoiding the presence of free Na^+ ions. As observed in Figure 16, both α and β-$NaGdF_4$ phases precipitated upon heat treatment at 600 °C for 1 h with a size of 4 and 13 nm, respectively. However, for longer heat treatments, up to 120 h, a relative decrease of the α phase, indicated by stars, is observed with respect to GC treated for only 1 h. Moreover, sharper peaks are observed after increasing the treatment time, thus indicating the formation of bigger crystals (more than 30 nm in size). Such behavior is quite different to that described for LaF_3, for which an increase of the treatment time increases neither the crystal size nor the crystal fraction. Therefore, it was suggested that LaF_3 crystals formation occurs as a fast precipitation when certain energy is given to the system, instead being a diffusion-limited process, as proposed by other authors. However, for $NaGdF_4$ OxGCs, a different crystallization mechanism occurs for polymorphous crystals;

in particular, the addition of alkaline earth elements could be responsible for the better crystallinity obtained by increasing the treatment time. In fact, the time-dependent diffusion of Na$^+$ ions could be relevant to achieve better crystallization.

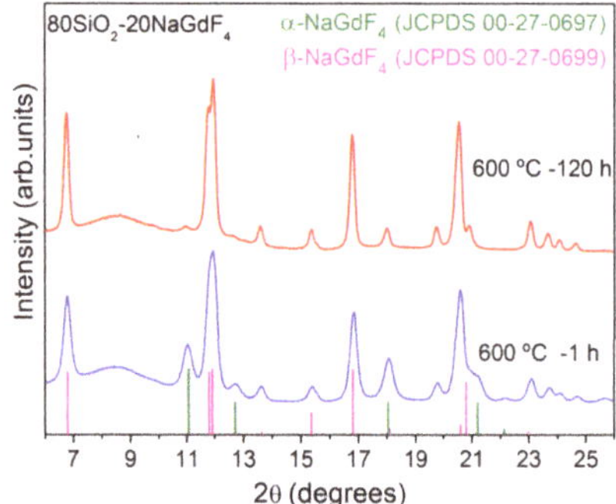

Figure 16. XRD of 80SiO$_2$-20NaGdF$_4$ GC treated at 600 °C for 1 and 120 h. The measurements were performed at the synchrotron SpLine BM25 of the ESRF.

Bartha et al. [114] studied NaYF$_4$ phosphors and observed that for heat treatment at 300 °C, both cubic and hexagonal NaYF$_4$ micro crystals appeared. However, by increasing the treatment temperature to 400–600 °C, only a pure hexagonal NaYF$_4$ phase was observed. The authors explained such behavior as an autocatalytic process where the initial cubic NaYF$_4$ phase played a catalytic role, causing its fast self-accelerated crystallization. The energy resulting from the disintegration process of the initial NaYF$_4$ crystals contributed to the growth of hexagonal NaYF$_4$ phase. This mechanism could also explain the crystallization of NaGdF$_4$ NCs in OxGCs but further investigation has still to be done.

Figure 17 shows a micrograph of 80SiO$_2$-20GdF$_3$ self-supported layer doped with 0.5 Eu^{3+} treated at 550 °C for 1 min. As for SiO$_2$-LaF$_3$ GCs, homogeneously distributed NCs are observed and no agglomerates or clusters are observed, as suggested by other authors for YF$_3$ NCs [57]. By a detailed analysis of the microstructures, both hexagonal and orthorhombic crystals were detected according to JCPDS. EDXS of Eu^{3+} doped 80SiO$_2$-20GdF$_3$ GC treated at 550 °C for 1 min revealed that Eu^{3+} ions are mostly concentrated into GdF$_3$ NCs. Therefore, dopant incorporation is very fast because even for a heat treatment as short as 1 min, most Eu^{3+} ions are already embedded in the crystal phase. A better explanation of dopant incorporation is given later when the results about XAS will be discussed.

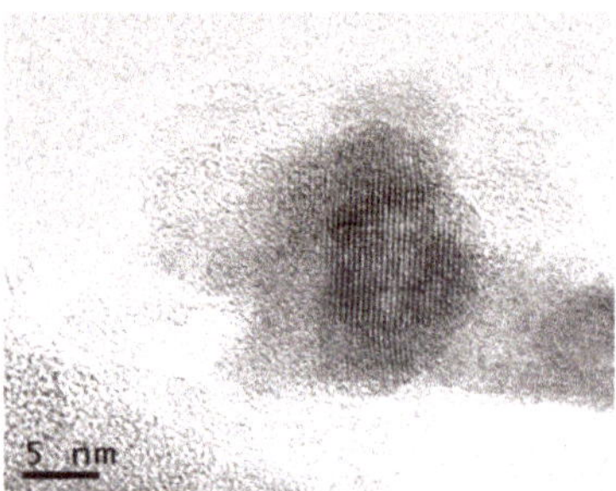

Figure 17. HRTEM of 80SiO$_2$-20GdF$_3$ self-supported layer treated at 550 °C for 1 min.

FTIR spectra of 80SiO$_2$-20GdF$_3$ xerogel and GC treated at 550 °C for 1 min are shown in Figure 18. Similar results are obtained with respect to the 80SiO$_2$-20LaF$_3$ composition given in Figure 12. For the xerogel sample TFA and acetate vibration bands are still present but disappear after heat treatment accompanied by LaF$_3$ crystallization. Similar features are obtained for SiO$_2$-NaGdF$_4$ compositions, thus indicating that a chemical reaction followed by NC precipitation is a common feature of all the studied compositions. However, differences in crystallization behavior are also observed by XRD and therefore the crystal growth can be dependent on crystal phase and synthesis conditions.

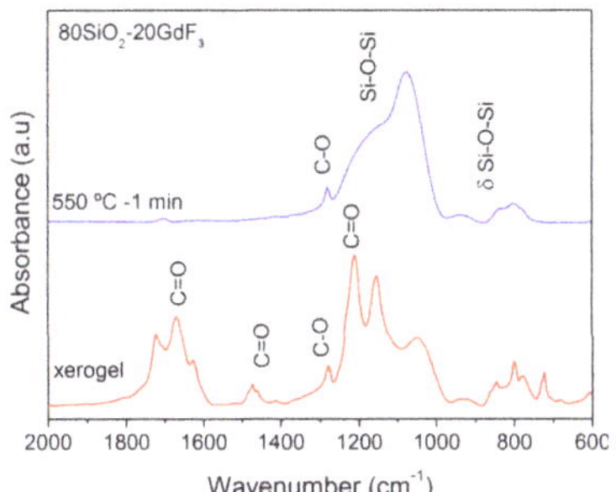

Figure 18. FTIR of 80SiO$_2$-20GdF$_3$ xerogel and GC treated at 550 °C for 1 min.

The last results that are shown in this paper deal with XAS spectra of 80SiO$_2$-20GdF$_3$ materials doped with Eu^{3+}. In particular, the spectra of the xerogel and GC sample are compared with the aim of gaining more insight into the Eu^{3+} environment. Indeed, better optical efficiencies are obtained when dopants are embedded into the NCs and knowledge of the real dopants fraction incorporated into the NCs is crucial to improve and optimize the luminescence emission of these materials. Even though a certain nominal dopant concentration is used, the real concentration into the fluoride NCs is rarely estimated, the only results being extrapolated from optical measurements [54]. Moreover, to estimate the effective concentration into the NCs, even with 10–20% error, knowledge of the crystal fraction is necessary and such values are seldom or never reported for SG OxGCs. Last year our group started acquiring data to obtain this relevant information, but a lot of work has still to be done. However, some interesting conclusions can already be drawn from the results shown below. Eu$_2$O$_3$ and EuF$_3$ spectra are shown in Figure 19a, together with their derivate curves in Figure 19b. The maximum of the derivate curves is 6977.9 and 6979.5 eV, for Eu$_2$O$_3$ and ErF$_3$, respectively. A lower energy is associated with less electronegativity and a lower field strength of the Ln^{3+} ion [115]. The uncertainty of the energy values is around 0.8 eV. Therefore, it can be confirmed that the two maxima are well separated from each other. Figure 20a,b show the results for a 80SiO$_2$ 20GdF$_3$ xerogel and GC samples treated at 550 °C for 1 min up to 8 h. In all cases, the absorption maximum is centered at ~6979 eV, practically the same value obtained for EuF$_3$ reference. Hence, even though by these measurements it is not possible to distinguish between a crystalline or amorphous environment, the crucial point is that a fluorine-rich environment is observed for all GCs and even for the xerogel sample. Such behavior could be explained considering that Eu^{3+} ions are still coordinated to the surrounding TFA ions in the xerogel sample, and, after heat treatment, GdF$_3$ crystals precipitate together with Eu^{3+} incorporation. These results are in agreement with Eu^{3+}-rich GdF$_3$ NCs observed by EDXS for 80SiO$_2$-20GdF$_3$ GC treated at 550 °C for 1 min.

Previous calculations performed for 80SiO$_2$-20LaF$_3$ samples doped with 0.5 Er^{3+} showed that almost 91% of Er^{3+} ions in GC samples are in a fluorine-rich environment, the effective concentration into the LaF$_3$ NCs thus being almost one order of magnitude higher than the nominal one. Similar results were obtained for MQ samples containing LaF$_3$ NCs and confirmed by PL results [116].

Finally, to conclude this section we show some optical results for 80SiO$_2$-20GdF$_3$ self-supported layers doped with 0.5 Eu^{3+}. Photoluminescence measurements of GC treated at 550 °C for 1 min showed well-resolved structure together with a narrowing of the Eu^{3+} with respect to the xerogel sample (Figure 21). Moreover, the R asymmetry ratio between the electric dipole transition (^{5}D$_0$-^{7}F$_2$) and the magnetic dipole transition (^{5}D$_0$-^{7}F$_1$) is reduced in GCs, thus indicating that Eu^{3+} ions are incorporated in the GdF_3 crystal phases. Moreover, the ET from Gd^{3+} to Eu^{3+} was observed in the GC sample and further supported by fluorescence decay lifetimes.

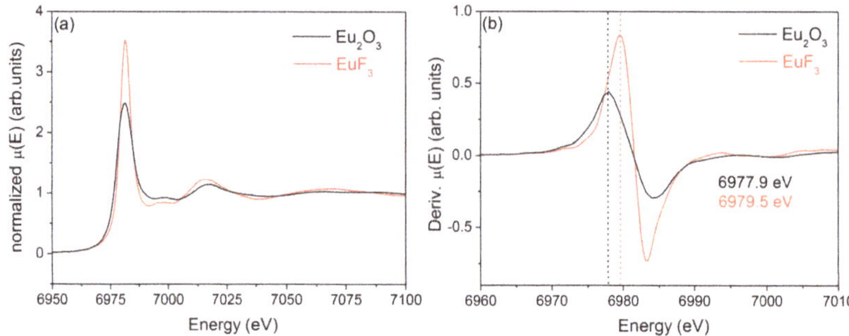

Figure 19. XAS (**a**) spectra and (**b**) derivate of Eu_2O_3 and EuF_3 reference samples. The values shown in (**b**) refer to the maximum of the derivate curves.

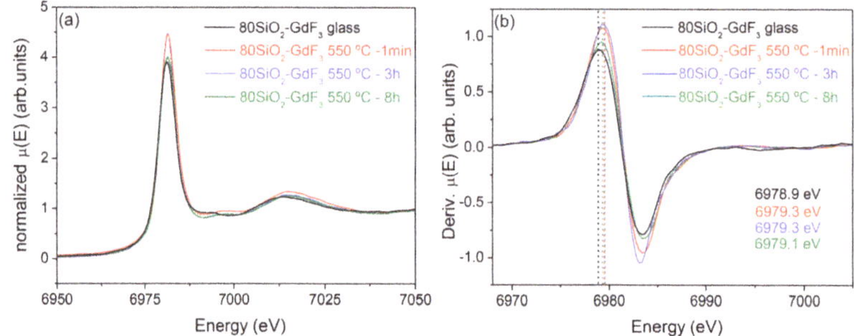

Figure 20. XAS (**a**) spectra and (**b**) derivate of $0.5Eu^{3+}$-doped $80SiO_2$-$20LaF_3$ xerogel and GC samples treated at 550 °C for 1 min up to 8 h. The values shown in (**b**) refer to the maximum of the derivate curves.

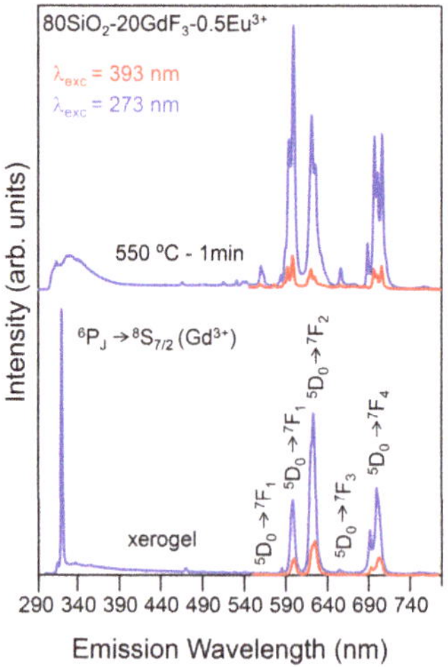

Figure 21. Emission spectra of Eu^{3+} and Gd^{3+} ions in xerogel (**bottom**) and GC (**top**) treated at 550 °C-1 min under excitation of Eu^{3+} at 393 nm (red spectra) and of Gd^{3+} at 273 nm (blue spectra).

4. Conclusions

Different research groups have studied the preparation of transparent GC materials using the MQ process and suitable control of the synthesis and heat treatment conditions. However, SG appears as a promising alternative to obtain innovative GCs with high fluorine content and high homogeneity at a lower temperature.

Most of the materials studied by SG are related with compositions $(100-x)SiO_2$-x $M_1F_2/M_1F_3/M_1M_2F_4/M_1M_2F_5$, containing a small nominal crystal fractions (x = 5–10 mol %), and are obtained using a similar synthesis procedure developed 20 years ago. Furthermore, the same SiO_2 precursor, TEOS or TMOS, is used in most works, and the precursor ratio rarely changes with respect to the first papers.

There is scarce information about the structural characterization of bulk and thin films; typical characterization is performed by means of XRD, HRTEM, and FTIR. However, most of the authors never calculated the real crystal fraction. Just one or two papers estimated the real active crystal content and concluded that for a nominal composition of 5 mol % of active phase, a 3 wt % of crystal phase is obtained after heat treatment. However, photoluminescence measurements of SG GCs showed very promising results. Ln^{3+} incorporation in fluoride NCs was demonstrated by several authors and produced an improvement in the optical properties (linear or non-) due to the low phonon energy of the crystal hosts, thus opening the way to use SG materials for photonic applications.

On the other side, our group introduced an important modification to the SG synthesis of OxGCs. First of all, LaF_3- and GdF_3-based compositions with a much higher content of active phase (up to 40 mol %) have been prepared for the first time, together with $NaGdF_4$ OxGCs. Moreover, new precursors and deep synthesis modification were performed by partial replacement of TEOS with MTES.

A chemical reaction followed by fast crystal precipitation was indicated as responsible for the crystallization mechanism of fluoride NCs, as opposed to conventional diffusion-controlled processes. LaF_3 NCs were demonstrated to be unstable with aging at crystallization or higher temperatures for a long treatment time, while for $NaGdF_4$ an increase in the crystal size was observed when the treatment time increased. Therefore, the evolution and behavior of the NCs depend on the synthesis and the crystal phase.

Rietveld refinement confirmed that a crystal fraction ~18 wt % is obtained for $80SiO_2$-$20LaF_3$ GC treated at 550 °C for 1 min, such a value being the highest reported to date for SG OxGCs.

HRTEM showed that homogeneously distributed fluoride NCs precipitate in the SiO_2 glass matrix after fast heat treatment at 550 °C for 1 min using TEOS/MTES as SiO_2 precursors. Moreover, EDXS confirmed dopants' incorporation in the fluoride NCs even after fast heat treatment at 550 °C for 1 min, suggesting that dopant incorporation occurs along with NC precipitation. Such results were also confirmed by XAS measurements, revealing a fluorine-rich environment even in the xerogel sample.

Optical measurements unambiguously showed dopant incorporation in low-phonon-energy fluoride NCs. Well-resolved Stark components and crystal-like spectra were obtained for GC samples, resulting in much higher emission intensities and more efficient ET process with respect to xerogel samples.

Considering all the results published up to now and the benefits offered by the SG method, we think that transparent SG OxGCs materials can be considered of great interest and promising for several photonic applications, but improvement of synthesis and processing is still necessary. In this sense, we hope that this paper will be useful for researchers working in this field.

Acknowledgments: This work was supported by MINECO under projects MAT2013-48246-C2-1-P, MAT2013-48246-C2-2-P, and MAT2017-87035-C2-1-P/-2-P (AEI/FEDER, UE) and Basque Country Government IT-943-16 and PPG17/07. The authors are grateful for access to the Spanish Beamline (SpLine) at the ESRF facilities in Grenoble to perform experiments MA-3350 and 25-01-1014. Jose Joaquín Velázquez also acknowledges MINECO for Grant FPDI-2013-16895.

Author Contributions: G.G., J.J.V., J.M., R.B., J.F., A.D. and Y.C. conceived and designed the experiments; G.G., J.J.V., R.B. and J.F. performed the experiments; G.G., J.J.V., R.B., J.F., A.D., Y.C. analyzed the data; G.G., J.J.V., J.M., R.B. and J.F. contributed reagents/materials/analysis tools; G.G., J.J.V., J.M., R.B., J.F., A.D. and Y.C. wrote the paper.

Conflicts of Interest: The authors declare no conflict of interest.

References

1. Feldmann, C.; Jüstel, T.; Ronda, C.R.; Schmidt, P.J. Inorganic luminescent materials: 100 years of research and application. *Adv. Funct. Mater.* **2003**, *13*, 511–516. [CrossRef]
2. George, N.C.; Denault, K.A.; Seshadri, R. Phosphors for solid-state white lightening. *Annu. Rev. Mater. Res.* **2013**, *43*, 481–501. [CrossRef]
3. Huang, X. Solid-state lighting: Red phosphor converts white LEDs. *Nat. Photonics* **2014**, *8*, 748–749. [CrossRef]
4. Huang, X.; Han, S.; Huang, W.; Liu, X. Enhancing solar cell efficiency: The search for luminescent materials as spectral converters. *Chem. Soc. Rev.* **2013**, *42*, 173–201. [CrossRef] [PubMed]
5. Eliseeva, S.V.; Bunzli, J.C.G. Lanthanide luminescence for functional materials and bio-sciences. *Chem. Soc. Rev.* **2010**, *39*, 189–227. [CrossRef] [PubMed]
6. Lee, H.U.; Park, S.Y.; Lee, S.C.; Choi, S.; Seo, S.; Kim, H.; Won, J.; Choi, K.; Kang, K.S.; Park, H.G.; et al. Black Phosphorus (BP) nanodots for potential biomedical applications. *Small* **2016**, *12*, 214–219. [CrossRef] [PubMed]
7. De Pablos-Martín, A.; Durán, A.; Pascual, M.J. Nanocrystallisation in oxyfluoride systems: Mechanisms of crystallisation and photonic properties. *Int. Mater. Rev.* **2012**, *57*, 165–186. [CrossRef]
8. Fedorov, P.P.; Luginina, A.A.; Popov, A.I. Transparent oxyfluoride glass ceramics. *J. Fluor. Chem.* **2015**, *172*, 22–50. [CrossRef]
9. De Pablos-Martín, A.; Ferrari, M.; Pascual, M.J.; Righini, G.C. Glass-ceramics: A class of nanostructured materials for photonics. *La Rivista del Nuovo Cimento* **2015**, *38*, 311–369. [CrossRef]
10. Wang, Y.; Ohwaki, J. New transparent vitroceramics codoped with Er^{3+} and Yb^{3+} for efficient frequency upconversion. *Appl. Phys. Lett.* **1993**, *63*, 3268–3270. [CrossRef]
11. Stevenson, A.J.; Serier-Brault, H.; Gredin, P.; Mortier, M. Fluoride materials for optical applications: Single crystals, ceramics, glasses, and glass–ceramics. *J. Fluor. Chem.* **2011**, *132*, 1165–1173. [CrossRef]
12. Lopez-Iscoa, P.; Salminen, T.; Hakkarainen, T.; Petit, L.; Janner, D.; Boetti, N.; Lastusaari, M.; Pugliese, D.; Paturi, P.; Milanese, D. Effect of partial crystallization on the structural and luminescence properties of Er^{3+}-doped phosphate glasses. *Materials* **2017**, *10*, 473. [CrossRef] [PubMed]
13. Höland, W.; Beall, G. *Glass-Ceramic Technology*, 3rd ed.; The American ceramic society: Westerville, OH, USA, 2002; pp. 15–18.
14. Augustyn, E.; Żelechower, M.; Stróż, D.; Chrapoński, J. The microstructure of erbium-ytterbium co-doped oxyfluoride glass-ceramic optical fibers. *Opt. Mater.* **2012**, *34*, 944–950. [CrossRef]
15. Reben, M.; Dorosz, D.; Wasylak, J.; Burtan-Gwizdala, B.; Jaglarz, J.; Zontek, J. Nd^{3+}-doped oxyfluoride glass ceramics optical fibre with SrF_2 nanocrystals. *Opt. Appl.* **2012**, *42*, 353–364. [CrossRef]
16. Krishnaiah, K.V.; Ledemi, Y.; Genevois, C.; Veron, E.; Sauvage, X.; Morency, S.; Soares de Lima Filho, E.; Nemova, G.; Allix, M.; Messaddeq, Y. Ytterbium-doped oxyfluoride nano-glass-ceramic fibers for laser cooling. *Opt. Mater. Express* **2017**, *7*, 1980–1994. [CrossRef]
17. Gorni, G.; Balda, R.; Fernández, J.; Ipparraguirre, I.; Velázquez, J.J.; Castro, Y.; Chen, G.; Sundararayan, M.; Pascual, M.J.; Durán, A. Oxyfluoride glass-ceramic fibers doped with Nd^{3+}: Structural and optical characterization. *CrystEngComm* **2017**, *19*, 6620–6629. [CrossRef]
18. Roberts, R.B.; Tainsh, R.J.; White, G.K. Thermal properties of Zerodur at low temperatures. *Cryogenics* **1982**, *22*, 566–568. [CrossRef]
19. Owens, G.J.; Singh, R.K.; Foroutan, F.; Alqaysi, M.; Han, C.M.; Mahapatra, C.; Kim, H.W.; Knowles, J.C. Sol-gel based materials for biomedical applications. *Prog. Mater. Sci.* **2016**, *77*, 1–79. [CrossRef]
20. Rywak, A.A.; Burlitch, J.M. Sol-gel synthesis of nanocrystalline magnesium fluoride: Its use in the preparation of MgF_2 films and MgF_2-SiO_2 composites. *Chem. Mater.* **1996**, *8*, 60–67. [CrossRef]

21. Rywak, A.A.; Burlitch, J.M. The crystal chemistry and thermal stability of sol-gel prepared fluoride-substituted talc. *Phys. Chem. Miner.* **1996**, *23*, 418–431. [CrossRef]
22. Fujihara, S.; Tada, M.; Kimura, T. Preparation and characterization of MgF_2 thin film by a trifluoroacetic acid method. *Thin Solid Films* **1997**, *304*, 252–255. [CrossRef]
23. Luo, W.; Wang, Y.; Bao, F.; Zhou, L.; Wang, X. Crystallization behavior of PbF_2-SiO_2 based bulk xerogels. *J. Non-Cryst. Solids* **2004**, *347*, 31–38. [CrossRef]
24. Del-Castillo, J.; Yanes, A.C.; Méndez-Ramos, J.; Tikhomirov, V.K.; Rodríguez, V.D. Structure and up-conversion luminescence in sol–gel derived Er^{3+}-Yb^{3+} co-doped SiO_2:PbF_2 nano-glass-ceramics. *Opt. Mater.* **2009**, *32*, 104–107. [CrossRef]
25. Del-Castillo, J.; Yanes, A.C.; Méndez-Ramos, J.; Tikhomirov, V.K.; Moshchalkov, V.V.; Rodríguez, V.D. Sol-gel preparation and white up-conversion luminescence in rare-earth doped PbF_2 nanocrystals dissolved in silica glass. *J. Sol-Gel Sci. Technol.* **2010**, *53*, 509–514. [CrossRef]
26. Szpikowska-Sroka, B.; Zur, L.; Czoik, R.; Goryczka, T.; Swinarew, A.S.; Żadło, M.; Pisarski, W.A. Long-lived emission from Eu^{3+}:PbF_2 nanocrystals distributed into sol-gel silica glass. *J. Sol-Gel Sci. Technol.* **2013**, *68*, 278–283. [CrossRef]
27. Szpikowska-Sroka, B.; Pawlik, N.; Zur, L.; Czoik, R.; Goryczka, T.M.; Pisarski, W.A. Effect of fluoride ions on the optical properties of Eu^{3+}:PbF_2 nanocrystals embedded into sol-gel host materials. *Mater. Chem. Phys.* **2016**, *174*, 138–142. [CrossRef]
28. Szpikowska-Sroka, B.; Pawlik, N.; Goryczka, T.; Pietrasik, E.; Bańczyk, M.; Pisarski, W.A. Lead fluoride β-PbF_2 nanocrystals containing Eu^{3+} and Tb^{3+} ions embedded in sol-gel materials: Thermal, structural and optical investigations. *Ceram. Int.* **2017**, *43*, 8424–8432. [CrossRef]
29. Yu, Y.; Chen, D.; Wang, Y.; Luo, W.; Zheng, Y.; Cheng, Y.; Zhou, L. Structural evolution and its influence on luminescence of SiO_2-SrF_2-ErF_3 glass ceramics prepared by sol-gel method. *Mater. Chem. Phys.* **2006**, *100*, 241–245. [CrossRef]
30. Yu, Y.; Wang, Y.; Chen, D.; Liu, F. Efficient upconversion luminescence of Er^{3+}:SrF_2-SiO_2-Al_2O_3 sol-gel glass ceramics. *Ceram. Int.* **2008**, *34*, 2143–2146. [CrossRef]
31. Zhou, L.; Chen, D.; Luo, W.; Wang, Y.; Yu, Y.; Liu, F. Transparent glass ceramic containing Er^{3+}:CaF_2 nano-crystals prepared by sol-gel method. *Mater. Lett.* **2007**, *61*, 3988–3990. [CrossRef]
32. Georgescu, S.; Voiculescu, A.M.; Matei, C.; Secu, C.E.; Negre, R.F.; Secu, M. Ultraviolet and visible up-conversion luminescence of Er^{3+}/Yb^{3+} co-doped CaF_2 nanocrystals in sol-gel derived glass-ceramics. *J. Lumin.* **2013**, *143*, 150–156. [CrossRef]
33. Jiang, Y.; Fan, J.; Jiang, B.; Mao, X.; Zhou, C.; Zhang, L. Structure and optical properties of transparent Er^{3+}-doped CaF_2-silica glass ceramic prepared by controllable sol-gel method. *Ceram. Int.* **2016**, *42*, 9571–9576. [CrossRef]
34. Chen, D.; Wang, Y.; Yu, Y.; Ma, E.; Zhou, L. Microstructure and luminescence of transparent glass ceramic containing Er^{3+}:BaF_2 nano-crystals. *J. Solid State Chem.* **2006**, *179*, 532–537. [CrossRef]
35. Secu, C.E.; Secu, M.; Ghica, C.; Mihut, L. Rare-earth doped sol-gel derived oxyfluoride glass-ceramics: Structural and optical characterization. *Opt. Mater.* **2011**, *33*, 1770–1774. [CrossRef]
36. Secu, C.E.; Bartha, C.; Polosan, S.; Secu, M. Thermally activated conversion of a silicate gel to an oxyfluoride glass ceramic: Optical study using Eu^{3+} probe ion. *J. Lumin.* **2014**, *146*, 539–543. [CrossRef]
37. Fujihara, S.; Kitta, S.; Kimura, T. Porous Phosphor thin films of oxyfluoride SiO_2-$BaMgF_4$: Eu^{2+} glass-ceramics prepared by sol-gel method. *Chem. Lett.* **2003**, *32*, 928–929. [CrossRef]
38. Kitta, S.; Fujihara, S.; Kimura, T. Porous SiO_2-$BaMgF_4$:Eu(II) glass-ceramic thin films and their strong blue photoluminescence. *J. Sol-Gel Sci. Technol.* **2004**, *32*, 263–266. [CrossRef]
39. Blasse, G.; van den Heuvel, G.P.M.; Van Dijk, T. Energy transfer from Gd^{3+} to Tb^{3+} and Eu^{3+}. *Chem. Phys. Lett.* **1979**, *62*, 600–602. [CrossRef]
40. Grzyb, T.; Runowski, M.; Lis, S. Facile synthesis, structural and spectroscopic properties of GdF_3:Ce^{3+}, Ln^{3+} (Ln^{3+} = Sm^{3+}, Eu^{3+}, Tb^{3+}, Dy^{3+}) nanocrystals with bright multicolor luminescence. *J. Lumin.* **2014**, *154*, 479–486. [CrossRef]
41. Pokhrel, M.; Mimun, L.C.; Yust, B.; Kumar, G.A.; Dhanale, A.; Tang, L.; Sardara, D.K. Stokes emission in GdF_3:Nd^{3+} nanoparticles for bioimaging probes. *Nanoscale* **2014**, *6*, 1667–1674. [CrossRef] [PubMed]
42. Fujihara, S.; Tada, M.; Kimura, T. Sol-gel processing of LaF_3 thin films. *J. Ceram. Soc. JPN* **1998**, *106*, 124–126. [CrossRef]

43. Fujihara, S.; Tada, M.; Kimura, T. Formation of LaF$_3$ microcrystals in sol-gel silica. *J. Non-Cryst. Solids* **1999**, *244*, 267–274. [CrossRef]

44. Tada, M.; Fujihara, S.; Kimura, T. Sol-gel processing and characterization of alkaline earth and rare-earth fluoride thin films. *J. Mater. Res.* **1999**, *14*, 1610–1616. [CrossRef]

45. Fujihara, S.; Kato, T.; Kimura, T. Influence of solution composition on the formation of SiO$_2$/LaF$_3$ composites in the sol-gel process. *J. Mater. Sci.* **2000**, *35*, 2763–2767. [CrossRef]

46. Fujihara, S.; Tada, M.; Kimura, T. Controlling factors for the conversion of trifluoroacetate sols into thin metal fluoride coatings. *J. Sol-Gel Sci. Technol.* **2000**, *19*, 311–314. [CrossRef]

47. Fujihara, S.; Kato, T.; Kimura, T. Sol-gel synthesis of silica-based oxyfluoride glass-ceramic thin films: incorporation of Eu^{3+} activators into crystallites. *J. Am. Ceram. Soc.* **2001**, *84*, 2716–2718. [CrossRef]

48. Ribeiro, S.J.L.; Araújo, C.C.; Bueno, L.A.; Gonçalves, R.R.; Messaddeq, Y. Sol-gel Eu^{3+}/Tm^{3+} doped transparent glass-ceramic waveguides. *J. Non-Cryst. Solids* **2004**, *348*, 180–184. [CrossRef]

49. Biswas, A.; Maciel, G.S.; Friend, C.S.; Prasad, P.N. Upconversion properties of a transparent Er^{3+}-Yb^{3+}co-doped LaF$_3$-SiO$_2$ glass-ceramics prepared by sol-gel method. *J. Non-Cryst. Solids* **2003**, *316*, 393–397. [CrossRef]

50. Yanes, A.C.; del-Castillo, J.; Méndez-Ramos, J.; Rodríguez, V.D.; Torres, M.E.; Arbiol, J. Luminescence and structural characterization of transparent nanostructured Eu^{3+}-doped LaF$_3$-SiO$_2$ glass-ceramics prepared by sol-gel method. *Opt. Mater.* **2007**, *9*, 999–1003. [CrossRef]

51. Velázquez, J.J.; Yanes, A.C.; del Castillo, J.; Méndez-Ramos, J.; Rodríguez, V.D. Optical properties of Ho^{3+}-Yb^{3+} co-doped nanostructured SiO$_2$-LaF$_3$ glass-ceramics prepared by sol-gel method. *Phys. Status Solidi A* **2007**, *204*, 1762–1768. [CrossRef]

52. Méndez-Ramos, J.; Velázquez, J.J.; Yanes, A.C.; del Castillo, J.; Rodríguez, V.D. Up-conversion in nanostructured Yb^{3+}-Tm^{3+} co-doped sol-gel derived SiO$_2$-LaF$_3$ transparent glass-ceramics. *Phys. Status Solidi A* **2008**, *205*, 330–334. [CrossRef]

53. Yanes, A.C.; Velázquez, J.J.; del Castillo, J.; Méndez-Ramos, J.; Rodríguez, V.D. Colour tuneability and white light generation in Yb^{3+}-Ho^{3+}-Tm^{3+} co-doped SiO$_2$-LaF$_3$ nano-glass-ceramics prepared by sol-gel method. *J. Sol-Gel Sci. Technol.* **2009**, *51*, 4–9. [CrossRef]

54. Velázquez, J.J.; Rodríguez, V.D.; Yanes, A.C.; del Castillo, J.; Méndez-Ramos, J. Increase in the Tb^{3+} green emission in SiO$_2$-LaF$_3$ nano-glass-ceramics by codoping with Dy^{3+} ions. *J. Appl. Phys.* **2010**, *108*, 113530–113536. [CrossRef]

55. Velázquez, J.J.; Rodríguez, V.D.; Yanes, A.C.; del Castillo, J.; Méndez-Ramos, J. Photon down-shifting by energy transfer from Sm^{3+} to Eu^{3+} ions in sol-gel SiO$_2$-LaF$_3$ nano-glass-ceramics for photovoltaics. *Appl. Phys. B* **2012**, *108*, 577–583. [CrossRef]

56. Dejneka, M.J. The luminescence and structure of novel transparent oxyfuoride glass-ceramics. *J. Non-Cryst. Solids* **1998**, *239*, 149–155. [CrossRef]

57. Luo, W.; Wang, Y.; Cheng, Y.; Bao, F.; Zhou, L. Crystallization and structural evolution of YF$_3$-SiO$_2$ xerogel. *Mater. Sci. Eng. B* **2006**, *127*, 218–223. [CrossRef]

58. Méndez-Ramos, J.; Santana-Alonso, A.; Yanes, A.C.; del Castillo, J.; Rodríguez, V.D. Rare-earth doped YF$_3$ nanocrystals embedded in sol-gel silica glass matrix for white light generation. *J. Lumin.* **2010**, *130*, 2508–2511. [CrossRef]

59. Santana-Alonso, A.; Méndez-Ramos, J.; Yanes, A.C.; del Castillo, J.; Rodríguez, V.D. White light up-conversion in transparent sol-gel derived glass-ceramics containing Yb^{3+}-Er^{3+}-Tm^{3+} triply-doped YF$_3$ nanocrystals. *Mater. Chem. Phys.* **2010**, *124*, 699–703. [CrossRef]

60. Yanes, A.C.; Santana-Alonso, A.; Méndez-Ramos, J.; del Castillo, J.; Rodríguez, V.D. Novel sol-gel nano-glass-ceramics comprising Ln^{3+}-Doped YF$_3$ nanocrystals: structure and high efficient UV up-conversion. *Adv. Funct. Mater.* **2011**, *21*, 3136–3142. [CrossRef]

61. Chen, D.; Wang, Y.; Yu, Y.; Huang, P. Structure and optical spectroscopy of Eu-doped glass ceramics containing GdF$_3$ nanocrystals. *J. Phys. Chem. C* **2008**, *112*, 18943–18947. [CrossRef]

62. Shan, Z.; Chen, D.; Yu, Y.; Huang, P.; Lin, H.; Wang, Y. Luminescence in rare earth-doped transparent glass ceramics containing GdF$_3$ nanocrystals for lighting applications. *J. Mater. Sci.* **2010**, *45*, 2775–2779. [CrossRef]

63. Yin, W.; Zhao, L.; Zhou, L.; Gu, Z.; Liu, X.; Tian, G.; Jin, S.; Yan, L.; Ren, W.; Xing, G.; Zhao, Y. Enhanced red emission from GdF$_3$:Yb^{3+}, Er^{3+} upconversion nanocrystals by Li$^+$ doping and their application for bioimaging. *Chem. Eur. J.* **2012**, *18*, 9239–9245. [CrossRef] [PubMed]

64. Fujihara, S.; Koji, S.; Kimura, T. Structure and optical properties of (Gd,Eu)F_3-nanocrystallized sol-gel silica films. *J. Mater. Chem.* **2004**, *14*, 1331–1335. [CrossRef]

65. Szpikowska-Sroka, B.; Zur, L.; Czoik, R.; Goryczka, T.; Żądło, M.; Pisarski, W.A. Ultraviolet-to-visible downconversion luminescence in solgel oxyfluoride glass ceramics containing Eu^{3+}:GdF_3 nanocrystals. *Opt. Lett.* **2014**, *39*, 3181–3184. [CrossRef] [PubMed]

66. Szpikowska-Sroka, B.; Pawlik, N.; Goryczka, T.; Pisarski, W.A. Influence of silicate sol-gel host matrices and catalyst agents on the luminescent properties of Eu^{3+}/Gd^{3+} under different excitation wavelengths. *RSC Adv.* **2015**, *5*, 98773–98782. [CrossRef]

67. Pawlik, N.; Szpikowska-Sroka, B.; Sołtys, M.; Pisarski, W.A. Optical properties of silica sol-gel materials singly- and doubly-doped with Eu^{3+} and Gd^{3+} ions. *J. Rare Earth* **2016**, *34*, 786–795. [CrossRef]

68. Kano, T.; Yamamoto, H.; Otomo, Y. $NaLnF_4$: Yb^{3+}, Er^{3+} (Ln: Y, Gd, La): Efficient green-emitting infrared-excited phosphors. *J. Electrochem. Soc.* **1972**, *119*, 1561–1564. [CrossRef]

69. Krämer, K.W.; Biner, D.; Frei, G.; Güdel, H.U.; Hehlen, M.P.; Lüthi, S.R. Hexagonal sodium yttrium fluoride based green and blue emitting upconversion phosphors. *Chem. Mater.* **2004**, *16*, 1244–1251. [CrossRef]

70. Liu, F.; Ma, E.; Chen, D.; Yu, Y.; Wang, Y. Tunable red-green upconversion luminescence in novel transparent glass ceramics containing Er: $NaYF_4$ nanocrystals. *J. Phys. Chem. B* **2006**, *110*, 20843–20846. [CrossRef] [PubMed]

71. De Pablos-Martín, A.; Méndez-Ramos, J.; del Castillo, J.; Durán, A.; Rodríguez, V.D.; Pascual, M.J. Crystallization and up-conversion luminescence properties of Er^{3+}/Yb^{3+}-doped $NaYF_4$-based nano-glass-ceramics. *J. Eur. Ceram. Soc.* **2015**, *35*, 1831–1840. [CrossRef]

72. Yanes, A.C.; Santana-Alonso, A.; Méndez-Ramos, J.; del Castillo, J.; Rodríguez, V.D. Yb^{3+}-Er^{3+} co-doped sol-gel transparent nano-glass-ceramics containing $NaYF_4$ nanocrystals for tuneable up-conversion phosphors. *J. Alloy. Compd.* **2009**, *480*, 706–710. [CrossRef]

73. Santana-Alonso, A.; Yanes, A.C.; Méndez-Ramos, J.; del Castillo, J.; Rodríguez, V.D. Sol-gel transparent nano-glass-ceramics containing Eu^{3+}-doped $NaYF_4$ nanocrystals. *J. Non-Cryst. Solids* **2010**, *356*, 933–936. [CrossRef]

74. Santana-Alonso, A.; Méndez-Ramos, J.; Yanes, A.C.; del Castillo, J.; Rodríguez, V.D. Up-conversion in sol-gel derived nano-glass-ceramics comprising $NaYF_4$ nano-crystals doped with Yb^{3+}, Ho^{3+} and Tm^{3+}. *Opt. Mater.* **2010**, *32*, 903–908. [CrossRef]

75. Méndez-Ramos, J.; Yanes, A.C.; Santana-Alonso, A.; del Castillo, J.; Rodríguez, V.D. Colour tuneability in sol-gel nano-glass-ceramics comprising Yb^{3+}-Er^{3+}-Tm^{3+} Co-Doped $NaYF_4$ nanocrystals. *J. Nanosci. Nanotechno.* **2010**, *10*, 1273–1277. [CrossRef]

76. Méndez-Ramos, J.; Yanes, A.C.; Santana-Alonso, A.; del Castillo, J. Highly efficient up-conversion and bright white light in RE co-doped KYF_4 nanocrystals in sol-gel silica matrix. *Chem. Phys. Lett.* **2013**, *555*, 196–201. [CrossRef]

77. Yanes, A.C.; Santana-Alonso, A.; Méndez-Ramos, J.; del Castillo, J. Structure and intense UV up-conversion emissions in RE^{3+}-doped sol-gel glass-ceramics containing KYF_4 nanocrystals. *Appl. Phys. B* **2013**, *113*, 589–596. [CrossRef]

78. Del Castillo, J.; Yanes, A.C.; Santana-Alonso, A.; Méndez-Ramos, J. Efficient dual-wavelength excitation of Tb^{3+} emission in rare-earth doped KYF_4 cubic nanocrystals dispersed in silica sol-gel matrix. *Opt. Mater.* **2014**, *37*, 511–515. [CrossRef]

79. Yanes, A.C.; del Castillo, J. Enhanced emission via energy transfer in RE co-doped SiO_2-KYF_4 nano-glass-ceramics for white LEDs. *J. Alloy. Compd.* **2016**, *658*, 170–178. [CrossRef]

80. Deng, D.; Xu, S.; Zhao, S.; Li, C.; Wang, H.; Ju, H. Enhancement of upconversion luminescence in Tm^{3+}/Er^{3+}/Yb^{3+}-codoped glass ceramic containing $LiYF_4$ nanocrystals. *J. Lumin.* **2009**, *129*, 1266–1270. [CrossRef]

81. Kawamura, G.; Yoshimura, R.; Ota, K.; Oh, S.Y.; Hakiri, N.; Muto, H.; Hayakawa, T.; Matsuda, A. A unique approach to characterization of sol-gel-derived rare-earth-doped oxyfluoride glass-ceramics. *J. Am. Ceram. Soc.* **2013**, *96*, 476–480. [CrossRef]

82. Secu, C.E.; Negrea, R.F.; Secu, M. Eu^{3+} probe ion for rare-earth dopant site structure in sol-gel derived $LiYF_4$ oxyfluoride glass-ceramic. *Opt. Mater.* **2013**, *35*, 2456–2460. [CrossRef]

83. Kawamura, G.; Yoshimura, R.; Ota, K.; Oh, S.Y.; Muto, H.; Hayakawa, T.; Matsuda, A. Extraction of Nd^{3+}-doped $LiYF_4$ phosphor from sol-gel-derived oxyfluoride glass ceramics by hydrofluoric acid treatment. *Opt. Mater.* **2013**, *45*, 1879–1881. [CrossRef]

84. Secu, M.; Secu, C.E. Up-conversion luminescence of Er^{3+}/Yb^{3+} co-doped $LiYF_4$ nanocrystals in sol-gel derived oxyfluoride glass-ceramics. *J. Non-Cryst. Solids* **2015**, *426*, 78–82. [CrossRef]

85. Del Castillo, J.; Yanes, A.C.; Abe, S.; Smet, P.F. Site selective spectroscopy in $BaYF_5$:RE^{3+} (RE = Eu, Sm) nano-glass-ceramics. *J. Alloy. Compd.* **2015**, *635*, 136–141. [CrossRef]

86. Del Castillo, J.; Yanes, A.C. Bright luminescence of Gd^{3+} sensitized RE^{3+}-doped SiO_2-$BaGdF_5$ glass-ceramics for UV-LEDs colour conversion. *J. Alloy. Compd.* **2017**, *695*, 3736–3743. [CrossRef]

87. Chiodini, N.; Paleari, A.; DiMartino, D.; Spinolo, G. SnO_2 nanocrystals in SiO_2 : A wide-band-gap quantum-dot system. *Appl. Phys. Lett.* **2022**, *81*, 1702–1704. [CrossRef]

88. Du, J.; Zhao, R.; Xie, Y.; Li, J. Size-controlled synthesis of SnO_2 quantum dots and their gas-sensing performance. *Appl Surf. Sci.* **2015**, *346*, 256–262. [CrossRef]

89. Nogami, M.; Enomoto, T.; Hayakawa, T. Enhanced fluorescence of Eu^{3+} induced by energy transfer from nanosized SnO_2 crystals in glass. *J. Lumin.* **2002**, *97*, 147–152. [CrossRef]

90. Chiodini, N.; Paleari, A.; Spinolo, G.; Crespi, P. Photorefractivity in SiO_2:SnO_2 glass-ceramics by visible light. *J. Non-Cryst. Solids* **2003**, *322*, 266–271. [CrossRef]

91. Yanes, A.C.; del Castillo, J.; Torres, M.; Peraza, J.; Rodríguez, V.D.; Méndez-Ramos, J. Nanocrystal-size selective spectroscopy in SnO^2: Eu^{3+} semiconductor quantum dots. *Appl. Phys. Lett.* **2004**, *85*, 2343–2345. [CrossRef]

92. Del Castillo, J.; Rodríguez, V.D.; Yanes, A.C.; Méndez-Ramos, J.; Torres, M.E. Luminescent properties of transparent nanostructured Eu^{3+} doped SnO_2–SiO_2 glass-ceramics prepared by the sol-gel method. *Nanotechnology* **2005**, *16*, S300–S303. [CrossRef]

93. Del Castillo, J.; Yanes, A.C.; Velázquez, J.J.; Méndez-Ramos, J.; Rodríguez, V.D. Luminescent properties of Eu^{3+}-Tb^{3+}-doped SiO_2-SnO_2-based nano-glass-ceramics prepared by sol-gel method. *J. Alloy. Compd.* **2009**, *473*, 571–575. [CrossRef]

94. Morais, E.A.; Ribeiro, S.J.L.; Scalvi, L.V.A.; Santilli, C.V.; Ruggiero, L.O.; Pulcinelli, S.H.; Messadeq, Y. Optical characteristics of Er^{3+}-Yb^{3+} doped SnO_2 xerogels. *J. Alloy. Compd.* **2002**, *344*, 217–220. [CrossRef]

95. Van Tran, T.T.; Si Bui, T.; Turrell, S.; Capoen, B.; Roussel, P.; Bouazaoui, M.; Ferrari, M.; Cristini, O.; Kinowski, C. Controlled SnO_2 nanocrystal growth in SiO_2-SnO_2 glass-ceramic monoliths. *J. Raman Spectrosc.* **2012**, *43*, 869–875. [CrossRef]

96. Van Tran, T.T.; Turrell, S.; Capoen, B.; Le Van, H.; Ferrari, M.; Ristic, D.; Boussekey, L.; Kinowski, C. Environment segregation of Er^{3+} emission in bulk sol-gel-derived SiO_2-SnO_2 glass ceramics. *J. Mater. Sci.* **2014**, *49*, 8226–8233. [CrossRef]

97. Gonçalves, R.R.; Messaddeq, Y.; Chiasera, A.; Jestin, Y.; Ferrari, M.; Ribeiro, S.J.L. Erbium-activated silica-zirconia planar waveguides prepared by sol-gel route. *Thin Solid Films* **2008**, *516*, 3094–3097. [CrossRef]

98. Gonçalves, R.R.; Guimarães, J.J.; Ferrari, J.L.; Maia, L.J.Q.; Ribeiro, S.J.L. Active planar waveguides based on sol-gel Er^{3+}-doped SiO_2-ZrO_2 for photonic applications: Morphological, structural and optical properties. *J. Non-Cryst. Solids* **2008**, *354*, 4846–4851. [CrossRef]

99. Suhaimi, N.F.M.; Rashid, S.N.M.; Junit, N.F.H.M.; Iznie Razakia, N.; Abd-Rahmana, M.K.; Ferrari, M. Effect of Zirconia in Er^{3+}-doped SiO_2-ZrO_2 for planar waveguide laser. In Proceedings of the 2012 IEEE 3rd International Conference on Photonics (ICP), Penang, Malaysia, 1–3 October 2012.

100. Dos Santos Cunha, C.; Ferrari, J.L.; de Oliveira, D.C.; Maia, L.J.Q.; Gomes, A.S.L.; Ribeiro, S.J.L.; Gonçalves, R.R. NIR luminescent Er^{3+}/Yb^{3+} co-doped SiO_2-ZrO_2 nanostructured planar and channel waveguides: Optical and structural properties. *Mater. Chem. Phys.* **2012**, *136*, 120–129. [CrossRef]

101. Ribeiro, S.J.L.; Messaddeq, Y.; Gonçalves, R.R.; Ferrari, M.; Montagna, M.; Aegerter, M.A. Low optical loss planar waveguides prepared in an organic-inorganic hybrid system. *Appl. Phys. Lett.* **2000**, *77*, 3502–3504. [CrossRef]

102. Jestin, Y.; Armellini, C.; Chiappini, A.; Chiaser, A.; Ferrari, M.; Goyes, C.; Montagna, M.; Moser, E.; Nunzi Conti, G.; Pelli, S.; et al. Erbium activated HfO_2 based glass-ceramics waveguides for photonics. *J. Non-Cryst. Solids* **2007**, *353*, 494–497. [CrossRef]

103. Ferrari, J.L.; Lima, K.O.; Maia, L.J.Q.; Ribeiro, S.J.L.; Gonçalves, R.R. Structural and spectroscopic properties of luminescent Er^{3+}-Doped SiO_2-Ta_2O_5 nanocomposites. *J. Am. Ceram. Soc.* **2011**, *94*, 1230–1237. [CrossRef]

104. Aquino, F.T.; Ferrari, J.L.; Ribeiro, S.J.L.; Ferrier, A.; Goldner, P.; Gonçalves, R.R. Broadband NIR emission in novel sol-gel Er^{3+}-doped SiO_2-Nb_2O_5 glass ceramic planar waveguides for photonic applications. *Opt. Mater.* **2013**, *35*, 387–396. [CrossRef]

105. Secu, M.; Secu, C.E.; Bartha, C. Crystallization and luminescence properties of a new Eu^{3+}-doped LaOCl nano-glass-ceramic. *J. Eur. Ceram. Soc.* **2016**, *36*, 203–207. [CrossRef]

106. Innocenzi, P.; Abdirashid, M.O.; Guglielmi, M. Structure and properties of sol-gel coatings from methyltriethoxysilane and tetraethoxysilane. *J. Sol-Gel Sci. Technol.* **1994**, *3*, 47–55. [CrossRef]

107. Ravel, B.; Newville, M. ATHENA, ARTEMIS, HEPHAESTUS: Data analysis for X-ray absorption spectroscopy using IFEFFIT. *J. Synchrotron Radiat.* **2005**, *12*, 537–541. [CrossRef] [PubMed]

108. Ma, Y.; Lee, H.R.; Tsuru, T. Study on preparation and hydrophobicity of MTES derived silica sol and gel. *Adv. Mater. Res.* **2012**, *535–537*, 2563–2566. [CrossRef]

109. Jia, Y.Q. Crystal Radii and effective ionic radii of the rare earth ions. *J. Sol. State Chem.* **1991**, *95*, 184–187. [CrossRef]

110. Gorni, G.; Pascual, M.J.; Caballero, A.; Velázquez, J.J.; Mosa, J.; Castro, Y.; Durán, A. Crystallization mechanism in sol-gel oxyfluoride glass-ceramics. *J. Non-Cryst. Solids* **2018**, accepted.

111. Gorni, G.; Balda, R.; Ferández, J.; Velázquez, J.J.; Pascual, L.; Mosa, J.; Durán, A.; Castro, Y. $80SiO_2$-$20LaF_3$ oxyfluoride glass ceramic coatings doped with Nd^{3+} for optical applications. *Int. J. Appl. Glass Sci.* **2017**, in press. [CrossRef]

112. Velázquez, J.J.; Mosa, J.; Gorni, G.; Balda, R.; Ferández, J.; Pascual, L.; Durán, A.; Castro, Y. Transparent SiO_2-GdF_3 sol-gel nano-glass ceramics for optical applications. *J. Sol-Gel Sci. Technol.* **2018**, under-review.

113. Chen, D.; Yu, Y.; Huang, P.; Wang, Y. Nanocrystallization of lanthanide trifluoride in an aluminosilicate glass matrix: Dimorphism and rare earth partition. *CrystEngComm* **2009**, *11*, 1686–1690. [CrossRef]

114. Bartha, C.; Secu, C.E.; Matei, E.; Secu, M. Crystallization kinetics mechanism investigation of sol-gel-derived $NaYF_4$:(Yb,Er) up-converting phosphors. *CrystEngComm* **2017**, *19*, 4992–5000. [CrossRef]

115. Agarwal, B.K.; Verma, L.P. A rule for chemical shifts of X-ray absorption edges. *J. Phys. C Solid State* **1970**, *3*, 535–537. [CrossRef]

116. Gorni, G.; Velázquez, J.J.; Mather, G.C.; Durán, A.; Chen, G.; Sundararaja, M.; Balda, R.; Ferández, J.; Pascual, M.J. Selective excitation in transparent oxyfluoride glass-ceramics doped with Nd^{3+}. *J. Eur. Ceram. Soc.* **2017**, *37*, 1695–1706. [CrossRef]

MDPI

St. Alban-Anlage 66

4052 Basel

Switzerland

Tel. +41 61 683 77 34

Fax +41 61 302 89 18

www.mdpi.com

Materials Editorial Office

E-mail: materials@mdpi.com

www.mdpi.com/journal/materials